THE ENVIRONMENT
A REVOLUTION
IN ATTITUDES

THE ENVIRONMENT

A REVOLUTION
IN ATTITUDES

Cornelia Blair

INFORMATION PLUS REFERENCE SERIES
Formerly published by Information Plus, Wylie, Texas

GALE GROUP

Detroit
New York
San Francisco
London
Boston
Woodbridge, CT

THE ENVIRONMENT: A REVOLUTION IN ATTITUDES

Cornelia Blair, *Author*

The Gale Group Staff:

Editorial: John F. McCoy, *Project Manager and Series Editor*; Michael T. Reade, *Series Associate Editor*; Jason M. Everett, *Series Assistant Editor*; Rita Runchock, *Managing Editor*; Luann Brennan, *Editor*; Thomas Carson, *Editor*; Andrew Claps, *Editor*; Kathleen Droste, *Editor*; Nancy Matuszak, *Editor*; Christy Wood, *Associate Editor*; Ryan McNeill, *Assistant Editor*; Jeffrey Telford, *Assistant Editor*

Image and Multimedia Content: Barbara J. Yarrow, *Manager, Imaging and Multimedia Content*; Robyn Young, *Project Manager, Imaging and Multimedia Content*; Dean Dauphinais, *Senior Editor, Imaging and Multimedia Content*; Kelly A. Quin, *Editor, Imaging and Multimedia Content*; Leitha Etheridge-Sims, Mary K. Grimes, and David G. Oblender, *Image Catalogers*; Pamela A. Reed, *Imaging Coordinator*; Randy Bassett, *Imaging Supervisor*; Robert Duncan, *Senior Imaging Specialist*; Dan Newell, *Imaging Specialist*; Christine O'Bryan, *Graphic Specialist*

Indexing: Jennifer Dye, *Indexing Specialist*; Lynne Maday, *Indexing Specialist*

Permissions: Maria Franklin, *Permissions Manager*; Margaret Chamberlain, *Permissions Specialist*; Julie Juengling, *Permissions Specialist*

Product Design: Michelle DiMercurio, *Senior Art Director*; Kenn Zorn, *Product Design Manager*

Production: Mary Beth Trimper, *Composition Manager*; Gary Leach and Carolyn Roney, *Typesetting Specialists*; NeKita McKee, *Buyer*; Dorothy Maki, *Manufacturing Manager*

ISBN 0-7876-5103-6 (set)
ISBN 0-7876-5142-7 (this volume)
ISSN 1532-270X (this volume)
Printed in the United States of America
10 9 8 7 6 5 4 3 2 1

TABLE OF CONTENTS

CHAPTER 1

People all over the world are beginning to understand that Earth's resources are limited and must be protected. Federal and state governments have created laws to protect these resources, which has an impact on world economies and businesses. Responsibility also falls to governments to protect its citizens from environmental hazards and crimes. The international community has held conferences and concluded treaties, impacting the environment. Public opinion and knowledge about environmental issues continues to evolve.

CHAPTER 2

Scientists agree that the Earth is getting warmer. Most scientists believe that the greenhouse effect accounts for this change, however, some are uncertain or disagree. Gases such as carbon dioxide, methane, nitrous oxide, and chlorofluorocarbons are seen as major causes of global warming and other changes. Many nations, as well as the American public, have committed themselves to reducing emissions of these gases.

CHAPTER 3

Ozone can be either harmful or protective to life on Earth, depending on where it is found. Human made chemicals, such as chlorofluro-carbons, destroy the protective layer of ozone found in the upper atmosphere. Without this protective layer, humans, plants, and animals are in danger. The international community plays an important role in protecting the ozone layer.

CHAPTER 4

The growth of industrialized nations and consumer societies has created a greater need for collection and disposal of refuse. Humans get rid of household and non-household waste in four ways: through landfills, incinerators, recycling, and composting. Hazardous and nuclear waste, on the other hand, requires even more special consideration. Federal and state governments must ensure that disposal and former industrial sites are safe for humans, animals, and plant-life.

CHAPTER 5

Air pollution can make people sick and damage the environment. Various gases and particles cause this pollution. Emissions from automobiles, airplanes, factories, and power plants are mostly to blame, although these industries are trying to address these problems through new technologies and alternative solutions. Governments also try to address air quality issues by passing legislation designed to clean up the air. The problem of air pollution is of a worldwide nature.

CHAPTER 6

Water is the most essential resource for sustaining all forms of life. Human alteration of the environment can have a devastating affect on the water supply. Legislation, such as the Clean Water Act, seeks to improve and protect the integrity of America's surface water, groundwater, oceans, coastal and drinking water. Nonetheless, water is polluted by many chemicals and contaminants.

CHAPTER 7

Burning fossil fuels, caused by natural occurrences and human activity, emit sulfate and nitrate into the atmosphere, leading to acid rain. The results of acid rain on humans and the environment can be harmful and destructive. A worldwide problem, politicians and environmental groups have sought to reduce the pollution that causes acid rain through legislation.

CHAPTER 8

Many of the substances released into the environment by modern, industrialized society are harmful to humans and other living creatures. Chemicals in pesticides and fertilizers are harmful to the enivornment. Large concentrations of toxins such as asbestos, chlorine, lead, radon, and others also pose a danger to the environment. This chapter also examines tobacco, the Gulf War syndrome, indoor air quality, noise pollution, and food safety.

CHAPTER 9

Without nature's ecological contribution, human, plant, and animal life could not exist. Forests, wetlands, and mountains are endangered by the activities of humans. The diversity of life on Earth is also at risk, with many species being destroyed. The extraction of renewable and nonrenewable resources has a detrimental effect on the environment. Local, national, and international efforts must be linked to deal with pressures on the environment.

PREFACE

The Environment: A Revolution in Attitudes is the latest volume in the ever-growing *Information Plus Reference Series*. Previously published by the Information Plus company of Wylie, Texas, the *Information Plus Reference Series* (and its companion set, the *Information Plus Compact Series*) became a Gale Group product when Gale and Information Plus merged in early 2000. Those of you familiar with the series as published by Information Plus will notice a few changes from the 1999 edition. Gale has adopted a new layout and style that we hope you will find easy to use. Other improvements include greatly expanded indexes in each book, and more descriptive tables of contents.

While some changes have been made to the design, the purpose of the *Information Plus Reference Series* remains the same. Each volume of the series presents the latest facts on a topic of pressing concern in modern American life. These topics include today's most controversial and most studied social issues: abortion, capital punishment, care for the elderly, crime, health care, the environment, immigration, minorities, social welfare, women, youth, and many more. Although written especially for the high school and undergraduate student, this series is an excellent resource for anyone in need of factual information on current affairs.

By presenting the facts, it is Gale's intention to provide its readers with everything they need to reach an informed opinion on current issues. To that end, there is a particular emphasis in this series on the presentation of scientific studies, surveys, and statistics. This data is generally presented in the form of tables, charts, and other graphics placed within the text of each book. Every graphic is directly referred to and carefully explained in the text. The source of each graphic is presented within the graphic itself. The data used in these graphics is drawn from the most reputable and reliable sources, in particular from the various branches of the U.S. government and from major independent polling organizations. Every effort was made to secure the most recent information available. The reader should bear in mind that many major studies take years to conduct, and that additional years often pass before the data from these studies is made available to the public. Therefore, in many cases the most recent information available in 2000 dated from 1997 or 1998. Older statistics are sometimes presented as well, if they are of particular interest and no more-recent information exists.

Although statistics are a major focus of the *Information Plus Reference Series* they are by no means its only content. Each book also presents the widely held positions and important ideas that shape how the book's subject is discussed in the United States. These positions are explained in detail and, where possible, in the words of those who support them. Some of the other material to be found in these books includes: historical background; descriptions of major events related to the subject; relevant laws and court cases; and examples of how these issues play out in American life. Some books also feature primary documents, or have pro and con debate sections giving the words and opinions of prominent Americans on both sides of a controversial topic. All material is presented in an even-handed and unbiased manner; the reader will never be encouraged to accept one view of an issue over another.

HOW TO USE THIS BOOK

The condition of the world's environment is a broad issue, and one of great concern to people worldwide. Ever since the late nineteenth century, mankind has developed an ever greater ability to alter the natural world, both deliberately and unintentionally. Many people fear that without proper restraint, mankind's actions could forever alter, or even eliminate, life on Earth. There are those, however, who feel that these fears are exaggerated, and that the substantial cost of environmental protection on business and industry should be reduced, or at least not expanded any further. The conflict between these two positions has serious environmental, political, and economic ramifications, both within the United States and between nations. This book examines the steps taken to protect the Earth's natural environment, and the controversies that surround them.

The Environment: A Revolution in Attitudes consists of ten chapters and three appendices. Each chapter is devoted to a particular aspect of environmental protection. For a summary of the information covered in each chapter, please see the synopses provided in the Table of Contents at the front of the book. Chapters generally begin with an overview of the basic facts and background information on the chapter's topic, then proceed to examine sub-topics of particular interest. For example, Chapter 4: Waste Disposal begins with an overview of how trash is created in the United States in the present day, how this differs from in the past, and why. The nature of the waste that Americans produce is examined. The chapter then discusses the primary methods through which waste is disposed of, such as landfills, recycling, and incinerators. The later part of the chapter examines in detail the special problems presented by hazardous waste. Readers can find their way through any chapter by looking for the section and sub-section headings, which are clearly set off from the text. Or, they can refer to the book's extensive index, if they already know what they are looking for.

Statistical Information

The tables and figures featured throughout *The Environment: A Revolution in Attitudes* will be of particular use to the reader in learning about this issue. These tables and figures represent an extensive collection of the most recent and important statistics on the environment. For example, the amounts of different kinds of pollutants found in air across the United States, the percentage of American lakes and streams which are unsafe to drink from or swim in, the amount of wetlands that are destroyed each year, and opinion polls on how willing Americans are to protect the environment even if doing so harms the economy. Gale believes that making this information available to the reader is the most important way in which we fulfill the goal of this book: To help readers understand the issues and controversies surrounding the environment, and reach their own conclusions about them.

Each table or figure has a unique identifier appearing above it, for ease of identification and reference. Titles for the tables and figures explain their purpose. At the end of each table or figure, the original source of the data is provided. The reader can also find the source information for all of the tables and figures gathered together in the Acknowledgments section.

In order to help readers understand these often complicated statistics, all tables and figures are explained in the text. References in the text direct the reader to the relevant statistics. Furthermore, the contents of all tables and figures are fully indexed. Please see the opening section of the index at the back of this volume for a description of how to find tables and figures within it.

In addition to the main body text and images, *The Environment: A Revolution in Attitudes* has three appendices. The first appendix is the Important Names and Addresses directory. Here the reader will find contact information for organizations that study endangered species, or play an important role in setting the policies that affect them. The second appendix is the Resources section, which is provided to assist the reader in conducting his or her own research. In this section, the author and editors of *The Environment: A Revolution in Attitudes* describe some of the sources that were most useful during the compilation of this book. The final appendix is this book's index. It has been greatly expanded from previous editions, and should make it even easier to find specific topics in this book.

COMMENTS AND SUGGESTIONS

The editor of the *Information Plus Reference Series* welcomes your feedback on *The Environment: A Revolution in Attitudes* Please direct all correspondence to:

Editor

Information Plus Reference Series

27500 Drake Rd.

Farmington Hills, MI, 48331-3535

ACKNOWLEDGMENTS

Permission to use the following quotes, photographs, illustrations, figures, charts and tables appearing in The Environment *were received from the following sources:*

Figures 4.1 and 10.8. Corbis Corporation. Reproduced by permission.

Figure 4.21. U.S. Department of Energy.

Figure 4.23. U.S. Department of Energy, Office of Civilian Radioactive Waste Management.

Figure 1.3. Environmental Defense Fund. Reproduced by permission.

Figure 5.18. General Motors. AP/Wide World. Reproduced by permission.

Tables 4.1 and 4.2. Courtesy of *Biocycle*, J.G. Press. Reproduced by permission.

Figure 1.6. U. S. National Aeronautics and Space Administration. Reproduced by permission.

Figure 10.12. Photograph by Roger Ressmeyer. Reproduced by permisssion.

Figure 10.9. Photograph by Keven Schafer. Corbis Corporation. Reproduced by permission.

Figure 1.5, Table 1.7. Courtesy of the *Wirthlin Report*, Wirthlin Worldwide, 1998. Reproduced by permission.

"Greenhouse Gas Profile," table 2.1, and "The Weather Forecast for 2050," table 2.6, 1990 from *The Greenhouse Trap*; "Human Population Growth," figure 1.1 1998 from *Conserving the World's Biodiversity*. Copyright 1984, 1990, 1998 by the World Resources Institute. Reproduced by permission.

CHAPTER 1

THE STATE OF THE ENVIRONMENT—AN OVERVIEW

We have not inherited the Earth from our fathers. We are borrowing it from our children.
—American Indian saying

Photographs from outer space impress upon the world that humanity shares one planet, and a small one at that. Earth is one ecosystem. There may be differences in race, nationality, religion, and language, but everyone resides on the same orbiting planet.

General concern about the environment is a relatively recent phenomenon. Two closely related factors explain the rising concern during the second half of the twentieth century: global industrialization following World War II and the worldwide population explosion. People once thought the life-sustaining resources of planet Earth were infinite; now they know they are finite and limited.

HISTORICAL ATTITUDES TOWARDS THE ENVIRONMENT

The Industrial Revolution

Humankind has always altered the environment around itself. For much of human history, however, these changes were fairly limited. The world was too vast and people too few to have more than a limited effect on the environment, especially since they had only primitive tools and technology to aid them. All of this began to change in the 1800s, however. First in Europe and then in America, powerful new machines, such as steam engines, were developed and put into use. These new technologies led to great increases in the amount and quality of goods that could be manufactured and the amount of food that could be harvested. As a result, the quality of life rose substantially and the population began to boom. The so-called Industrial Revolution was underway.

At the same time the Industrial Revolution enabled people to live better in other ways, it increased the rate of pollution. For many years pollution was thought to be an insignificant side effect of growth and progress. In fact, at one time, people looked on the smokestacks belching black soot as a healthy sign of economic growth. The reality was that pollution, along with the increased demands for natural resources and living space that resulted from the Industrial Revolution, was beginning to have a significant effect on the environment.

Twentieth Century America

For much of the early twentieth century, Americans accepted pollution as an inevitable cost of economic progress. After World War II, however, more and more incidents made people aware of environmental problems caused by human activities. Los Angeles "smog," a smoky haze of pollution that formed like a fog in the city, contributed a new word to the language. Swimming holes became so polluted they were poisonous. Still, little action was taken.

In the 1960s, environmental awareness began to increase, partly in response to the 1962 publication of Rachel Carson's book *Silent Spring*, which exposed the toll of the chemical pesticide DDT on bird populations. Other signs of the drastic effects of pollution on the environment became harder and harder to ignore. For example, in 1969, the Cuyahoga River near Cleveland, Ohio, burst into flames due to pollutants in the water.

Environmental protection rapidly became very popular with the public, particularly with the younger generation. *The New York Times,* on November 30, 1969, carried a lengthy article reporting on the astonishing increase in environmental interest. The article noted that concern about the environmental crisis was especially strong on college campuses, where it was threatening to become even more of an issue than the Vietnam War.

FACTORS CONTRIBUTING TO ENVIRONMENTAL ACTIVISM. In 1969, according to Opinion Research of

Princeton, New Jersey, only 1 percent of Americans polled expressed concern for the environment. By 1971 fully one-quarter reported that protecting the environment was important. What motivated Americans to this new awareness? Among the likely factors were:

- An affluent economy and increased leisure time

- The emergence of an "activist" upper middle class—college-educated, affluent, concerned, and youthful

- The rise of television, an increasingly aggressive press, and advocacy journalism (supporting specific causes)

- An advanced scientific community with increasing funding, new technology, and vast communication capabilities

EARTH DAY AND THE BIRTH OF ENVIRONMENTAL PROTECTION. The idea for Earth Day evolved over a period beginning in the early 1960s. Nationwide "teach-ins" were being held on campuses across the nation to protest the Vietnam War. Democratic Senator Gaylord Nelson of Wisconsin, troubled by the apathy of American leaders toward the environment, announced that a grass-roots demonstration on behalf of the environment would be held in the spring of 1970 and invited everyone to participate. On April 22, 1970, 20 million people participated in massive rallies on American campuses and in large cities. Earth Day would go on to develop into an annual event, still being celebrated by the year 2000.

With public opinion loudly expressed by the Earth Day demonstrations, Congress and President Richard Nixon passed a series of unprecedented laws in 1970, creating the Environmental Protection Agency (EPA), an organization devoted to setting limits for water and air pollutants and to investigating the environmental impact of proposed federally funded projects. In the years that followed, many more environmental laws were passed setting basic rules for interaction with the environment. Most notable among these laws were the Clean Air Act (91-604, 1970), the Clean Water Act (PL 92-500, 1972), the Endangered Species Act (PL 93-205, 1973), the Safe Drinking Water Act (PL 93-523, 1974), and the Resource Conservation and Recovery Act (PL 95-510, 1976).

THE STATUS OF ENVIRONMENTAL ISSUES IN THE UNITED STATES AT THE END OF THE TWENTIETH CENTURY. Since the 1970s, the state of the environment has continued to be a major political issue of interest to many Americans. Many activist organizations, such as Greenpeace and the Natural Resources Defense Council, have come into being. Virtually every state has established one or more agencies charged with protecting the environment. Many universities and colleges offer programs in environmental education. Billions of dollars are spent every year by state and federal governments for environ-

mental protection and improvement. These efforts have, in turn, led to many improvements in the state of the environment. Many of the most dangerous chemicals that once polluted the air and water have either been banned or their emissions into the environment greatly reduced. Yet other environmental problems have arisen or worsened since 1970, such as the possibility of global warming and the depletion of Earth's natural resources. International conferences addressing these issues have produced mixed results.

Some studies have suggested that concern about the environment declined in the 1990s. To explain this, many people point to the fact that obvious dangers, such as rivers on fire and belching smokestacks, have seen substantial improvement. Those dangers that remain, such as global warming and ozone depletion in the atmosphere, are largely invisible, and the public may not as easily accept or be concerned about their existence.

Another factor in the decline in environmental activism is money. Many of the cheapest and easiest environmental problems to fix were resolved in the 1970s, 1980s, and 1990s. Most of the problems that remained to be dealt with at the end of the century were so large or complicated that it was believed that tremendous amounts of money would have to be spent before even modest improvements would be seen. Many Americans, especially those who felt their jobs were threatened by environmental regulations, questioned whether these increased costs were worth the relatively small benefits they would provide. Even among those who wanted to see further environmental improvements, environmental issues were competing for funds with other pressing human needs such as AIDS, homelessness, and starvation in many parts of the world.

THE ROLE OF POPULATION IN THE ENVIRONMENTAL EQUATION

Earth's population is believed to have grown more from 1950 through 2000 than it did during the previous four million years. (See Figure 1.1.) For centuries, deaths largely offset births, resulting in slow population growth. Beginning around 1950, high birth rates in developing countries, coupled with a reduction in death rates and reduced infant mortality (which led to an overall lengthening of the life span), dramatically impacted population growth. Between 1950 and 1990 the global population doubled from 2.5 billion to more than 5.5 billion people. Another billion people are estimated to have been added between 1990 and 2000. Furthermore, Earth's population is expected to nearly double one more time before stabilizing.

The growing number of people increases demands on natural resources. More people require more food, fuel, clothing, and other necessities for life. All of these must be supplied from the planet's resources and from the sun's energy. These facts—combined with the realization that the earth's resources are limited, not infinite—pose serious questions about rapid population growth. Can the world's resources support its population and maintain the environment, or will human needs overwhelm Earth's capacity to provide?

Most scientists believe that population growth will eventually cease. As a nation's economy develops and its standards of living rise, its population generally stabilizes. This "demographic shift" has already been observed in highly industrialized countries such as Japan and the nations of Western Europe where the fertility rate is generally at or below replacement level. Similarly, while most developing nations are still experiencing population growth, the rate of growth has peaked and is now on the decline everywhere except in Africa. The United Nations projects that this slow decline in growth rates will continue until all regions of Earth have developed high standards of living, experienced a demographic shift, and had their populations stabilize. By the time this occurs, however, Earth's population is expected to have reached 12 billion or more.

It is not known if Earth's resources can support a population of 12 billion. Some feel that there is evidence that population growth is already pressing, or has exceeded, the capacity of natural resources in many areas. Environmentalists and others warn that, without conservation, resource protection, and drastic measures to curb population, humankind and planet Earth might suffer serious harm. Other analysts do not share these opinions, however. They point to the fact that, at many times throughout history, there have been those who claimed that industrial development could not be sustained. For example, at one time it was feared that, as the world ran out of whale oil for lamps, great cities would be left in darkness. Others predicted that when the earth's supply of coal was exhausted, the industrial economy would collapse. In these and other cases, new technologies averted the expected disasters.

THE IMPACT OF ENVIRONMENTAL PROTECTION ON THE U.S. ECONOMY

How Government Regulations Work

Since federal and state governments began actively protecting the environment in the 1970s, they have acted primarily by creating rules (called regulations) that say how Americans can affect the environment around them. In order to get people and organizations to comply with these regulations, the government fines, imprisons, or otherwise punishes those who violate them.

Most of these rules are aimed at controlling the behavior of businesses and industries, as their behaviors are much easier to monitor and control than those of indi-

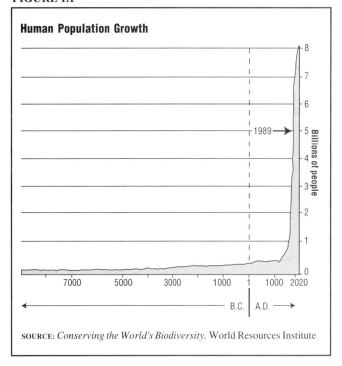

FIGURE 1.1

Human Population Growth

SOURCE: *Conserving the World's Biodiversity*. World Resources Institute

vidual citizens. So, for example, to reduce air pollution, the government might regulate the lawn mower industry by telling it not to make lawn mowers that release more than a certain amount of air pollution. This is much easier for the government than the alternative: checking how much pollution is released when an individual mows his or her lawn, and punishing that person if it is too much.

Attitudes of Business Towards Environmental Regulation

All environmental regulations interfere with how businesses would otherwise operate. They force businesses to design their products differently, install special machinery in their factories, or even stop certain activities entirely. In the most extreme cases, entire industries might be shut down, such as when the government determined that the chemical DDT, once widely used as a pesticide, was too hazardous to human health to be used at all.

The changes required to comply with environmental regulations almost always result in a business making less profit than it otherwise would. This is especially true for industries that extract natural resources, such as mining and logging; for industries that produce lots of pollution, such as electrical power generation; and for industries whose products are potentially hazardous, such as the chemical industry. Compliance with governmental regulations is a very significant cost item for some industries and, according to Department of Commerce estimates, may represent 10 percent or more of their total costs.

It is not surprising that businesses and business leaders object to the increased costs that environmental

TABLE 1.1

Environmental Industry Growth

REVENUE ($ BILLIONS) & GROWTH (AVERAGE ANNUAL)

EBI's MARKET SEGMENTS	1996	1998	2000	AVG. GROWTH (95-2000)
Services				
Analytical Services	1.2	1.2	1.2	0.6% →
Wastewater Treatment Works	24.0	25.9	27.8	3.7% ↗
Solid Waste Management	33.9	35.8	37.4	2.5% ↗
Hazardous Waste Management	6.0	5.6	5.1	-3.8% ↘
Remediation/Industrial Services	8.6	8.8	8.8	0.6% →
Consulting & Engineering	15.2	15.6	15.8	1.2% ↗
Equipment				
Water Equipment & Chemicals	17.5	19.0	20.4	4.1% ↗
Instruments & Information Systems	3.1	3.4	3.7	4.2% ↗
Air Pollution Control	15.7	16.2	12.5	2.2% →
Waste Management Equipment	12.0	12.2	12.5	0.6% →
Process & Prevention Technology	0.8	1.0	1.3	10.0% ↗
Resources				
Water Utilities	26.4	28.2	29.9	3.2% ↗
Resource Recovery	14.3	15.6	16.6	3.6% ↗
Energy Sources	2.4	2.7	3.1	6.0% ↗
Total Industry	**181.1**	**199.4**	**210.7**	
Average Growth Rate	1.2%	2.7%	2.2%	

SOURCE: *U.S. Environmental Industry, 1998.* Environmental Business International, Inc.

regulations bring. Some business leaders claim that environmental regulations will make their businesses unprofitable by driving up prices and production costs, forcing them to close plants and lay off workers if not shut down entirely. As early as the 1970s, these sentiments were finding support among politicians and the general population, and an anti-regulatory movement developed (see below).

DOES ENVIRONMENTAL PROTECTION DESTROY JOBS? By the end of the twentieth century the large-scale layoffs that some business people predicted had not come to pass. Michael Renner, in *State of the World 2000* (Worldwatch Institute, W.W. Norton Company, New York, 2000), states that job loss due to environmental regulation has been relatively limited. According to Renner, at least as many people have gained jobs due to environmental regulations as have lost them. He points out that environmental regulations have led to the creation of an entirely new industry that earns its profits by helping other businesses comply with regulations, mostly by helping them deal with pollution.

Although the number of people who lose their jobs may be outnumbered by those who gain new jobs, some individuals are certainly hurt by environmental regulations. Renner contends that policy changes intended to protect the environment must have a clear and pre-determined schedule. This way workers will know in advance what is expected of them, what jobs will be in demand, and can seek training to get them into those positions.

Even for those who accept this positive view of the overall effects of environmental regulation on business, it is unquestionable that some businesses and their employees are badly hurt by environmental regulations. As long as this remains the case, attempts to expand environmental regulation will certainly be met with opposition from those businesses and people who stand to lose their livelihood because of it.

The Business of Environmental Protection

In order for the United States and other nations to meet their environmental goals, an environmental protection industry has emerged. Its activities include pollution control, waste management, cleanup of contaminate sites, pollution prevention, recycling, and production of energy from alternative sources (solar and wind energy).

According to Environmental Business International, Inc. (EBI), a private organization that offers business and market information to the environmental industry and is the source of the most comprehensive data available on the industry, the United States market for pollution control technologies and other environmental services topped $210 billion in 2000. This was an increase of 5.7 percent over 1998. (See Table 1.1.) The majority of the companies in the environmental industry are in the public sector, providing potable (drinkable) water and wastewater treatment services. Among the private sector firms, the largest portion provides solid waste management services to communities. The environmental industry is generating many new jobs, both in traditional industries such as steel and iron production, and in emerging industries such as wind energy.

According to EBI, developed nations represent about 90 percent of the entire pollution control industry; the United States alone accounts for approximately 40 percent of the industry. Pollution control markets are, however, growing rapidly in other countries. Asian countries (excluding Japan) are projected to account for more than 7 percent of the world total by 2001, making it the fastest growing market worldwide. Plagued by heavy pollution, China is rapidly increasing its investments in pollution control and prevention.

Expansion in Asian markets will greatly depend on the region's economic status. If the economy worsens, it may lead governments to give the environment less priority. Eastern Europe and Russia are examples. They have a legacy of abusive environmental practices that increase their environmental needs but a lack of funding that limits an otherwise huge market for cleanup and new technology.

The Office of Technology Policy of the U.S. Department of Commerce, in *The U.S. Environmental Industry: Meeting the Challenge* (1998), found that the United States environmental industry compared favorably, in terms of revenue, to such well-established industries as paper and allied products, petroleum refining, and aerospace. Its revenues are nearly as large as the motor vehicle and car body industries. The environmental industry employed more people than many well-established industries—it was larger than the chemicals and allied products industry, the paper industry, or the motor vehicles and car bodies industry. The report noted that the environmental industry is in a transition phase. In the past it focused on remedial cleanup; in the future it will be focused more upon prevention.

THE ANTI-REGULATORY MOVEMENT

Ever since the mid-1980s, dissatisfaction with government regulation has grown. In 1994 the newly elected Republican-controlled Congress attempted to strike down a wide variety of federal regulations, including environmental regulations that they considered overly burdensome. Bills were introduced to relax regulations established under the Clean Water Act, Endangered Species Act, the Superfund toxic-waste cleanup program, the Safe Drinking Water Act, and other environmental statutes. Much of that legislation ultimately failed to pass. However, congressional budget cuts for the agencies responsible for carrying out these acts meant that many of the laws were not strongly enforced.

Several factors contributed to this reaction to federal regulation. During the early days of the environmental era the United States was experiencing a post-World War II economic boom, leading Americans to regard regulatory costs as sustainable. During the 1970s and 1980s, as economic growth slowed, wages stagnated and Americans became uncertain about the future. An increasing number of Americans started to question the costs of environmental protection. They began to pay more attention to the business leaders and politicians who had claimed since the beginning of the environmental protection movement that regulations would hurt the economy and cost people jobs. Although the national economy improved tremendously in the 1990s, it did not eliminate people's concerns about the potential negative effects of environmental regulations on the economy.

The impact of environmental regulations on private property use has also played a very important role in the change in attitudes. The Endangered Species Act and the wetlands provisions of the Clean Water Act spurred a grassroots "private property rights" movement. Many people became concerned that these acts, and other legislation, would allow the government to "take" or devalue properties without compensation. For example, if federal regulations prohibited construction on a plot of land that was protected by law, then the owner of that land often felt that the government was unfairly limiting the use of his or her property. At the very least, the owner wanted government compensation for decreasing the monetary value of the land.

Finally there are those who feel that environmental regulation by the government, while not necessarily bad, has gone too far. Many people feel that the federal government has overstepped its authority and should allow state and local governments to make their own rules on environmental issues. Similarly some people feel that existing regulations are too strict and should be relaxed in order to generate economic growth.

LITIGATION AND ENVIRONMENTAL POLICY

The courts have been an important forum for developing environmental policy because they allow citizens to challenge complex environmental laws and to affect the decision-making process. Both supporters and opponents of environmental protection have successfully used the courts to change environmental policy and law. Successful challenges can force the legislature to change laws or even have the law suspended as unconstitutional. A lawsuit can also be filed to seek compensation for harm to a person, property, or an economic interest. Sometimes lawsuits have prompted the creation of entirely new laws such as the federal Superfund law (1980) and the Toxic Substances Control Act (1976). Even the threat of a lawsuit, and the bad publicity it can bring, is sometimes enough to get a business or the government to change its behavior.

There are many different situations under which an individual or organization can go to court over environmental laws and regulations. One common occurrence is for an individual or group to sue the government in order to block a law or regulation from going into effect. For example, when the government halted logging in Northwest forests because of threats to endangered owls, logging companies fought to halt enforcement of those protections because that would decrease the industry's income and cause the loss of jobs.

Some lawsuits are filed not to block an environmental law or regulation from going into effect but because the claimants feel that the government owes them compensation for the negative effects of the law. For example, in 1986, David Lucas bought two residential lots on a South Carolina barrier island. He planned to build houses on these lots, just as had been done on other nearby lots. At the time he bought the land this was entirely legal but, in 1988, South Carolina passed the Beachfront Management Act. Designed to protect the state's beaches from erosion, it prohibited new construction on land in danger of erosion, which included the land Lucas owned.

Lucas went to court claiming that the Beachfront Management Act had violated the Fifth Amendment of the Constitution of the United States by preventing him from building on his property. The Fifth Amendment states, among other things, that no person's "private property shall be taken for public use, without just compensation." Lucas argued that preventing him from building on his property was equivalent to taking it, and so the government of South Carolina had to compensate him for it. On June 29, 1992, the United States Supreme Court, in *Lucas v. South Carolina* (60 LW 4842), agreed with Lucas in a 7–2 decision, and South Carolina was ordered to compensate him.

Situations in which supporters of environmental protection have filed lawsuits are when they feel the government is not properly enforcing the law. Environmental groups like the Sierra Club and Greenpeace have sued the government on many occasions to compel it to officially recognize certain species as endangered.

ENVIRONMENTAL JUSTICE—AN EVOLVING ISSUE

All Americans deserve clean air, pure water, land that is safe to live on, and food that is safe to eat. Although there has been some significant progress made, some communities continue to bear a disproportionate burden of pollution.
—Carol M. Browner, U.S. EPA Administrator

The so-called environmental justice issue stems from concerns that racial minorities are disproportionately subject to environmental hazards. The EPA defines environmental justice as "the fair treatment and meaningful involvement of all people regardless of race, color, natural origin, or income with respect to the development, implementation, and enforcement of environmental laws, regulations and policies."

The environmental justice movement gained national attention in 1982 when a demonstration took place against the building of a hazardous waste landfill in Warren County, North Carolina, a county with a predominantly black population. A resulting 1983 congressional study found that, in three of four landfills surveyed, blacks made up the majority of the population living nearby and at least 26 percent of the population in those communities was below the poverty level. In 1987 the United Church of Christ published a nationwide study, *Toxic Waste and Race in the United States*, reporting that race was the most significant factor among the variables tested in determining locations of hazardous waste facilities.

A 1990 EPA report (*Environmental Equity: Reducing Risk for All Communities*) concluded that racial minorities and low-income people bore a disproportionate burden of environmental risk. These groups were exposed to lead, air pollutants, hazardous waste facilities, contaminated fish, and agricultural pesticides in far greater frequencies than the general population.

In 1992 the Environmental Protection Agency (EPA) established the Office of Environmental Justice to address environmental impacts affecting minority and low-income communities. In 1994 President Clinton issued Executive Order 12898 (*Federal Actions to Address Environmental Justice in Minority Populations and Low-Income Populations*), requiring federal agencies to develop a comprehensive strategy for including environmental justice in their decision-making.

Examples of environmental injustice are claims that:

- Low-income Americans, especially minorities, may be more likely than other groups to live near landfills, incinerators, and hazardous waste facilities.

- Low-income and black children often have higher than normal levels of lead in their blood.

- Greater proportions of Hispanic and black Americans than whites live in communities that fail to meet air quality standards.

- Higher percentages of hired farm workers in the United States are minorities. It has been estimated that more than 300,000 farm workers may suffer pesticide-related illnesses each year.

- Low-income and minority fishermen who use fish as their sole source of protein are generally not well informed about the risk of eating contaminated fish from certain lakes, rivers, and streams.

In Harm's Way

In 1997 residents of Kennedy Heights, a Houston, Texas, neighborhood of about 1,400 people, complained about a variety of illnesses such as cancer, tumors, lupus (an autoimmune disease), and rashes. They eventually discovered that their homes, built 30 years ago, were sitting on top of a number of oil pits that had been abandoned in the 1920s. Some tests of the municipal water found traces of crude oil, and residents believed oil sludge from the pits had seeped into the water supply and damaged their health. To make matters worse, their homes lost virtually all their resale value once it became known that the land was contaminated.

The predominantly black homeowners had not been told that their property sat over an abandoned oil dump. They blamed Chevron Oil Company which owned the property for a time after acquiring it from Gulf Oil. Chevron denied contamination could have caused any illness. Accusations of environmental racism mounted, attracting ever-wider attention. Reverend Jesse Jackson visited the site in 1997 to call for a boycott of Chevron products. In March 1999 Chevron agreed to pay $8 million into a fund for distribution to Kennedy Heights claimants.

In 1998 some of the residents of Chester, Pennsylvania, sued the Pennsylvania Department of Environmental

Protection. In their case (*James M. Seif v. Chester Residents Concerned Citizens et al.*), they claimed that the department had violated Title IV of the Civil Rights Act of 1964 (PL 88-352) by issuing a solid waste permit for a facility in a minority community in Chester called Chester Heights. The residents claimed the waste permit violated their civil rights because, while only two permits had been issued in adjoining non-minority communities, five such permits had been approved for their community. Although the case was dismissed when the treatment facility withdrew its application, the suit demonstrated the possibility of claiming environmental discrimination under Title IV. It should be noted, however, that few environmental racism cases have been successful because it is difficult to prove that racial discrimination is the cause of particular environmental policies.

"NEW" CRIME—ENVIRONMENTAL CRIME

As recently as 20 years ago, very few Americans understood that harming the environment could be considered a "crime." The American public began to recognize the seriousness of environmental crimes. Most Americans believe that damaging the environment is a serious crime and that corporate officials should be held responsible for environmental offenses committed by their firms. Even though the immediate consequences of an offense may not be obvious or severe, environmental crime is a serious problem and does have victims—the cumulative costs in damage to the environment and the toll to humans in illness, injury, and death can be considerable.

Law enforcement agencies generally believe that successful criminal prosecution—even the threat of it—is the best deterrent to environmental crime. Under the dual sovereignty doctrine, both state and federal governments can independently prosecute environmental crimes without violating the double jeopardy or due process clauses of the American Constitution.

Federal criminal enforcement has grown from a misdemeanor penalty for dumping contaminants into waterways without a permit to a felony for clandestine (secret) dumping. Several federal laws now include criminal penalties. Companies, their officials, and staff can be prosecuted for knowingly violating any one of a number of crimes. Such crimes include transporting hazardous waste to an unlicensed facility, storing and disposing of hazardous waste without a permit, failing to notify of a hazardous substance release, falsifying documents, dumping into a wetland, or violating air quality standards.

In 1997 the EPA levied fines of $264.4 million and referred 704 cases to the Justice Department. That year federal prosecutors filed 207 civil and 446 criminal enforcement actions in U.S. district courts for violations of environmental (64 percent) or wildlife (36 percent)

FIGURE 1.2

Federal Environmental and Wildlife Protection Acts

Environmental protection acts

Clean Air Act — to prevent the deterioration of air quality. (42 U.S.C. §§ 7401-7491)

Clear Water Act — to regulate the sources of water pollution. (33 U.S.C. §§ 1251-1370)

Comprehensive Environmental Response, Compensation, and Liability (CERCLA) — to address problem of abandoned hazardous waste sites. (42 U.S.C. §§ 9601-9675)

Resource Conservation and Recovery Act (RCRA) — to protect human health and the environment from dangers associated with waste management. (42 U.S.C. §§ 6901-6992)

Toxic Substances Control Act (TSCA) — to regulate chemical substances in which the public or environment may become exposed. (15 U.S.C. §§ 2601-2671)

Act to Prevent Pollution by Ships (APPS) — to address the discharge of harmful substances into the oceans. (33 U.S.C. §§ 1901-1950)

Emergency Planning and Community Right to Know Act (EPCRA) — to protect the environment from pollution. (42 U.S.C. §§ 11001-11050)

Wildlife acts

Endangered Species Act (ESA) — to conserve the various species of fish, wildlife, and plants facing extinction. (16 U.S.C. §§ 1531(b))

Bald and Golden Eagle Protection Act (BGEPA) — to provide a program for the conservation of bald and golden eagles. (16 U.S.C. §§ 668)

Migratory Bird Treaty Act (MBTA) —to protect migratory birds during their nesting season. (16 U.S.C. §§ 707)

Lacey Act — to control the trade in of exotic fish, wildlife, and plants into the United States. (16 U.S.C. §§ 3372)

SOURCE: *Federal Enforcement of Environmental Laws 1997.* Bureau of Justice Statistics

---

laws. Figure 1.2 shows the federal environmental and wildlife protection acts.

In 1997 almost half of defendants charged violated the Clean Water Act, followed by the Resource Conservation and Recovery Act (RCRA) and the Clean Air Act. (See Table 1.2.) A monetary award or settlement was paid in 74 cases; the average award was about $2.5 million. Cases involving the RCRA produced the largest settlements—an average of $5.4 million. (See Table 1.3.)

In 1997 federal officials reported that the sale of contraband Freon (a refrigerant used in air conditioning systems) had become more profitable than cocaine. (In Mexico, which shares a 2,000-mile border with the United States, Freon is still legal to manufacture and export but it is banned in the United States.) A court case involving Freon brought the first-ever U.S. felony conviction based on environmental crimes.

Since the 1970s, environmental laws have become more complicated. The strictness of those laws may have

TABLE 1.2

TABLE 1.3

Federal Environmental Enforcement Actions Initiated, by Type of Violation and Enforcement Action, 1997

Type of violation	Total	Type of enforcement action		
		Administrative	Civil[a]	Criminal[b]
Total	4,129	3,427	207	446
Environmental protection	3,842	3,427	204	211
Clean Air Act	457	391	35	31
Clean Water Act	1,812	1,642	62	108
RCRA	505	423	23	59
CERCLA	391	305	82	4
TSCA	191	185	0	6
Other	486	481	2	3
Wildlife	287	—	3	235

—Statistics not available.

[a]Represents filing by U.S. Attorneys in U.S. district court only. Statistics describing administrative actions for wildlife and conservation offenses were not available.

[b]Criminal actions include only those offenses classified as felonies or Class A misdemeanors.

SOURCE: *Federal Enforcement of Environmental Laws 1997.* Bureau of Justice Statistics

Monetary Award/Settlement in U.S. District Court Cases Charging a Civil Violation of Federal Environmental Law, 1997

Type of violation	Cases with a monetary award/settlement		
	Total	Mean	Median
Total	74	$2,454,447	$287,500
Environmental protection	74	2,454,447	287,500
Clean Air Act	15	520,039	275,000
Clean Water Act	24	2,877,190	98,927
CERCLA	22	1,815,722	440,000
RCRA	12	5,402,087	279,800
Other	1	—	—
Wildlife	0	—	—

—Not calculated, too few cases.

SOURCE: *Federal Enforcement of Environmental Laws 1997.* Bureau of Justice Statistics

contributed to increasing the incidence of environmental violations. First, many businesses have found compliance increasingly expensive, and many are simply avoiding the costs even if it means violating the law. These businesses consider the penalties just another "cost of doing business." Second, businesses and their legal counsel are becoming increasingly "savvy" in avoiding prosecution through the use of dummy corporations, intermediaries, and procedural techniques.

EXPOSING POLLUTERS TO PUBLIC PRESSURE

In 1997 the Clinton Administration and the EPA initiated the Sector Facility Indexing Project (SFIP), a project to inform the public about industries that pollute the environment. The latest in an ambitious campaign to expand "right-to-know" initiatives, SFIP publishes "pollution profiles" on-line (http://es.epa.gov/oeca/sfi/) on five industrial sectors—oil production, steel, other metals, autos, and paper—in hopes of exposing polluters to possible pressure from an informed public. The profiles include the number of federal inspections, episodes of noncompliance, penalties imposed in recent years, releases of pollutants, pollution spills, injuries and deaths, toxicity of chemicals released, ratios of pollution to production, and demographics (racial and income breakdowns) of the population within a three-mile radius of a plant. As of 2000, SFIP profiled approximately 640 individual facilities.

THE INTERNATIONAL RESPONSE TO ENVIRONMENTAL PROBLEMS

Environmental issues have never been neatly bound by national borders. Activities taking place in one country often have an impact on the environment of other countries, if not that of the entire Earth. In fact, many of the most important aspects of environmental protection involve things which are not located within any particular country, such as the oceans, or belong to no one, such as the atmosphere. In an attempt to deal with these issues, the international community has held a number of conferences and concluded numerous treaties.

A First Step—The Stockholm Conference

In 1972 the United Nations met in Stockholm, Sweden, for a conference on the environment. Delegates from 113 countries met, each reporting the state of his or her nation's environment—forests, water, farmland, and other natural resources. The countries represented essentially fell into two groups. The industrialized countries were primarily concerned about how to protect the environment by preventing pollution and overpopulation and conserving of natural resources. The less developed nations were more concerned about problems of widespread hunger, disease, and poverty they all faced. They did consider the environment very important however, and were willing to protect it as long as doing so did not have a major negative economic impact on their citizens.

By the end of the two-week meeting, the delegates had agreed that the human environment had to be protected, even as industrialization proceeded in the less developed countries. They established the United Nations Environment Program (UNEP), which included Earthwatch, a program to monitor changes in the physical and biological resources of the earth. The most important outcome of the conference was awareness of Earth's ecology as a whole. For the first time in global history, the environmental problems of both rich and poor nations were put in perspective. General agreement emerged to protect natural resources, encourage family planning and population control, and protect against the negative effects of industrialization.

Some Difficulties Facing International Environmental Protection

Since the 1972 conference, hundreds of environmental treaties have been signed. From this one might assume that great progress has been made, but this is not truly the case. Most experts believe that international cooperation is not keeping pace with the world's ever-growing interdependence and the rapidly deteriorating condition of much of the environment. Carbon dioxide levels are at record highs, water shortages exist around the world, fisheries are becoming depleted, and many scientists are warning that large numbers of species are becoming extinct. The reason for this is that, while nations agree on the fact that the environment must be safeguarded, they disagree sharply on the issue of what role each nation should play in protecting it.

Less developed nations are generally unwilling to alter their laws and economy to end environmentally destructive behaviors because a shift to environmentally friendly behaviors is too expensive, they claim, for their economies to handle. Yet the richer, industrialized, nations generally refuse to alter their own behavior unless the less developed nations do so as well. Their reason is not so much the cost of change but they believe it is unfair that the less developed nations want them to carry most of the burden of environmental protection.

The less developed nations respond by pointing out that the industrialized nations became rich by using the very same practices they now want the less developed nations to stop. They claim it is unfair to be expected to limit their economic development in ways that the industrialized nations themselves never would have done.

This is a difficult disagreement but not an impossible one to resolve. When both sides are willing to compromise, agreements can be reached. These compromises usually require the industrialized nations to make bigger changes in their behavior, and to help the less developed nations change without too negative an impact on their economy.

Even agreements like these face many obstacles, however. Environmental agreements seldom include a means of enforcement but rely instead on each signing country to keep its word. Faced with the actual, immediate costs of implementing environmental agreements, many countries eventually back down from their commitments.

The 1992 Earth Summit

The 1992 Earth Summit in Rio de Janeiro is an example of a conference where compromises were made and agreements reached but little change actually resulted.

Mounting global concern for the environment prompted the United Nations to convene a summit meeting in Rio de Janeiro in 1992. Approximately 180 governments participated, making it one of the largest and most important environmental summits ever. As with environmental summits in the past, the conference was split between industrialized and developing nations.

The main accomplishments of the Earth Summit were pacts on global warming (see Chapter 2) and biodiversity (see Chapter 9). President George Bush signed the global-warming treaty for the United States. President Bill Clinton signed the biodiversity treaty in 1994. These agreements came about largely because the industrialized nations also agreed to commit 0.7 percent of their gross national products to assist developing countries by the year 2000.

Problems arose soon after the summit ended, however. Participating countries submit annual reports to the 53-nation United Nations Commission on Sustainable Development (CSD), a standing body set up to implement the Rio agreements. The CSD concluded in 1994 that most countries were failing to provide the money and expertise necessary to implement the plans set at Rio. Chairman of the CSD, Klaus Toepfer of Germany, reported that the world's efforts to finance the goals had fallen "significantly short of expectations and requirements."

By 1996 a number of national governments, including the United States, had prepared plans for environmental protection and submitted them to the CSD. Hundreds of municipalities had also written plans of action. The CSD once again found, however, that other issues had crowded out environmental concerns. As developed and less developed nations alike worried about the potential effects of implementing the Rio agreements, they found reasons to delay implementation and reduce funding for those that had been implemented.

THE WORLD TRADE ORGANIZATION. The World Trade Organization (WTO) is an international organization whose purpose is to encourage free trade between its members. Most of the world's nations are members. Although the WTO was officially founded in 1995, it is the result of decades of international cooperation under the General Agreement on Tariffs and Trade (GATT). The WTO continues to administer the free trade system established under GATT.

One of the primary missions of the WTO is to eliminate barriers to free trade. Doing so can have a negative effect on environmental protection, however, because laws designed to protect the environment often have the effect of restricting trade. If the WTO finds that a member nation is restricting trade in violation of GATT, then other members are permitted to raise their tariffs (import taxes) on goods from that nation until the barriers to trade are eliminated. Most of the time nations quickly change their laws to eliminate barriers to trade, rather than suffer high taxes.

Since it has forced many environmental laws to be weakened over the years, the WTO is greatly disliked by many environmentalists in the United States. Also, there are organizations that think its power over internal U.S. affairs is too great. When the WTO met in Seattle, Washington, in 1999, tens of thousands of activists protested in the city, including environmental activists. This massive protest succeeded in overshadowing the WTO meeting itself, and drew public attention to the problems, environmental and otherwise, with free trade organizations. This was due in no small part to the violent rioting and property damage caused by some protesters.

Americans are not the only ones who take issue with some of the WTO's actions regarding the environment. For example, Europeans opposed to genetically modifying food, a procedure in wide-spread use in the United States by the year 2000, wanted restrictions placed on the sale of U.S. food in Europe, but such restrictions would violate the GATT and invite retaliation by the United States.

THE NORTH AMERICAN FREE TRADE AGREEMENT. The North American Free Trade Agreement (NAFTA), signed in 1994, is another major free trade agreement with the potential to negatively impact environmental protection in the United States. Members are the United States, Canada, and Mexico, and its purpose is to eliminate trade barriers between these three countries. This includes most tariffs, investment restrictions, and import quotas. While its scope is much smaller than the WTO, NAFTA's impact on the three member countries is even greater than that of GATT.

A significant difference between NAFTA and GATT is that NAFTA is the first treaty of its kind ever to be accompanied by an environmental protection agreement. To discourage countries from weakening environmental standards in the name of increasing foreign trade, the United States, Canada, and Mexico signed the North American Agreement on Environmental Cooperation (NAAEC). Under NAAEC, a member country can be challenged if it or one of its states fails to enforce its environmental laws. A challenge can be brought by one of the member nations, or any interested party (such as an environmental protection group) can petition the NAAEC commission. If the commission finds a member country is showing a "persistent pattern of failure...to enforce its environmental law" then it may fine that country. If the fines are not paid, the other members are permitted to suspend NAFTA benefits in an amount not exceeding the amount of the assessed fine.

Even with NAAEC, some U.S. environmentalists and state officials feared that NAFTA could result in the weakening of numerous humane laws and the reversal of 30 years of advances in animal protection and environmental

TABLE 1.4

Opinion Regarding the Seriousness of Various Social Problems
Next, I am going to read a list of problems currently facing our country. For each one please tell me how serious of a problem you consider it to be for our country -- extremely serious, very serious, somewhat serious, or not serious at all. [RANDOM ORDER: A-G]

	Extremely serious	Very serious	Somewhat serious	Not serious at all	No opinion
A. Hunger and homelessness					
2000 Apr 3-9	24%	42	30	4	*
B. Crime and violence					
2000 Apr 3-9	33%	49	17	1	*
C. Environmental problems					
2000 Apr 3-9	17%	38	39	5	1
D. Poor health care					
2000 Apr 3-9	29%	38	28	5	*
E. Drug use					
2000 Apr 3-9	38%	45	15	2	*
F. Racial conflict					
2000 Apr 3-9	18%	33	43	5	1
G. Illegal immigration					
2000 Apr 3-9	15%	30	45	8	2

SOURCE: Gallup Poll, 4/17/2000. Gallup Organization, 2000

cleanup. In response, Congress provided more protection for state laws and included more environmental language than in any previous trade agreement. The implementing legislation for NAFTA in the United States allows states much input and requires notification of actions that may affect them. In addition, during NAFTA discussions, the Border Environment Cooperation Commission (BECC) and the North American Development Bank (NADBank) were created. Independent of NAFTA itself, these agencies are intended to ensure that policy discussions are open and fairly enforced, to consider allegations that a country is not enforcing environmental laws, to help communities finance environmental infrastructures, and to resolve disputes, particularly those that cross borders.

Despite all these measures designed to make sure that NAFTA does not trample on environmental protection, environmentalists still see the need for concern. They point out that United States laws designed to protect certain animals could be challenged as barriers to free trade under NAFTA. They also point to the increased pollution in Mexico and along its border with the United States that has resulted from the increase in trade between those two countries.

PUBLIC OPINION ON THE ENVIRONMENT

A Decline in Concern in the 1990s

An April 2000 Gallup Poll found that 83 percent of Americans agreed with the environmental movement, including 43 percent who "strongly agreed." Those views had remained essentially the same over the preceding decade. When respondents were asked to rate the seriousness of a variety of problems facing the nation, however,

TABLE 1.5

Opinion Regarding the Seriousness of Various Environmental Problems

Percentage expressing a "great deal" of concern

	2000	1999	1989	Since 1989 Change
The loss of tropical rain forests	51%	49%	42%	+9
The "greenhouse effect" or global warming	40	34	35	+5
Contamination of soil and water by radioactivity from nuclear facilities	52	48	54	-2
Damage to the earth's ozone layer	49	44	51	-2
Air pollution	59	52	63	-4
Contamination of soil and water by toxic waste	64	63	69	-5
Pollution of rivers, lakes and reservoirs	66	61	72	-6
Ocean and beach pollution	54	50	60	-6
The loss of natural habitat for wildlife	51	51	58	-7
Acid rain	34	29	41	-7
Pollution of drinking water	72	68	--	--
Extinction of plant and animal species	45	--	--	--
Urban sprawl and loss of open spaces	42	--	--	--

SOURCE: Gallup Poll, 4/17/2000. Gallup Organization, 2000

TABLE 1.6

Opinion Regarding Impact of Various Social Movements

	Great deal/ Moderate amount (combined) %	A great deal %	A moderate amount %
1. Civil Rights Movement	85	50	35
2. Women's Rights Movement	82	42	40
3. Gun-Control Movement	74	38	36
4. Abortion Rights Movement	75	36	39
5. Environmental Movement	76	30	46
6. Consumers Rights Movement	67	21	46
7. Gay and Lesbian Rights Movement	59	24	35
8. Animal Rights Movement	50	15	35

SOURCE: Gallup Poll, 4/18/2000. Gallup Organization, 2000

they ranked the environment well below crime and violence, drugs, hunger and homelessness, and health care. (See Table 1.4.)

The poll went on to measure attitudes about 13 specific environmental issues, ten of which have been studied since 1989. Public concern had either declined or remained the same for eight of the issues. The greatest decline in concern was for acid rain, loss of wildlife habitat, and water pollution. Concern rose for the loss of tropical rain forests and global warming. (See Table 1.5.) The leveling off of concern in 2000, compared to the peak level of 1989, may be due to decreased visibility of environmental problems and less media attention to environmental issues.

According to the 2000 Gallup Poll, Americans seemed to feel that more has been done about the environment during the 1990s than during the 1980s. When asked how much progress has been made in addressing certain issues, 26 percent said "a great deal of progress" and 64 percent said "only some progress." In 1990 only 14 percent felt that a "great deal of progress" had been made, and 63 percent said "only some progress."

When Americans were asked to rate the impact that various social movements have had on the nation's poli-

cies, 76 percent of respondents rated the environmental movement as having "a great deal" or "a moderate amount" of success. They ranked only the civil rights (85 percent) and women's rights (82 percent) movements as having more impact, although the abortion rights (75 percent) and gun-control movements (74 percent) were statistically even with the environment. The movements cited as having less impact on public policy were consumer rights (67 percent), gay and lesbian rights (59 percent), and animal rights (50 percent). (See Table 1.6.)

Americans do many things in varying degrees to protect the environment. When questioned about those behaviors, 90 percent recycled, 83 percent tried to reduce energy and water usage and avoided environmentally harmful products, and 73 percent claimed they bought environmentally beneficial products. Questions about direct support for environmental groups and causes found that 40 percent reported contributing to environmentalist groups, 31 percent had signed petitions for environmentalist causes, and 20 percent had attended a meeting about an environmental issue. Twenty-eight percent claimed they have voted or worked for a candidate because of that candidate's position on environmental issues. Other activities included contacting a public official about an issue (18 percent) or participating actively in an environmental group (15 percent).

Relatively few people reported engaging in forms of "green" consumerism, such as complaining to a business about products or policies that harm the environment (13 percent) or trading stocks based on the environmental record of a company (9 percent).

Generations Agree on "Green Issues"

A 2000 poll conducted for the Environmental Defense Fund, an environmental protection organization, compared the environmental attitudes of young adults and those of the baby boom generation. Although the pollsters anticipated significant differences between the

FIGURE 1.3

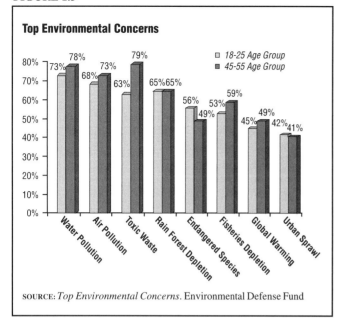

Top Environmental Concerns

SOURCE: *Top Environmental Concerns*. Environmental Defense Fund

FIGURE 1.4

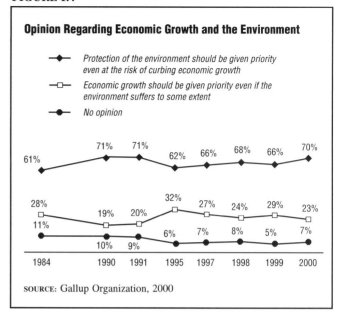

Opinion Regarding Economic Growth and the Environment

SOURCE: Gallup Organization, 2000

generations, the survey found much similarity. When asked what they considered the top environmental concerns, both generations ranked them similarly in order of importance. Toxic waste, water pollution, air pollution, and rain forest depletion were rated by both groups as the most pressing problems. (See Figure 1.3.)

The Environment or the Economy?

Much debate centers on whether protecting the environment or economic growth is more important. Since 1989, Gallup has asked Americans to choose between protecting the environment, even at the risk of curbing the economy, and economic growth, even if the environment suffers to some extent. The public has always claimed to favor the environmental alternative. In 1984, 61 percent favored protection of the environment, with 28 percent choosing economic growth and 11 percent expressing no opinion. By 2000, 70 percent of Americans selected the environmental position, 23 percent favored economic growth, and 7 percent had no opinion. (See Figure 1.4.)

The poll also found there were differences by age and political persuasion. Support for environmental protection was 71 percent of those aged 18 to 49 years, 51 percent of those 65 and over, 51 percent of political conservatives, and 81 percent of political liberals.

According to Wirthlin Worldwide, a Virginia-based marketing research and consulting company, another relationship may exist between the environment and the economy. In its report *Environmental Support Softens amid Economic Uncertainty* (McLean, Virginia, 1998), Wirthlin noted that environmental support in the United States generally moves in accord with the economy. The report compared unemployment rates with the data shown in Figure

1.5 and found a statistically significant inverse relationship between unemployment rates and support for environmental protection. The report concluded that when the economy is healthy, public concern for the environment runs high; when the economy is bad, the environment becomes a much less important issue in people's minds.

Polling American Consumers

A 1996 Hartman Group study (*Food and the Environment: A Consumer's Perspective*) of 2,000 household grocery buyers found that most Americans want Earth-sustainable products (agriculture that uses organic growing techniques to limit chemical use on crops). The study found that 45 percent were open to the idea of buying such products and 7 percent were eager to buy them. Many people who were only minimally involved with environmentalism claimed they would choose an Earth-sustainable product if given a reason, an opportunity, and a competitive price.

The 1998 Wirthlin report also asked Americans which industries caused or solved environmental problems. Respondents ranked chemical and oil and gas industries as the worst "causers" of pollution. The computer and natural gas industries ranked best. (See Table 1.7.)

What American Voters Say

According to an April 2000 survey conducted for the League of Conservation Voters (LCV), 64 percent of United States voters viewed environmental issues as very or somewhat important. Furthermore, 78 percent of respondents claimed they favored candidates who vote for the environment over those who vote for fewer government regulations. An overwhelming 92 percent said that environmental law-breakers should pay fines, while

FIGURE 1.5

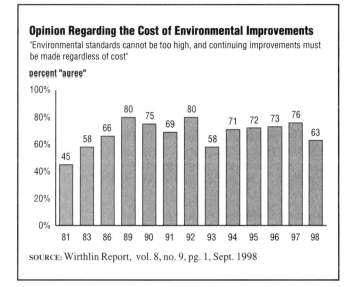

Opinion Regarding the Cost of Environmental Improvements

"Environmental standards cannot be too high, and continuing improvements must be made regardless of cost"

percent "agree"

SOURCE: Wirthlin Report, vol. 8, no. 9, pg. 1, Sept. 1998

TABLE 1.7

Trends in the Environmental Image of Selected Industries, 1993–1998
FIVE YEAR TREND

	INDUSTRY	1993 score*	1996 score*	1998 score*
best	computers	25	11	15
	natural gas	11	-9	7
	electrical utilities	n/a	n/a	-11
	furniture	-29	-19	-15
	steel	n/a	-40	-21
	forest/paper products	-28	-36	-34
	plastics	-38	-36	-35
	nuclear power	-42	-50	-43
	fast food	-69	-43	-45
	automotive	n/a	-50	-56
	oil/gasoline	-59	-69	-65
worst	chemicals	-59	-75	-71

* difference score = % who say this industry has SOLVED environmental problems minus % who say this industry has CAUSED environmental problems. Higher scores are better. "NEITHER" responses are not reflected in the scores.

SOURCE: Wirthlin Report, vol. 8, no. 9, pg. 2, Sept. 1998

89 percent believed that businesses that pollute should pay higher fees for dumping permits to help pay for environmental law enforcement.

More than 60 percent perceived pro-environment candidates as trustworthy and responsible. Those positive traits were not attributed to candidates who favored reducing environmental regulations on business. Deb Callahan, president of the LCV Education Fund, noted that the environment can be a key issue in political campaigns, because environmental concern is perceived by many voters to reflect positively on the character of candidates.

An April 2000 Gallup Poll asked respondents to rate the importance of environmental issues in deciding how they would vote in the 2000 presidential election. Only 26 percent reported environmental protection as an "extremely important" issue to their vote, although an additional 40 percent said it was "very important." Table 1.8 shows the importance of various issues to American voters in the 2000 race. The environment ranked about the same as gun control and tax reductions. It was less important to most Americans than education, health care, crime, Social Security, family values, and the economy, but ahead of foreign affairs, abortion, and campaign finance reform.

How Reliable are Polls on Environmental Issues?

Some experts suggest that opinion polls are an unreliable guide to how voters actually feel about environmental issues. Although polls of Americans indicate that concern for environmental issues is substantial, this same level of concern does not manifest itself when it comes to actual voting and purchasing decisions. Some observers suggest that people often claim in polls that they are interested in environmental issues because they are trying to give the pollster the answer that he/she wants to hear.

TABLE 1.8

Opinion Regarding Importance of Selected Issues in Vote for President

(Extremely + Very Important)

1. Education	89%
2. Health care	82
3. Crime	81
4. Social Security	80
5. Family values	79
6. The economy	77
7. Gun control	69
8. Environmental protection	66
9. Tax reductions	63
10. Foreign affairs	57
11. Abortion	53
12. Campaign finance reform	40

SOURCE: Gallup Poll, 4/17/2000. Gallup Organization, 2000

In other words, they are giving what they think is the "right" answer. In actuality, respondents may be more interested in other issues and, in the voting booth, may vote other than their poll answers would indicate.

ENVIRONMENTAL EDUCATION

Lacking Basic Knowledge

In 1997 the National Environmental Education and Training Foundation (NEETF), a private non-profit organization, conducted its sixth annual study of American adults' knowledge of environmental issues. The study, conducted by Roper Starch Worldwide, found that Americans supported environmental protection but were generally ignorant on the leading causes of pollution, energy sources, and the hazards in some household products.

FIGURE 1.6

The Earth, as seen from a U.S. satellite. (*U. S. National Aeronautics and Space Administration.*)

The study found that, although 72 percent of Americans say that water quality regulation has not gone far enough, about the same number do not know that the leading cause of water pollution is water runoff from farmland, parking lots, lawns, and streets. Twice as many respondents thought factories were the leading water polluters than knew it comes from runoff. Only one-third of respondents knew that most electricity is produced by burning coal, natural gas, and oil, while half thought that electricity was produced mostly by hydropower (which accounts for only 12 percent of energy in the United States). This may explain why lawmakers have difficulty engaging the public in solutions to global warming and air quality.

When the study asked if environmental education should be taught in schools, 95 percent of respondents said yes. This included those who did not do well on tests of environmental knowledge. Only 2 percent actively opposed teaching environmental education in the schools.

TEACHING ABOUT THE ENVIRONMENT IN SCHOOLS

Many states require schools to incorporate environmental concepts, such as ecology, conservation, and environmental law, into many subjects in all grade levels. Some even require special training in environmentalism for teachers. Since 1992, the EPA, with congressional authority to spend a total of $13 million on environmental education, gave grants to about 1,200 such projects.

Although this mandating of environmental education pleases environmentalists, and studies have shown that most Americans support environmental education, some people have protested. These critics claim most environmental education in the schools is based on flawed information, biased presentations, and questionable objectives. Critics also say it leads to brainwashing and pushing a regulatory mindset on students. Some critics contend that, at worst, impressionable children are being trained to believe that the environment is in immediate danger of catastrophe because of consumption, economic growth, and free-market capitalism.

NASA'S MISSION TO PLANET EARTH

In the past three decades, humans have changed the way they think about the planet. Missions to other planets have found those planets interesting and diverse but also sterile. Earth is unique; as far as is known, it is the only planet capable of sustaining life.

While environmental change is certain, many things remain unknown. Among those things is how humans alter Earth. One way to study the effects of humans on Earth as a whole is from space. Since 1991, NASA has undertaken a comprehensive program—Mission to Planet Earth (MTPE)—to study the planet as an environmental system. By using satellites and other tools focused on Earth, scientists hope to expand what is known about how natural processes affect humans and how humans may, in turn, impact the planet.

Mission to Planet Earth has three components: a series of Earth-observing satellites, an advanced data system, and teams of scientists—a total of 1,795 researchers—who study the data collected. Key areas of study include clouds, water and energy cycles, oceans, atmospheric chemistry, land surface, water and ecosystem processes, glaciers and polar ice sheets, and the solid earth. (Figure 1.6 shows Earth as photographed from a satellite.)

Phase I of MTPE was comprised of satellites, Space Shuttle missions, and various land- and airborne-based studies. Phase II began April 15, 1999, (delayed from 1998 because of technical problems) with the launch of Landsat 7, the centerpiece of MTPE, from Vandenberg Air Force Base in California. The first Earth Observing System (EOS) satellite, Landsat 7's goal is to take an estimated 250 photographs of Earth each day. These will be available not only for national security purposes but also to private sector, commercial, and academic segments of society to help answer pressing environmental issues in all these areas.

CHAPTER 2

THE GREENHOUSE EFFECT AND CLIMATE CHANGE

CLIMATE AND HUMAN EVOLUTION

Earth scientists are in the midst of a revolution in their understanding of how climate change occurred in the past. Researchers are discovering the geological and astronomical forces that have changed the planet's environment from hot to cold, wet to dry, and back again, over hundreds of millions of years. Dramatic climate change is nothing new for planet Earth. The climate of the past 10,000 years, during which civilization developed, is a mere blip in a much bigger history of climate change. In fact, Earth's climate will most certainly continue to go through dramatic changes—even without the influence of human activity. Most scientists believe that, if not for extreme climate shifts some 65 million years ago, most of the animals on Earth today, including humans, would probably not be here.

Astronomical cycles have caused the climate to fluctuate between long periods of cold lasting 50,000 to 80,000 years and shorter periods of warmth lasting about 10,000 years. Some scientists believe that these geologically rapid shifts between warm and cold were the catalyst behind human evolution, forcing humans to adapt physiologically and socially in order to cope with the changing climate.

THE WORLD CLIMATE

Earth's climate is a delicate balance of energy inputs, chemical processes, and physical phenomena. Temperatures on Venus are too hot for the human body; on Mars, a person would instantly freeze to death. This difference in temperature is due to the varying composition of each planet's atmosphere. All three planets receive huge quantities of solar energy, but the amount radiated back into space as heat depends upon the atmospheric composition of the particular planet. Some gases, such as carbon dioxide (CO_2) and methane (CH_4), absorb and maintain heat in the same way that glass traps heat in a greenhouse.

These gases in Earth's atmosphere allow temperatures to build up, keeping our planet warm and habitable. That is why the increased buildup of these and other gases caused by pollution is often called the "greenhouse effect." (See Figure 2.1.)

It is necessary, however, to distinguish between the "natural" and a possible "enhanced" greenhouse effect. The natural greenhouse effect creates a climate in which life can exist, causing the mean temperature of Earth's surface to be about 33 degrees warmer than it would be if natural greenhouse gases were not present. Without this process, Earth would be frigid and uninhabitable. An "enhanced" greenhouse effect, sometimes called the anthropogenic effect, refers to the possible increase in the temperature of Earth's surface due to human activity.

The scorching heat of Venus is the result of an atmosphere that is composed largely of carbon dioxide. The atmosphere of Earth, however, is nitrogen- and oxygen-based and contains only 0.03 percent carbon dioxide. This percentage has varied little over the past million years, permitting a stable climate favorable to life. The blanket of air enveloping Earth moderates its temperature and sustains life.

A Revolutionary Idea

Earth's atmosphere was first compared to a glass vessel in 1827 by the French mathematician Jean-Baptiste Fourier. In the 1850s British physicist John Tyndall tried to measure the heat-trapping properties of various components of the atmosphere. By the 1890s scientists had concluded that the great increase in combustion in the Industrial Revolution had the potential to change the atmosphere's load of carbon dioxide. In 1896 the Swedish chemist Svante Arrhenius made the revolutionary suggestion that human activities could actually disrupt this delicate balance. He theorized that the rapid increase in the use of coal that came with the Industrial

FIGURE 2.1

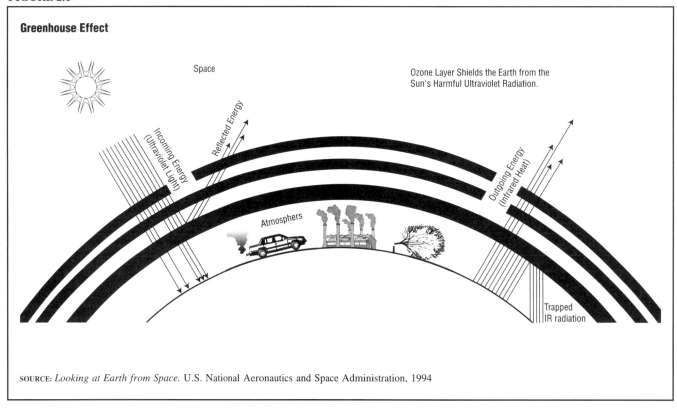

Greenhouse Effect

Space

Ozone Layer Shields the Earth from the
Sun's Harmful Ultraviolet Radiation.

Incoming Energy
(Ultraviolet Light)

Reflected Energy

Outgoing Energy
(Infrared Heat)

Atmosphers

Trapped
IR radiation

SOURCE: *Looking at Earth from Space.* U.S. National Aeronautics and Space Administration, 1994

TABLE 2.1

Greenhouse Gas Profile

Greenhouse Gas	Sources	Lifespan
Carbon di-oxide (CO_2)	Fossil fuels, deforestation, soil destruction	500 years
Methane (CH_4)	Cattle, biomass, rice paddies, gas leaks, mining, termites	7-10 years
Nitrous oxide (N_2O)	Fossil fuels, soil cultivation, deforestation	140-190 years
Chlorofluoro-carbons (CFC 11 and 12)	Refrigeration, air conditioning, aerosols, foam blowing, solvents	65-110 years
Ozone	Photo-chemical processes	Hours to days in upper troposphere (1 hour in up-per strato-sphere)

SOURCE: *The Greenhouse Trap: What We're Doing to the Atmosphere and How We Can Slow Global Warming.* World Resources Institute, 1990

Revolution could increase carbon dioxide concentrations and cause a gradual rise in temperatures. For almost six decades, his theory stirred little interest.

In 1957, studies at the Scripps Institute of Oceanography in California suggested that, indeed, half the carbon dioxide released by industry was being permanently trapped in the atmosphere. The studies showed that atmospheric concentrations of carbon dioxide in the previous 30 years were greater than in the previous two centuries and that the gas had reached its highest level in 160,000 years.

Findings in the 1980s and 1990s provided more disturbing evidence. Scientists detected increases in other, even more potent gases that contribute to the greenhouse effect, notably chlorofluorocarbons (CFC 11 and 12), methane, nitrous oxide (N_2O), and halocarbons (CFCs, methyl chloroform, HCFCs). (See Table 2.1.) Atmospheric concentrations of these gases from the 1700s through the 1900s has increased drastically. (See Figure 2.2.) Overall, total emissions of these gases in the United States alone increased about 10 percent from 1990 to 1998, although the rate of increase from 1997 to 1998 (0.2 percent), was down from the 1.2 percent average annual growth rate of the 1990s. (See Table 2.2.)

Global warming was only acknowledged as an international problem at the end of the last century. At the world's first ecological summit, the 1972 Stockholm Conference, climate change was not even listed among the threats to society.

FIGURE 2.2

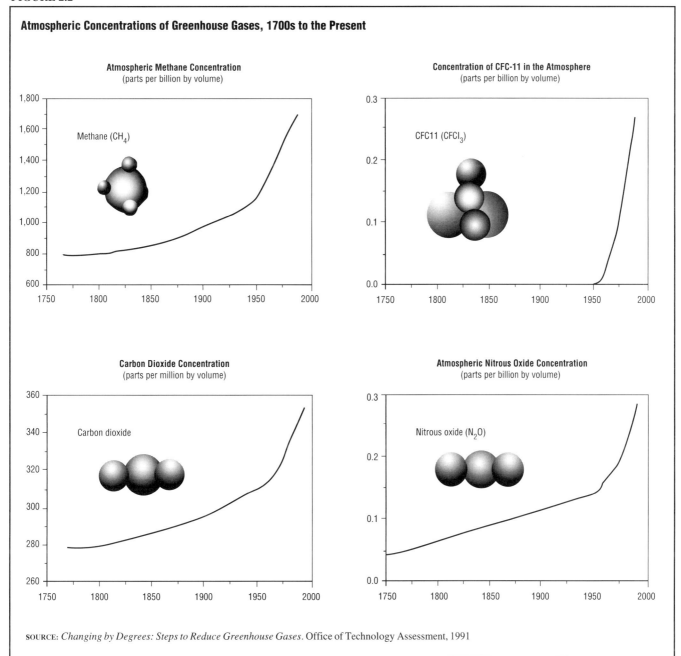

Atmospheric Concentrations of Greenhouse Gases, 1700s to the Present

Atmospheric Methane Concentration
(parts per billion by volume)

Methane (CH$_4$)

Concentration of CFC-11 in the Atmosphere
(parts per billion by volume)

CFC11 (CFCl$_3$)

Carbon Dioxide Concentration
(parts per million by volume)

Carbon dioxide

Atmospheric Nitrous Oxide Concentration
(parts per billion by volume)

Nitrous oxide (N$_2$O)

SOURCE: *Changing by Degrees: Steps to Reduce Greenhouse Gases.* Office of Technology Assessment, 1991

Is the Earth Getting Warmer?

As of the year 2000, however, even experts were not sure whether the world had already experienced human-induced climate change. Scientists have been unable to provide a definitive answer because they do not know how much the global climate has varied on its own in the relatively recent past (about 1,000 years). Temperature records based on thermometers go back only about 150 years. Investigators have turned, therefore, to "proxy" (indirect) means of measuring past temperatures. These methods include chemical evidence of climactic change contained in fossils, corals, ancient ice, and growth rings in trees.

Scientists do know, however, that humans are putting increasing amounts of gases, including carbon dioxide, into the atmosphere. (See Figure 2.3.) These gases are building up and could possibly trap energy from the sun. No one is certain how this accumulation affects Earth's climate.

RECORD HIGHS. In April 2000 the National Oceanic and Atmospheric Administration (NOAA) announced that the first quarter of 2000 was the warmest such three-month period in the United States during the past 106 years of record keeping. The average worldwide temperature in January, February, and March 2000—41.7 degrees Fahrenheit—was one degree higher than the previous record set in 1990. The nine-month period from June 1999 to March 2000 was the hottest similar period on record.

TABLE 2.2

U.S. Greenhouse Gas Emissions, by Gas, 1990–1998
(Million Metric Tons of Gas)

Gas	1990	1991	1992	1993	1994	1995	1996	1997	P1998
Carbon Dioxide	4,939.0	4,886.0	4,972.9	5,091.5	5,169.7	5,220.5	5,395.6	5,464.9	5,483.4
Methane	30.2	30.5	30.6	29.9	30.0	30.2	29.3	29.3	28.8
Nitrous Oxide	1.2	1.2	1.2	1.2	1.3	1.3	1.2	1.2	1.2
Halocarbons and Other Gases									
CFC-11, CFC-12, CFC-113	0.2	0.2	0.1	0.1	0.1	0.1	0.1	0.0	0.0
HCFC-22	0.1	0.1	0.1	0.1	0.1	0.1	0.1	0.1	0.1
HFCs, PFCs, and SF_6	*	*	*	*	*	*	*	*	*
Methyl Chloroform	0.2	0.2	0.1	0.1	0.1	*	*	*	*
Criteria Pollutants That Affect Climate									
Carbon Monoxide	86.8	88.6	85.5	85.6	89.5	80.7	82.3	79.2	NA
Nitrogen Oxides	21.2	21.3	21.6	21.8	22.1	21.5	21.3	21.4	NA
Nonmethane VOCs	18.9	19.0	18.6	18.8	19.4	18.6	17.4	17.3	NA

*Less than 50,000 metric tons of gas.

P = preliminary data. NA = not available.

Note: Data in this table are revised from the data contained in the previous EIA report, *Emissions of Greenhouse Gases in the United States 1997*, DOE/EIA-0573(97) (Washington, DC, October 1998).

SOURCE: *Emissions of Greenhouse Gases in the United States 1998*. Energy Information Administration

In 1998, Drs. Michael E. Mann and Raymond S. Bradley of the University of Massachusetts at Amherst, and Dr. Malcolm K. Hughes of the University of Arizona at Tucson, surveyed proxy evidence of temperatures in the Northern Hemisphere since 1400. They discovered that the twentieth century was the warmest century of the past 600 years and that the warmest years in that entire period were 1990, 1995, and 1997. They concluded that the warming trend of the past few years seems to be closely connected to the emission of greenhouse gases by humans. Some experts, however, question whether studies of proxy evidence will ever be reliable enough to yield valuable information on global warming.

Three international agencies have compiled long-term data on surface temperatures—the British Meteorological Office in Bracknell, U.K., the National Climatic Data Center in Asheville, North Carolina, and the NASA Goddard Institute for Space Studies in New York. Temperature measurements from these organizations reported that the 1990s were the warmest decade of the century and the warmest decade since humans began measuring temperatures in the mid-nineteenth century. The average global surface temperature was approximately 1 degree Fahrenheit warmer than a century ago, and this rise has increased more rapidly in the past two decades.

While some scientists are uncertain whether the greenhouse effect accounts for the change, most believe that it is the most likely explanation. Climate models suggest the potential for a warming from 2 to 6 degrees over the next one hundred years, warmer than Earth has been for millions of years.

However, some scientists observe that major climate events should be viewed in terms of thousands of years, not just a century. A record of only the past century may indicate, but not prove, that a major change has occurred. Is it caused by greenhouse gases, or is it natural variability? While many experts believe it is not possible to conclude that the warming is caused by greenhouse gases emitted by human activity, the rising temperature is roughly on track with that of computer models programmed to predict the course of greenhouse warming.

FURTHER SIGNS. In 1990 the first Intergovernmental Panel on Climate Change (IPCC—see below) noted several early signs of actual climate change. The average warm-season temperature in Alaska has risen nearly 3 degrees Fahrenheit in the last 50 years. Glaciers have generally receded and have become thinner on average by about 30 feet in the last 40 years. There is about 5 percent less sea ice in the Bering Sea than in the 1950s; and permafrost is thawing causing the ground to subside, opening holes in roads, producing landslides and erosion, threatening roads and bridges, and causing local floods. Ice cellars in northern villages have thawed and become useless. More precipitation now falls as rain than snow, and the snow melts faster, causing more running and standing water. While this could be natural variability, it is the kind of change expected of global warming, that is, the Arctic will warm more than the global average.

SOME POSSIBLE POSITIVE EFFECTS. The IPCC also identified several possible positive consequences of global warming. Agriculture in the northern United States and southern Canada, on the West Coast, and in interior parts of the West, could benefit—as could the evergreen forests

FIGURE 2.3

FIGURE 2.4

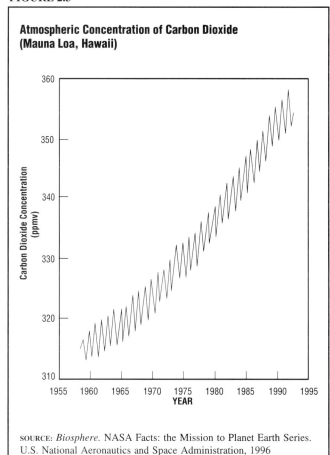

Atmospheric Concentration of Carbon Dioxide (Mauna Loa, Hawaii)

SOURCE: *Biosphere.* NASA Facts: the Mission to Planet Earth Series. U.S. National Aeronautics and Space Administration, 1996

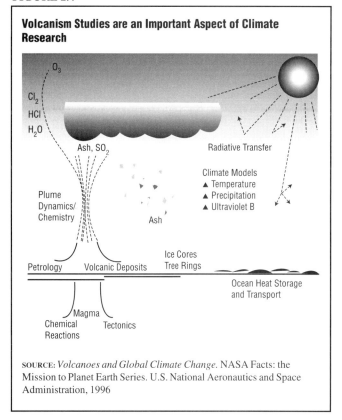

Volcanism Studies are an Important Aspect of Climate Research

SOURCE: *Volcanoes and Global Climate Change.* NASA Facts: the Mission to Planet Earth Series. U.S. National Aeronautics and Space Administration, 1996

of the West Coast. Milder winters could reduce the number of cold-weather deaths, as well as the cost of snow clearance and heating. Northern waters could remain open longer for navigation; the Arctic Ocean might become ice-free, opening a new trade route between Europe and Asia.

The Effect of Volcanic Activity

Volcanic activity, such as the 1991 eruption of the Mount Pinatubo volcano in the Philippines, can temporarily offset recent global warming trends. Volcanoes spew vast quantities of particles and gases into the atmosphere, including sulfur dioxide that combines with water to form tiny supercooled droplets. The droplets create a long-lasting global haze that reflects and scatters sunlight, reducing energy from the sun and preventing its rays from heating Earth, which causes the planet to cool. (See Figure 2.4.)

This also occurred in 1982, when the El Chichon volcano in Mexico depressed global temperatures for about four years. The 1991 Pinatubo eruption was about twice the size of the El Chichon eruption. NASA reported that satellite sensors measured the sulfur dioxide cloud from Mount Pinatubo at 15 million tons, about twice the size of the one emitted by El Chichon. NASA found that the haze of sulfur from the eruption reflected enough sunlight

to cool the earth by about one degree Fahrenheit, as was predicted by computer models.

In 1815 a major eruption of the Tambora volcano in Indonesia produced serious weather-related disruptions, such as crop-killing summer frosts in the United States and Canada. It became known as the "year without a summer." On the other hand, for several years following the Tambora eruption, people around the world commented about the beautiful sunsets.

In addition to effects on the weather, some scientists are concerned that volcanic eruptions may contribute to further destruction of the stratospheric ozone layer. (See Chapter 3.)

Clouds—A Key Variable in Climate Change

Clouds may hold a key to understanding climate change. Although we see clouds virtually every day, surprisingly little is known about them—where they occur, their role in energy and water transfer, and their ability to reflect solar heat. Earth's climate tries to maintain a balance between the energy that reaches Earth from the sun and the energy that radiates back from Earth into space. Scientists refer to this as Earth's "radiation budget." The components of Earth's system are the planet's surface, atmosphere, and clouds. (See Figure 2.5.)

Different parts of Earth have different capacities to reflect solar energy. Oceans and rain forests reflect only a

FIGURE 2.5

Clouds and Their Effect on Global Temperature

The shortwave rays from the sun are scattered in a cloud. Many of the rays return to space. The resulting "cloud albedo forcing," taken by itself, tends to cause a cooling of the Earth.

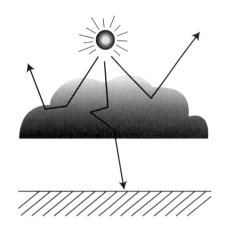

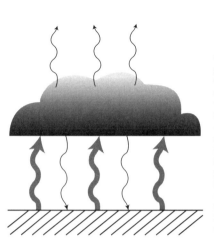

Longwave rays emitted by the Earth are absorbed and reemitted by a cloud, with some rays going to space and some going to the surface. Wavy arrows indicate longwave rays (distinguished from straight arrows, which indicate shortwave rays, as in the previous figure), and thicker arrows indicate more energy. The resulting "cloud greenhouse forcing," taken by itself, tends to cause a warming of the Earth.

SOURCE: *Clouds and the Energy Cycle.* NASA Facts: the Mission to Planet Earth Series. U.S. National Aeronautics and Space Administration, 1996

small portion of the sun's energy. Deserts and clouds, on the other hand, reflect a large portion of solar energy. A cloud reflects more radiation back into space than the surface would in the absence of clouds. An increase in cloudiness can also act like the panels on a greenhouse roof.

The Mission to Planet Earth program (see Chapter 1) includes numerous scientific studies of clouds, and much new information has been gained:

- The effect of clouds on climate depends on the balance between the incoming solar radiation and the absorption of Earth's outgoing radiation.

- Low clouds have a cooling effect because they are optically thicker and reflect much of the incoming solar radiation out to space.

- High thin cirrus clouds have a warming effect because they transmit most of the incoming solar radiation while also trapping some of Earth's radiation and radiating it back to the surface.

- Deep convective clouds have neither a warming nor a cooling effect because their reflective and absorptive abilities cancel one another.

Another Possible Climate Culprit—Solar Cycles

Scientists have known for centuries that the sun is not a completely steady source of energy; it has seasons, storms, and rhythms of activity with sunspots and flares appearing in cycles of roughly 11 years. Some scientists contend that these factors play a role in climate change on Earth. Some

research, though sketchy and controversial, suggests that the sun's variability could account for some, if not all, of global warming to date. The biggest correlation occurred centuries ago—between 1640 and 1720—when sunspot activity fell sharply and Earth cooled about two degrees Fahrenheit. (The sun is brighter when sunspots appear and dimmer when they disappear.) The sun is now approaching a stormy period in its 11-year cycle, promising a wealth of new data that might help explain the phenomenon.

CARBON DIOXIDE

Carbon dioxide, a naturally occurring component of Earth's atmosphere, is generally considered the major cause of global warming. (See Figure 2.6.) The Energy Information Administration of the U.S. Department of Energy (DOE) reported in 1998 that carbon dioxide accounted for approximately 81 percent of greenhouse gas emissions in the United States. (See Figure 2.7.)

The burning of fossil fuels by industry and motor vehicles is, by far, the leading source of carbon dioxide in the atmosphere, accounting for 99 percent of the nation's emission of greenhouse gases. In 1998 fossil fuel combustion totaled almost 1.5 million metric tons of carbon equivalents (MMTCE), up from 1.3 MMTCE in 1990. (See Table 2.3.)

As populations and economies expand, they use ever-greater amounts of fossil fuels. The United States, with only 5 percent of the world's population, accounts for 25 percent of the world energy use, making it the most carbon-intensive country.

FIGURE 2.6

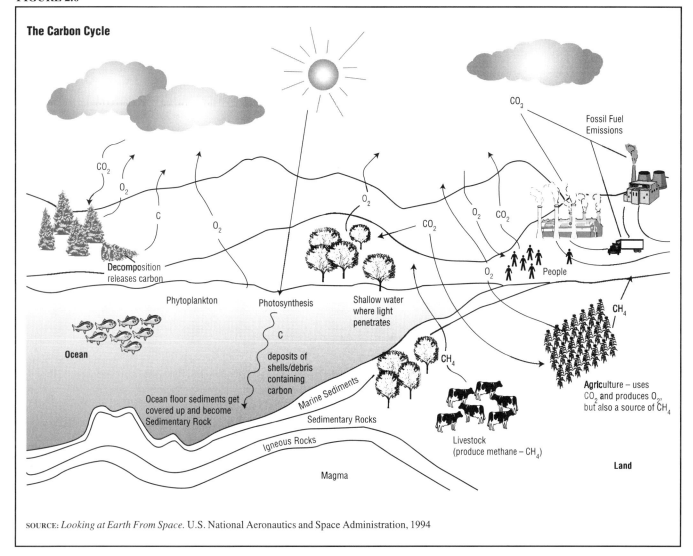

The Carbon Cycle

SOURCE: *Looking at Earth From Space.* U.S. National Aeronautics and Space Administration, 1994

In 1995, according to the Clinton Administration's *Climate Change Program* (2000), 73 percent of world carbon dioxide emissions from human activities came from the developed world. Eastern Europe and the former Soviet Union combined accounted for 27 percent of world emissions; the United States 22 percent, Western Europe 17 percent, and Asia 7 percent. Of the 27 percent that came from the developing world, China contributed 11 percent, followed by Latin America (4 percent), Africa and the Middle East (3 percent each), and other Asian nations (6 percent). The Administration predicts that, by 2035, the developing world will account for 50 percent of carbon dioxide emissions. Of that, China will contribute 17 percent, other Asian countries 14 percent, Africa 8 percent, Latin America 6 percent, and the Middle East 5 percent. (See Figure 2.8.)

The Role of the Forests as Carbon Sinks

Forests act as "sinks" (repositories), absorbing and storing carbon. Trees naturally absorb and neutralize carbon dioxide, although scientists do not agree on the extent to which forests can soak up excess amounts. The increas-

FIGURE 2.7

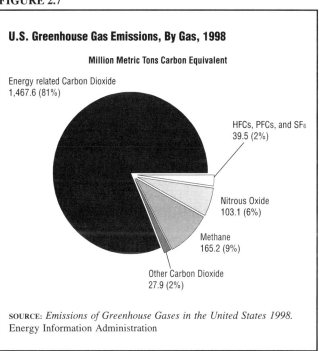

U.S. Greenhouse Gas Emissions, By Gas, 1998

Million Metric Tons Carbon Equivalent

Energy related Carbon Dioxide
1,467.6 (81%)

HFCs, PFCs, and SF₆
39.5 (2%)

Nitrous Oxide
103.1 (6%)

Methane
165.2 (9%)

Other Carbon Dioxide
27.9 (2%)

SOURCE: *Emissions of Greenhouse Gases in the United States 1998.* Energy Information Administration

TABLE 2.3

U.S. Sources of Carbon Dioxide Emissions and Sinks, 1990–1998 (Million Metric Tons of Carbon Equivalents)

Source	1990	1991	1992	1993	1994	1995	1996	1997	1998
Fossil Fuel Combustion	1,320.1	1,305.8	1,330.1	1,361.5	1,382.0	1,392.0	1,441.3	1,460.7	1,468.2
Cement Manufacture	9.1	8.9	8.9	9.4	9.8	10.0	10.1	10.5	10.7
Natural Gas Flaring	2.5	2.8	2.8	3.7	3.8	4.7	4.5	4.2	3.9
Lime Manufacture	3.0	3.0	3.1	3.1	3.2	3.4	3.6	3.7	3.7
Waste Combustion	2.8	3.0	3.0	3.1	3.1	3.0	3.1	3.4	3.5
Limestone and Dolomite Use	1.4	1.3	1.2	1.1	1.5	1.9	2.0	2.3	2.4
Soda Ash Manufacture and Consumption	1.1	1.1	1.1	1.1	1.1	1.2	1.2	1.2	1.2
Carbon Dioxide Consumption	0.2	0.2	0.2	0.2	0.2	0.3	0.3	0.4	0.4
Land-Use Change and Forestry (Sink)[a]	(316.4)	(316.3)	(316.2)	(212.7)	(212.3)	(211.8)	(211.3)	(211.1)	(210.8)
International Bunker Fuels[b]	32.2	32.7	30.0	27.2	26.7	27.5	27.9	29.9	31.3
Total Emissions	**1,340.3**	**1,326.1**	**1,350.4**	**1,383.3**	**1,404.8**	**1,416.5**	**1,466.2**	**1,486.4**	**1,494.0**
Net Emissions (Sources and Sinks)	**1,023.9**	**1,009.8**	**1,034.2**	**1,170.6**	**1,192.5**	**1,204.7**	**1,254.9**	**1,275.3**	**1,283.2**

[a] Sinks are only included in net emissions total. Estimates of net carbon sequestration due to land-use change and forestry activities exclude non-forest soils, and are based partially upon projections of forest carbon stocks.

[b] Emissions from International Bunker Fuels are not included in totals.

Note: Totals may not sum due to independent rounding.

SOURCE: *Draft - U.S. Greenhouse Gas Emissions and Sinks: 1990-1998.* U.S. Environmental Protection Agency

FIGURE 2.8

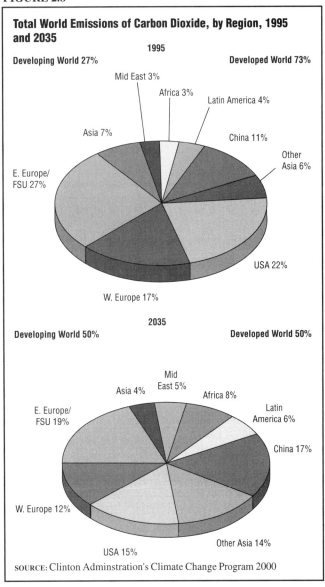

Total World Emissions of Carbon Dioxide, by Region, 1995 and 2035

1995
Developing World 27%
Developed World 73%
Mid East 3%
Africa 3%
Latin America 4%
Asia 7%
China 11%
E. Europe/FSU 27%
Other Asia 6%
USA 22%
W. Europe 17%

2035
Developing World 50%
Developed World 50%
Asia 4%
Mid East 5%
Africa 8%
E. Europe/FSU 19%
Latin America 6%
China 17%
W. Europe 12%
Other Asia 14%
USA 15%

SOURCE: Clinton Administration's Climate Change Program 2000

FIGURE 2.9

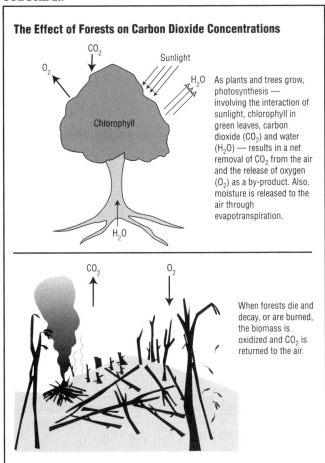

The Effect of Forests on Carbon Dioxide Concentrations

CO_2 Sunlight O_2 H_2O Chlorophyll H_2O

As plants and trees grow, photosynthesis — involving the interaction of sunlight, chlorophyll in green leaves, carbon dioxide (CO_2) and water (H_2O) — results in a net removal of CO_2 from the air and the release of oxygen (O_2) as a by-product. Also, moisture is released to the air through evapotranspiration.

CO_2 O_2

When forests die and decay, or are burned, the biomass is oxidized and CO_2 is returned to the air.

SOURCE: *Biosphere.* NASA Facts: the Mission to Planet Earth Series. U.S. National Aeronautics and Space Administration, 1996

TABLE 2.4

U.S. Sources of Methane Emissions, 1990–1998
(Million Metric Tons of Carbon Equivalents)

Source	1990	1991	1992	1993	1994	1995	1996	1997	1998
Landfills	58.2	58.1	59.1	59.6	59.9	60.5	60.2	60.2	58.8
Enteric Fermentation	32.7	32.8	33.2	33.7	34.5	34.9	34.5	34.2	33.7
Natural Gas Systems	33.0	33.4	33.9	34.6	34.3	34.0	34.6	34.1	33.6
Manure Management	15.0	15.5	16.0	17.1	18.8	19.7	20.4	22.1	22.9
Coal Mining	24.0	22.8	22.0	19.2	19.4	20.3	18.9	18.8	17.8
Petroleum Systems	7.4	7.5	7.2	6.9	6.7	6.7	6.5	6.5	6.3
Rice Cultivation	2.4	2.3	2.6	2.4	2.7	2.6	2.4	2.6	2.7
Stationary Sources	2.2	2.3	2.4	2.3	2.3	2.4	2.5	2.2	2.2
Mobile Sources	1.5	1.5	1.5	1.5	1.5	1.4	1.4	1.4	1.3
Wastewater Treatment	0.9	0.9	0.9	0.9	0.9	0.9	0.9	0.9	0.9
Petrochemical Production	0.3	0.3	0.3	0.4	0.4	0.4	0.4	0.4	0.4
Agricultural Residue Burning	0.2	0.2	0.2	0.1	0.2	0.2	0.2	0.2	0.2
Silicon Carbide Production	+	+	+	+	+	+	+	+	+
International Bunker Fuels	+	+	+	+	+	+	+	+	+
Total	**177.8**	**177.6**	**179.4**	**178.6**	**181.5**	**184.0**	**183.0**	**183.7**	**180.9**

+ Does not exceed 0.05 MMTCE

* Emissions from International Bunker Fuels are not included in totals.

Note: Totals may not sum due to independent rounding.

SOURCE: *Draft - U.S. Greenhouse Gas Emissions and Sinks: 1990-1998*. U.S. Environmental Protection Agency

ing levels of carbon dioxide in the atmosphere might conceivably be tolerated in Earth's normal carbon dioxide cycle if not for the additional complicating factor of deforestation. The burning of the Amazon rain forests and other forests has had a twofold effect: the immediate release of large amounts of carbon dioxide into the atmosphere from the fires, and the loss of trees to neutralize the carbon dioxide in the atmosphere. (See Figure 2.9.)

The Role of the Oceans as Carbon Sinks

Oceans have a profound effect on climate change, both because of their huge capacity to store heat and because they can moderate levels of atmospheric gases that regulate global temperatures. Covering more than 70 percent of Earth and holding 97 percent of the water on the planet's surface, oceans function as huge reservoirs of heat. Ocean currents transport this stored heat and dissolved gases so that different areas of the world serve as either sources or "sinks" for these components. While scientists know a great deal about oceanic and air circulation, they are less certain about the ocean's ability to store additional carbon dioxide or about how much heat it will absorb.

The top eight feet of the oceans hold as much heat as Earth's entire atmosphere. As ocean waters circulate globally, heat is transferred from low altitudes to high altitudes, from north to south, and vertically from surface to deep oceans and back. But how is this heat apportioned? If heat is able to circulate through the entire oceanic depth range, the process could take centuries and the world's oceans could serve to buffer or delay global warming. Researchers are working to determine that possibility, but—at present—it remains one of many unanswered questions.

The ocean is, by far, the largest reservoir of carbon in the carbon cycle. It holds approximately 50 times more carbon than the atmosphere and 20 times more than the terrestrial reservoir (land). Oceanographers and ecologists disagree over the carbon cycle/climate connection and over the ocean's capacity to absorb carbon dioxide. Some scientists believe that the oceans can absorb one to two billion tons of carbon dioxide a year, about the amount the world emitted in 1950. Until scientists can more accurately determine how much carbon dioxide can be buffered by ocean processes, the extent and speed of disruption in the carbon supply remains unclear.

METHANE

Methane is an important component of greenhouse emissions, second only to carbon dioxide. (See Table 2.2.) While there is less methane in the atmosphere, scientists estimate that it may be 21 times more effective at trapping heat in the atmosphere than carbon dioxide. Over the past two centuries, methane's concentration in the atmosphere has more than doubled, and scientists generally attribute those increases to human sources, such as landfills, natural gas systems, agricultural activities, coal mining, and wastewater treatment. According to the Environmental Protection Agency (EPA), in 1998, methane emissions totaled almost 181 MMTCE, up from almost 178 MMTCE in 1990. (See Table 2.4.)

NITROUS OXIDE

Nitrous oxide (N_2O) is formed by natural biological sources and by a number of human activities. Although nitrous oxide makes up a much smaller portion of green-

TABLE 2.5

U.S. Sources of Nitrous Oxide Emissions
(Million Metric Tons of Carbon Equivalents)

Source	1990	1991	1992	1993	1994	1995	1996	1997	1998
Agricultural Soil Management	75.3	76.3	78.2	77.3	83.5	80.4	82.4	84.2	83.9
Mobile Sources	13.8	14.6	15.7	16.5	17.1	17.4	17.5	17.3	17.2
Nitric Acid	4.9	4.9	5.0	5.1	5.3	5.4	5.6	5.8	5.8
Stationary Sources	3.7	3.6	3.7	3.8	3.8	3.9	4.0	4.0	4.1
Manure Management	3.4	3.6	3.5	3.7	3.8	3.7	3.8	3.9	4.0
Human Sewage	2.0	2.0	2.0	2.0	2.1	2.1	2.1	2.1	2.2
Adipic Acid	5.0	5.2	4.8	5.2	5.5	5.5	5.7	4.7	2.0

SOURCE: *Draft-U.S. Greenhouse Gas Emissions and Sinks: 1990-1998.* U.S. Environmental Protection Agency

house gases than carbon dioxide, like methane, it is much more (perhaps 310 times) powerful than carbon dioxide at trapping heat. (See Table 2.5 for activities that produced nitrous oxide from 1990 through 1998.) In 1998 emissions of nitrous oxide totaled 83.9 MMTCE, up from 75.3 MMTCE in 1990 but down slightly from 84.2 MMTCE in 1997.

CHLOROFLUOROCARBONS

Most scientists believe that chlorofluorocarbons (CFCs), an important class of modern industrial chemicals, are responsible for much of the increased greenhouse effect experienced during the 1980s. The United States is the leading producer of CFCs. CFCs are also responsible for depletion of the ozone layer in the stratosphere, which shields the earth from deadly ultraviolet radiation. (See Chapter 3.) Therefore, efforts to limit their use are better developed than for other greenhouse gases.

Beginning in the 1970s the United States and many other nations started banning the use of CFCs in aerosol sprays. In 1987 leaders of many world nations met in Montreal and agreed to cut output by 50 percent by the year 2000. In 1989, 82 nations signed the Helsinki Declaration, pledging to phase out five CFCs. Also, for the first time, a large number of third-world countries were actively engaged in the agreement.

EFFECTS OF A WARMING CLIMATE

Rising Sea Level

Some observers compare global warming to nuclear war in its potential to disrupt human and environmental systems. While some sources dispute the occurrence of human-induced climate change, if temperature increases were to take place, substantial changes would occur on Earth's surface. If average temperatures rise 1.5 to 4.5 degrees Celsius by 2030, the global sea level could rise 20–140 centimeters (8–56 inches). The Climate Institute in Washington, DC, forecasts a further rise of eight inches by 2030 and 26 inches by 2100. This rise would be caused by the expansion of seawater as it is warmed, as well as by melting glaciers and ice caps. The National Climatic Data Center reports that sea levels have risen by 10 inches in the last century.

More than half the population of the United States lives within 50 miles of a coastline. A rising sea level would narrow or destroy beaches, flood wetland areas, and either submerge or force costly fortification of shoreline property. Higher water levels would increase storm damage and many coastal cities worldwide would be flooded. Some islands, such as the Philippines, have already seen encroachment. Residents of the Maldive Islands in the Indian Ocean, which lie on average three feet above sea level, are already erecting artificial defenses, such as breakwaters made of concrete, to fend off rising seas.

Some 70 million people in low-lying areas of Bangladesh could be displaced by a one-meter (39.4 inches) rise. Such a rise would also threaten Tokyo, Osaka, and Nagoya in Japan, as well as coastal China and the Atlantic and Gulf Coasts of the United States. The rising waters would also intrude on inland rivers threatening fresh water supplies, and increasing the salt content of groundwater as the sea intrudes on freshwater aquifers (naturally occurring underground water reservoirs). Much of the increased rainfall would come not in steady, gentle rains favored by farmers, but from heavy storms and flooding. In 2000 some experts believed that trend had already begun.

Increased Heat

Warmer temperatures may also increase the evaporation rate, thereby increasing atmospheric water vapor and cloud cover, which in turn may affect regional rainfall patterns. A warmer climate would likely shift the rainbelt of the middle latitudes toward the poles as water-laden air of the tropics travels further toward the poles before the moisture condenses as precipitation. This would shift patterns of rainfall around the world. Wetter, more violent weather is projected for some regions. Forests, which are adapted to a narrow temperature and moisture range, would particularly be threatened by climate shifts.

The opposite problem—too little water—could worsen in arid areas such as the Middle East and parts of Africa. Some experts have suggested that the greenhouse effect and global warming are mild terms for a coming era that may be marked by heat waves that may make some regions virtually uninhabitable. Frequent droughts could plague North America and Asia, imperiling food production, as agriculture is particularly vulnerable to the effects of climate change. Most experts believe Africa would be the most vulnerable to climate change because its economy depends largely on rain-fed agriculture, and many farmers are too poor and ill equipped to adapt. Australia and Latin America could also be subject to severe drought.

The Effect of Increased Temperatures on Humans

Extra heat alone would be enough to kill some people. Some deaths would occur directly from heat-induced strokes and heart attacks. Air quality also deteriorates as temperatures rise. Hot, stagnant air contributes to the formation of atmospheric ozone, the main component of smog, which damages human lungs. Poor air quality can also aggravate asthma and other respiratory diseases. Increasing ultraviolet rays can increase the incidence of skin cancers, diminish the function of the human immune system, and cause eye problems such as cataracts. Higher temperatures and added rainfall could create ideal conditions for the spread of a host of infectious diseases by insects, including mosquito-borne malaria, dengue fever, and encephalitis.

NASA has begun using computer calculations in an attempt to predict the effects of global warming on some of the major cities in the United States. (See Table 2.6.)

Decreasing Biological Diversity

Biological diversity is also predicted to suffer from global warming. Loss of forests, tundra, and wetlands could irrevocably damage ecosystems. Some plants and animals that live in precise, narrow bands of temperature and humidity, like monarch butterflies and the edelweiss flower which grows in the Alps, may find their habitats wiped out altogether. Rising seas would cover coastal mangrove swamps causing the loss of many species, including the Bengal tiger. Plants and animals of the far north, like the polar bear and the walrus, would die out for lack of an acceptably cold environment. Many species cannot migrate rapidly enough to cope with climate change at the projected rate. Opportunistic species, such as weeds, often out-compete trees and other plants more valued by humans. Pests, such as certain insects, may survive where other species cannot.

The U.S. Forest Service believes that Eastern hemlock, yellow birch, beech, and sugar maple forests would gradually shift their ranges northward by 300 to 600 miles but would be severely limited by the warming and

TABLE 2.6

The Weather Forecast for 2050, Selected U.S. Cities

City	Days over 90°F		Days over 100°F	
	Today	2050	Today	2050
Washington, DC	36	87	1	12
Omaha, NE	37	86	3	21
New York, NY	15	48	0	4
Chicago, IL	16	56	0	6
Denver, CO	33	86	0	16
Los Angeles, CA	5	27	1	4
Memphis, TN	65	145	4	42
Dallas, TX	100	162	19	78

SOURCE: *The Greenhouse Trap: What We're Doing to the Atmosphere and How We Can Slow Global Warming.* World Resources Institute, 1990

would largely die out, along with the wildlife they shelter. Studies by the World Wildlife Fund International report that more than half the world's parks and reserves could be threatened by climate change. These include the Florida Everglades, Yellowstone National Park, the Great Smoky Mountains, and Redwood National Park in California. The EPA warned in 1988, "If current trends continue, it is likely that climate may change too quickly for many natural systems to adapt." (See Chapter 9 for more information on biological diversity.)

EL NIÑO

For centuries fishermen in the Pacific ocean off the coast of South America have known about the phenomenon called El Niño. Every three to five years, during December and January, fish in those waters virtually vanish, bringing fishing to a standstill. Fishermen gave this occurrence the name "El Niño," which means "the Child," because it occurs around the celebration of the birth of the Christ child. Although originating in the Pacific, the effects of El Niño are felt around the world. Computers, satellites, and improved data gathering have found that El Niño has been responsible for drastic climate change.

An El Niño occurs because of interactions between atmospheric winds and sea surfaces. In normal years, trade winds blow from east to west across the eastern Pacific. They drag the surface waters westward across the ocean, causing deeper, cold waters to rise to the surface. This "upwelling" of deep ocean waters carries nutrients from the bottom of the ocean that feeds fish populations in the upper waters. In an El Niño the westward movement of waters weaken, causing the upwelling of deep waters to cease. The resulting warming of the ocean waters further weakens trade winds and strengthens El Niño. Without upwelling, nutrients from deep waters are lacking, depleting fish populations. (See Figure 2.10.) The warm waters that normally lie in the western waters of the Pacific shift eastward. This turbulence creates weather conditions east-

FIGURE 2.10

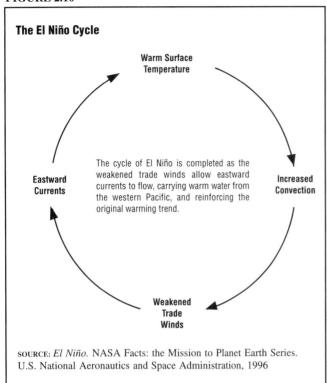

The El Niño Cycle

Warm Surface
Temperature

Eastward
Currents

The cycle of El Niño is completed as the weakened trade winds allow eastward currents to flow, carrying warm water from the western Pacific, and reinforcing the original warming trend.

Increased
Convection

Weakened
Trade
Winds

SOURCE: *El Niño.* NASA Facts: the Mission to Planet Earth Series. U.S. National Aeronautics and Space Administration, 1996

FIGURE 2.12

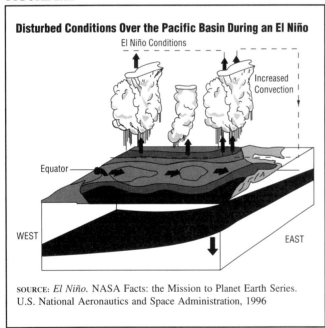

Disturbed Conditions Over the Pacific Basin During an El Niño

El Niño Conditions

Increased
Convection

Equator

WEST

EAST

SOURCE: *El Niño.* NASA Facts: the Mission to Planet Earth Series. U.S. National Aeronautics and Space Administration, 1996

FIGURE 2.11

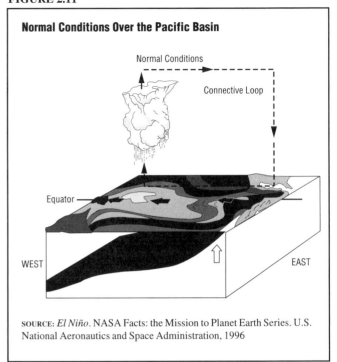

Normal Conditions Over the Pacific Basin

Normal Conditions

Connective Loop

Equator

WEST

EAST

SOURCE: *El Niño.* NASA Facts: the Mission to Planet Earth Series. U.S. National Aeronautics and Space Administration, 1996

ward—towering cumulus clouds reaching high into the atmosphere with strong up and down (vertical or convective) forces and the weakening of normal east-to-west trade winds. (See Figures 2.11 and 2.12.)

The worldwide effects of El Niño can include torrential rains, tornadoes, hurricanes, mudslides, flash flooding, beach and cliff erosion, sewage spills, drought, increased snowfall, and disruption in the marine food chain. Such weather events often affect regional energy and economic markets. The El Niño of 1982–83 is estimated to have caused $13.6 billion in damage and killed 2,000 people around the world. In 1997 and 1998, El Niño–related storms in California caused more than 9,000 people to seek federal disaster assistance for property losses. An estimated $3.6 million in aid was distributed to approximately 1,689 people. The American Fisheries Society estimated that 1990s storms have been the most devastating for marine life in more than a century.

GENERAL CIRCULATION MODELS

The science of global atmospheric change is still in its infancy. Most of our images of the world's environmental future must come from computer and mathematical models. These computer models are crude, imperfect representations of the real world and of the future world. They cannot indicate where changes will be the worst. Nor can they accurately calculate the heat reflected by clouds or absorbed by the ocean. As a result, scientists disagree over whether the forecasts of global warming are reliable.

Predictions by Computer Models

The most highly developed tools now available to project climatic changes are complex computer models called general circulation models (GCMs). (See Figure 2.13.) Such models are critical to the study of climate change, and they are costly. Five government agencies operate such projects: the DOE, NASA, the National Science Foundation (NSF), the NOAA, and the EPA.

FIGURE 2.13

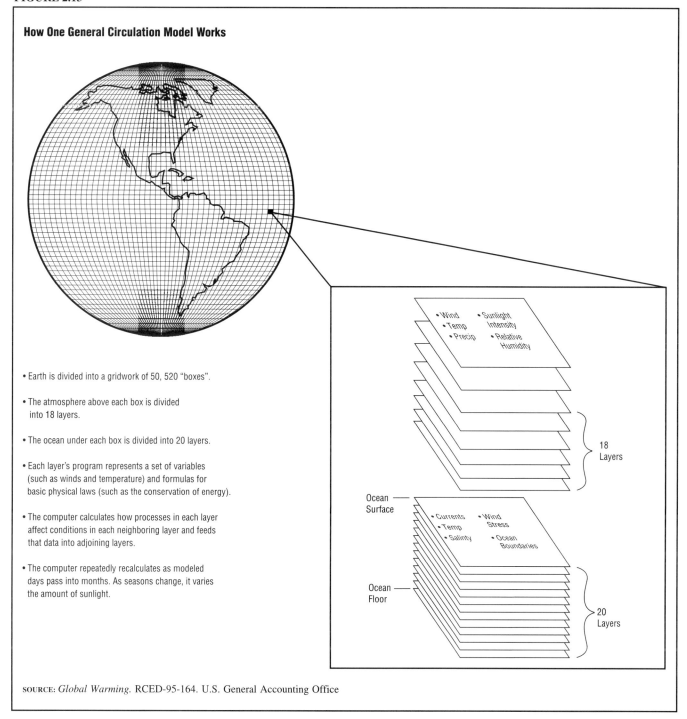

How One General Circulation Model Works

- Earth is divided into a gridwork of 50, 520 "boxes".

- The atmosphere above each box is divided into 18 layers.

- The ocean under each box is divided into 20 layers.

- Each layer's program represents a set of variables (such as winds and temperature) and formulas for basic physical laws (such as the conservation of energy).

- The computer calculates how processes in each layer affect conditions in each neighboring layer and feeds that data into adjoining layers.

- The computer repeatedly recalculates as modeled days pass into months. As seasons change, it varies the amount of sunlight.

• Wind • Sunlight
• Temp   Intensity
• Precip • Relative
            Humidity

18 Layers

Ocean Surface

• Currents • Wind
• Temp   Stress
• Salinty   • Ocean
             Boundaries

Ocean Floor

20 Layers

SOURCE: *Global Warming*. RCED-95-164. U.S. General Accounting Office

"FLAWED" MODELS. Although computer models have become more accurate during the past few decades, important uncertainties limit their predictive abilities. The most commonly cited flaw in climate computer models is the "cold start error." Even the most powerful computers are limited in their ability to store and analyze the vast quantity of data required to accurately simulate the global climate. Modelers have tried to overcome these limitations by introducing assumptions into their models that deliberately oversimplify some operations to free the GCMs' capacity for more critical operations. For example, model-ers have assumed that the ocean was not warmed by emissions of greenhouse gases before 1985. Although this assumption increases GCM capacity, it increases the uncertainty of the computer's predictions because the ocean will reach its capacity to absorb emissions sooner. Scientists do not know, however, how much or by how long the predictions are distorted because of this error.

Patrick Michaels, with the Center for the Study of American Business at Washington University, claims in *Global Deception: The Exaggeration of the Global*

TABLE 2.7

Highlights of the IPCC Scientific Assessment of Climate Change

The IPCC is certain that:
- There is a natural greenhouse effect that already keeps the Earth warmer than it would otherwise be.
- Emissions resulting from human activities are substantially increasing the atmospheric concentrations of the greenhouse gases.

The IPCC calculates with confidence that:
- Atmospheric concentrations of the long-lived gases (CO_2, N_2O, and the CFCs) adjust only slowly to changes in emissions. Continued emissions of these gases at present rates would commit us to increased concentrations for centuries ahead.
- The longer emissions continue to increase at present-day rates, the greater reductions would have to be for concentrations to stabilize at a given level.
- Immediate reductions (on the order of 60%) in emissions of long-lived gases (CO_2, N_2O, and the CFCs) from human activities would be required to stabilize their concentrations at today's levels; methane would require a 15 to 20% reduction.

Based on current model results, the IPCC predicts that:
- Under the IPCC Business-As-Usual Scenario[a], global mean temperature will increase about 0.3°C per decade (with an uncertainty range of 0.2 to 0.5 °C per decade), reaching about 1 °C above the present value by 2025 and 3 °C before the end of the 21st century .
- Land surfaces will warm more rapidly than the ocean, and high northern latitudes will warm more than the global mean in winter.
- Global mean sea level will rise about 6 cm per decade over the next century, rising about 20 cm by 2030 and 65 cm by the end of the 21st century.

All predictions are subject to many uncertainties with regard to the timing, magnitude, and regional patterns of climate change, due to incomplete understanding of:
- sources and sinks of greenhouse gases,
- clouds,
- oceans, and
- polar ice sheets.

The IPCC judgment is that:
- Global mean surface air temperature has increased by about 0.45 °C (with an uncertainty range of 0.3 to 0.6 °C) over the last 100 years, with the five globally averaged warmest years occurring in the 1980s.
- The size of this warming is broadly consistent with predictions of climate models, but it is also of the same magnitude as natural climate variability. Thus, the observed temperature increase could be largely due to natural variability; alternatively, this variability and other human factors could have offset a still larger human-induced greenhouse warming. The unequivocal detection of the enhanced greenhouse effect from observations is not likely for a decade or more.

[a]Assumes that emissions of all greenhouse gases continue at 1990 levels.

SOURCE: *Changing by Degrees: Steps to Reduce Greenhouse Gases.* Office of Technology Assessment, 1991

Warming Threat (1997), that the global warming crisis has been greatly overstated. He notes that newer GCM models, which are more physically realistic than the older models, tend to forecast less warming rather than more.

THE INTERNATIONAL PANEL ON CLIMATE CHANGE—A SCIENTIFIC CONSENSUS?

Climate-change science has been developing rapidly, leading observers to recognize the complexity of the issues. For example, new findings on the role of local air pollution in ozone depletion have lowered computer projections of the rate of global warming. In order to understand the issue of climate change and any possible global warming, scientists worldwide were called together in 1990 to exchange information. They formed the Intergovernmental Panel on Climate Change (IPCC) sponsored jointly by the United Nations Environmental Programme (UNEP) and the World Meteorological Organization. The panel's mission was to advise the parties to the 1992 Global Warming Treaty (see below). With contributions by nearly 400 of the world's climate experts from 25 nations, the panel consolidated the best available knowledge.

The IPCC report was the most comprehensive summary of climate-change science to date. With regard to various statements about climate change, the scientists responded based on how sure they were of global warming: "virtually certain" (nearly unanimous agreement and no credible alternative), "very probably" (roughly a 90

percent likelihood of occurring), "probable" (a two out of three chance of happening), and "uncertain" (evidence does not support). Some consensus emerged among many of those scientists that greenhouse gases do produce climate change, and the IPCC was certain that there is a natural greenhouse effect and that emissions resulting from human activities are contributing to the greenhouse gases. (See Table 2.7.)

A Landmark Judgment—The 1995 IPCC Report

In 1995 the IPCC reassessed the state of knowledge about climate change. The panel reaffirmed its earlier conclusions and updated its forecasts, predicting that, if no further action is taken to curb emissions of greenhouse gases, temperatures will increase 1.44 degrees to 6.3 degrees Fahrenheit by 2100. The panel concluded that the evidence suggested a human influence on global climate. The cautiously worded statement was a compromise following intense discussions. Nonetheless, it was a landmark conclusion because, until then, the panel had always said that global warming and climate changes could have been the result of natural variability. Despite acrimonious debate and challenges to the report, the conclusions have largely held up to the scientific scrutiny of many of the top specialists in climate science.

The 2000 Update

A preliminary version of the IPCC's 2001 report went even further than the organization's 1995 statement. It concluded that humans have "discernibly" affected Earth's climate and that the surface of the planet will likely warm between 2 and 9 degrees Fahrenheit by the end of the twenty-first century. The update was more definitive about global warming, declaring that, over the past 150 years, it is "exceptional and unlikely to be solely natural in origin."

SOME RESEARCHERS STILL QUESTION THE THEORY OF CLIMATE CHANGE

Dr. Richard Lindzen, professor of meteorology at the Massachusetts Institute of Technology, in the *Swiss Review of World Affairs* (July 1994), observed that a solid scientific foundation is lacking from claims that global warming is imminent. He believes that Earth's climate depends on more than the carbon dioxide content. Instead, Earth cools by atmospheric movements upward and poleward rather than only by radiation; and climate changes since 1850 are indistinguishable from natural variability. He further claims that no large-scale model includes all the known major factors in the water vapor cycle. In 1997 Dr. Lindzen added that the natural system has built-in resilience and that the observed changes are insignificant and not urgent.

Among the claims of critics of global climate warming are:

- The increase in the earth's surface temperature during the past 150 years is less than the best existing climate models can explain. As the models improve, however, they predict less and less warming.

- Climate has been known to change dramatically within a relatively short period without any human influence.

- Temperature readings already showed increased temperatures before carbon dioxide levels rose significantly (before 1940).

- Natural variations in climate may exceed any human-caused climate change.

- Some of the increase in temperatures can be attributed to sunspot activity.

- If warming should occur, it will not stress Earth; it may even have benefits, such as for agriculture, and may delay the next ice age.

- Reducing emissions will raise energy prices, reduce Gross Domestic Product (GDP), and produce job losses.

- While clouds are crucial to climate predictions, so little is known about them that computer models cannot produce accurate predictions.

THE RESPONSE OF THE NATIONS

Many industrialized countries have committed themselves to stabilizing or reducing carbon dioxide emissions. The Bush Administration (1989–92) opposed precise deadlines for carbon dioxide limits, arguing that the extent of the problem was too uncertain to justify painful economic measures. When President Clinton took office in 1993, he joined the European Community in calling for overall emissions to be stabilized at 1990 levels by the year 2000, but this goal was not met. Saudi Arabia and several other oil-producing nations have resisted the setting of new targets for emissions because they fear this will reduce the demand for their oil. On the other hand, many environmentalists believe that reducing pollution depends on increasing energy efficiency and gradually switching from fossil fuels to renewable energy, an idea most industrialized countries have been slow to embrace.

China

China's ongoing economic revolution may lead that nation to become the world's largest contributor to global warming. China is burning increasing amounts of coal and is expected to become the single greatest creator of greenhouse gases in the coming decades. China's leaders indicate that they are well aware that coal-burning causes pollution, but the Chinese government has made it clear that it will not sacrifice development for the environment. Chinese leaders argue that it is unfair to impose constraints on China or developing nations when Western countries have been willfully polluting the environment for more than a century, and China has only recently become a significant offender. Critics of China's argument note that China itself would be harmed by global warming, which could dry up crops, shrink water supplies, and cause the flooding of major coastal cities.

Island States

In 1992 representatives of 37 island nations, which make up the Alliance of Small Island States, drafted an agreement to present to the Earth Summit. These nations, including Cyprus and Malta (in the Mediterranean) and the Caribbean Islands, fear their existence is threatened by the rising sea level. They claim they will be the first victims of global warming, becoming a whole new category of environmental refugees. Tourist beaches are shrinking, dikes are being erected to protect reclaimed land, and some islands have already been evacuated. Since they are so vulnerable, these countries feel an urgency not felt by the northern industrial countries and the larger developing nations. Nonetheless, larger countries may have similar incentives to stabilize sea levels. A three-foot rise in ocean levels would render an estimated 72 million people homeless in China, 11 million in Bangladesh, and 8 million in Egypt.

A Global Warming Treaty

In 1992, 143 countries approved a United Nations global warming treaty in Rio de Janeiro, Brazil, that committed them to reduce the amount of greenhouse gases emitted into the atmosphere. Many environmentalists criticized the treaty as too weak because it did not establish specific targets that governments must meet. The treaty did not include specific targets mainly because then-President George Bush, representing the United States, refused to accept them. President Clinton signed the treaty in 1994, although adherence to the environmental measures has largely been disappointing even among those nations that originally signed.

Supporters of a global warming strategy advocate limiting the emissions of the four main greenhouse gases: carbon dioxide, methane, nitrous oxide, and CFCs. They also recommend a gradual transition away from fossil fuels, which currently provide about three-quarters of the world's energy. (See Chapter 10.)

In 1993, the United States, under the United Nations Framework Convention on Climate Change, released *The Climate Change Action Plan*, detailing the nation's response to climate change. The plan included a set of measures by both government and the private sector to lay a foundation for the nation's participation in world response to the climate challenge.

The actions called for under the *Action Plan* would reduce emissions for all greenhouse gases to 1990 levels by 2000. However, since the time these projections were prepared and the *Action Plan* was published, the economy grew at a more robust rate than anticipated, which led to increased emissions. Furthermore, the U.S. Congress did not provide full funding for the actions contained in the plan.

In 1995, 120 parties to the global warming treaty met in Berlin in what is known as the Berlin Mandate to determine the success of existing treaties and to embark on discussions of emissions after 2000. Differences persisted along north/south lines, with developing countries making essentially a moral argument for requiring more of the richer nations. They pointed out that the richer nations are responsible for most of the pollution. The Berlin talks essentially failed to endorse binding timetables for reductions in greenhouse gases.

The Kyoto Protocol—Rich versus Poor Countries

In December 1997 delegates from 166 countries met in Kyoto, Japan, at the United Nations Climate Change Conference to negotiate actions to reduce global warming. The task was more complicated and difficult than was envisioned in 1995, when parties to the 1992 Rio treaty on climate change decided that stronger action was needed. What was originally envisioned as a matter of deciding on a reduction target and a timetable for industrialized countries once again broadened into a contentious debate between developed and developing countries as to the proper role of each.

Some developed nations, including the United States, wanted to require all countries to reduce their emissions. However, developing countries felt the industrialized nations had caused, and were still causing, most global warming and expected the industrialized world to bear the brunt of economic sacrifices to clean up the environment. Even within the industrialized community, the European Union criticized the United States for lagging behind in reducing emissions, as they had previously pledged.

Different targets for different countries, tailored to their economic and social circumstances, emerged as a possible way to get around the impasse between nations. The final compromise, signed by the parties, was for the industrialized nations to cut emissions by an average of 5.2 percent between 2008 and 2012. The United States agreed to reduce emissions by 7 percent, European countries by 8 percent, and Japan by 6 percent. The treaty also set up an emission trading system that would allow countries that exceed their pollution limits to purchase on an open market "credits" from countries that pollute less. This provision was viewed as necessary to U.S. congressional approval. The developing nations feared that such a trading system would allow rich countries to buy their way into compliance rather than make unpopular emissions cuts. Enforcement mechanisms were not agreed to, nor did developing nations commit to binding participation. Individual countries had until March 1999 to ratify the treaty. Most nations waited to see what the United States would do. The U.S. Congress never ratified the treaty; therefore, in 2000, international support for the Protocol had largely evaporated.

THE UNITED STATES TAKES ACTION

During the late twentieth and early twenty-first centuries, the United States global warming program was coordinated through the Committee on Earth and Environmental Science (CEES) within the Office of Science and Technology Policy of the White House. Some 18 federal agencies were represented in the multi-year, multi-billion dollar research program, nine of which received the bulk of funding. NASA received 66 percent, the NSF 10 percent, and the remainder was shared by the Smithsonian Institute and a half dozen federal agencies, including the EPA, and the Departments of Agriculture, Commerce, Defense, Energy, and the Interior.

Even though the United States had a comprehensive global warming program in place, Congress was reluctant to take steps to reduce emissions. However, the Clinton Administration implemented some policies that did not require congressional approval. In 1999 President Clinton made a number of Executive Orders and began several new initiatives. These included The Climate Change Technology Initiative (CCTI), a five-year, $6.3 billion program of tax incentives and investments focusing on improving

TABLE 2.8

Opinion Regarding the Seriousness of Global Warming

Do you think that global warming will pose a serious threat to you or your way of life in your lifetime? Do you think that global warming will pose a serious threat to your children or the next generation of Americans in their lifetime?

	Your lifetime	Your children's
Yes	25%	65%
No	69	27
No opinion	6	8
	100%	100%

SOURCE: Gallup Poll, 12/02/1997. Gallup Organization, 2000

TABLE 2.9

Effect of Increased Energy Prices on Opinion Regarding Reducing Global Warming

Would you, personally, be willing or not willing to have the United States take steps to reduce global warming if costs for gasoline or electricity went up a great deal?

Yes, willing	44%
Not willing	48
Depends(vol.)	4
No opinion	4
	100%

SOURCE: Gallup Poll, 12/02/1997. Gallup Organization, 2000.

TABLE 2.10

Effect of Increased Unemployment on Opinion Regarding Reducing Global Warming

Would you, personally, be willing or not willing to have the United States take steps to reduce global warming if unemployment went up a great deal?

Yes, willing	34%
Not willing	54
Depends(vol.)	5
No opinion	7
	100%

SOURCE: Gallup Poll, 12/02/1997. Gallup Organization, 2000

energy efficiency and renewable energy technologies; The Wind Powering America initiative, seeking to supply 5 percent of the nation's electricity through wind technology by 2020; and a Brightfields initiative aimed at using former industrial sites contaminated with toxic waste for producing pollution-free solar energy. Executive Order 13134 aimed at coordinating federal efforts to develop technology to grow the economy while simultaneously solving some environmental problems, such as converting crops, trees, and other "biomass" into fuels, power, and products. Executive Order 13123 required all federal government agencies to reduce greenhouse gas emissions below 1990 levels by 2010. The federal government was the largest energy consumer in the nation, and the Clinton Administration wanted it to serve as an example to the nation's businesses and consumers, who ultimately could reap benefits from energy improvements.

PUBLIC OPINION

A November 1997 *New York Times* poll found that Americans were more willing than the government to take steps to counter the threat of global warming. Forty-nine percent of respondents said they believed global warming was a result of greenhouse gases; 16 percent thought it was due to normal climate fluctuations. Eight percent said it was due to both greenhouse gases and normal climate change, while 26 percent had no answer, and 1 percent said neither. When asked if they would be willing to invest in new appliances and insulation to cut household emissions of greenhouse gases, 47 percent said they would, 21 percent indicated they would not, 10 percent thought "it depends," and 4 percent said they already were doing so. Sixty-five percent of those polled claimed they would opt to take steps now to cut emissions of greenhouse gases; only 15 percent preferred to wait until many countries agreed to take steps together. The remainder had no answer.

When asked if they thought the weather had recently been normal or "stranger than normal," 67 percent thought it had been "stranger than normal," while 31 percent said it was normal, and 1 percent had no opinion. However, those who perceived the weather was unusual differed on the cause—34 percent had no idea, 17 percent said El Niño, 13 percent thought ozone/global warming, and 11 percent said pollution. Another 8 percent thought the unusual weather was caused by "God," and 7 percent said nature.

However, a November 1997 Gallup Poll suggested that, despite their concern about climate change, Americans were unlikely to support strict new limits on fossil fuel emissions that might result from the Kyoto Conference. By a 58 percent to 32 percent margin, Americans opposed committing to any treaty that would hold the United States to stricter energy standards than other large nations of the world. Although 69 percent of respondents did not think global warming would be a threat to their own lives, 65 percent said it would be a problem in their children's lifetime. (See Table 2.8.) Even so, 48 percent said they were unwilling to take steps to reduce global warming if costs for energy went up, and 44 percent claimed to be willing to pay higher energy costs to reduce global warming. (See Table 2.9.) An even greater percentage (54 percent) were unwilling to take steps to reduce global warming if unemployment would rise as a result. (See Table 2.10.)

Some Industry Support of Concerns

In the early 1990s, industries were virtually unanimous in claiming there was no cause for serious concern about global warming and in rejecting any agreement obligating countries to cut emissions of heat-trapping gases. Some change, however, was noted in talks held in Bonn, Germany, in 1997. British Petroleum, the world's third-largest oil company, broke ranks with other fossil-fuel producers by announcing that it believed there was enough scientific evidence to warrant concern about whether human activity is changing Earth's climate.

CHAPTER 3
A HOLE IN THE SKY: OZONE DEPLETION

EARTH'S PROTECTIVE LAYER

What Is Ozone?

Ozone is a colorless, gaseous form of oxygen found in Earth's atmosphere, primarily at altitudes below 30 miles. (See Figure 3.1.) It is an unstable molecule of three atoms of oxygen produced naturally in the middle and upper stratosphere. In the present-day stratosphere, the natural balance of the ozone layer has been altered, particularly by the introduction of human-made chlorofluorocarbons (CFCs).

Ozone can be either harmful or protective to life on Earth, depending on where it is found. Ozone in the stratosphere is important to life processes on Earth. At altitudes above 15 miles it acts as a shield to protect Earth from the sun's harmful ultraviolet radiation. In the lower ranges of the atmosphere (troposphere) ozone is one of several air pollutants that endanger life on Earth. Ironically, destruction of the ozone layer at the upper levels could increase the amount at the earth's surface. Ozone is the primary component in smog. As more radiation reaches the ground, the photochemical process that creates smog intensifies. Smog retards crop and tree growth, impairs health, and limits visibility. In addition, the decline in stratospheric ozone is thought to increase hydrogen peroxide in the stratosphere, contributing to acid rain.

Ozone Balance in the Stratosphere

"Ozone depletion" refers to the loss of stratospheric ozone as a result of the interaction between the sun's ultraviolet rays and regular oxygen molecules. Ultraviolet radiation causes the oxygen molecule to split and then recombine, producing ozone. Chlorine, which comes from the breakdown of CFCs, and substances such as halon and methyl chloroform, destroy this ozone. (See Figure 3.2.)

One way to understand this process is to compare it to a leaking bucket. As long as water is poured into a bucket

FIGURE 3.1

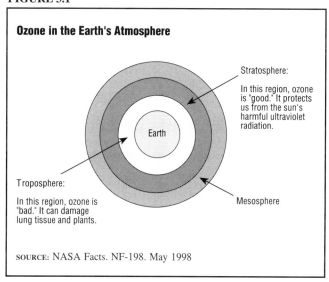

Ozone in the Earth's Atmosphere

Stratosphere: In this region, ozone is "good." It protects us from the sun's harmful ultraviolet radiation.

Earth

Troposphere: In this region, ozone is "bad." It can damage lung tissue and plants.

Mesosphere

SOURCE: NASA Facts. NF-198. May 1998

at the same rate that it leaks out, the amount of water remains constant. Similarly, as long as ozone is created at the same rate it is destroyed, the amount of ozone remains balanced. Scientists have found, however, that—since the 1970s—human activities have been altering that balance.

Ultraviolet radiation in the upper atmosphere breaks down CFC molecules, causing the destruction of ozone. Human production of certain chemicals, such as CFCs, adds to this destruction. Therefore, ozone loss exceeds ozone creation, as in the case of a leaking bucket forming extra leaks. In colder areas of the planet, such as Antarctica, where cloud and ice particles are present, reactions that hasten this ozone destruction also occur on the surface of ice particles.

Monitoring Ozone

In the mid-1970s scientists first began to speculate that the ozone layer was rapidly being destroyed by reactions high in the atmosphere involving chlorine- and

FIGURE 3.2

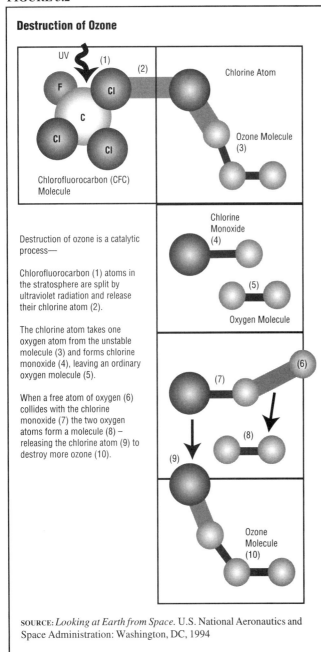

Destruction of Ozone

UV (1) (2)

F Cl

C

Cl Cl

Chlorofluorocarbon (CFC)
Molecule

Chlorine Atom

Ozone Molecule
(3)

Destruction of ozone is a catalytic
process—

Chlorofluorocarbon (1) atoms in
the stratosphere are split by
ultraviolet radiation and release
their chlorine atom (2).

The chlorine atom takes one
oxygen atom from the unstable
molecule (3) and forms chlorine
monoxide (4), leaving an ordinary
oxygen molecule (5).

When a free atom of oxygen (6)
collides with the chlorine
monoxide (7) the two oxygen
atoms form a molecule (8) –
releasing the chlorine atom (9) to
destroy more ozone (10).

Chlorine
Monoxide
(4)

(5)

Oxygen Molecule

(6)

(7)

(8)

(9)

Ozone
Molecule
(10)

SOURCE: *Looking at Earth from Space*. U.S. National Aeronautics and
Space Administration: Washington, DC, 1994

However, they are unsure how much of the loss is due to human activity and how much is the result of fluctuations in natural cycles.

The Measurement of Stratospheric Ozone

From the 1920s to the 1970s ozone was measured from the ground. Instruments around the globe measured the amount of ultraviolet radiation getting through the atmosphere at each location. From these measurements, scientists calculated the concentration of ozone in the atmosphere above that site. This data, however, did not provide an adequate picture of ozone levels around the globe. Since the late 1970s scientists have measured ozone levels from above the earth with satellites, sophisticated space instruments, aircraft, and balloons. As part of the National Aeronautics and Space Administration (NASA)'s Mission to Planet Earth program, numerous missions have been launched to study ozone. Table 3.1 shows the missions scheduled through 2002.

The Primary Culprit: CFCs

Although a number of chemicals can destroy stratospheric ozone, CFCs are the main offenders. (See Table 3.2.)When CFCs were invented in 1930, they were welcomed as chemical wonders. Discovered by Thomas Midgley Jr., they were everything the refrigeration industry needed at the time—nontoxic, nonflammable, noncorrosive, stable, and inexpensive. Their artificial cooling provided refrigeration for food and brought comfort to warm climates. The compound was originally marketed under the trademark Freon.

Over time, new formulations were discovered and the possibilities for use seemed endless. CFCs could be used as coolants in air conditioners and refrigerators, as propellants in aerosol sprays, in certain plastics such as polystyrene, in insulation, in fire extinguishers, and as cleaning agents. World production doubled every five years through 1970 and another growth spurt occurred in the 1980s as new uses were discovered—primarily as a solvent to clean circuit boards and computer chips.

Worldwide, aerosols still represent the largest use of CFCs, accounting for 25 percent of the total. CFCs are extremely stable; it is this stability that allows them to float intact through the troposphere (the layer of air nearest the earth's surface) and into the ozone layer. They reach the stratosphere after six to eight years. Once there, they can survive for up to 150 years. CFCs do not degrade in the lower atmosphere but, upon entering the stratosphere, they encounter the sun's ultraviolet radiation and eventually break down into chlorine, fluorine, and carbon. Many scientists believe it is the chlorine that damages the ozone layer.

In 1950 the worldwide production of CFCs was approximately 42,000 tons annually; in 1988, it peaked at

bromine-containing industrial chemicals. In 1984 British scientists at Halley Bay in Antarctica measured the ozone in the air column above them and discovered an "ozone hole." Measurements indicated ozone levels about 50 percent lower than they had been in the 1960s. Indeed, computer models revealed a hole larger than the United States and higher than Mount Everest.

Scientists conducted satellite, aircraft, balloon, and ground measurements to explain the cause. Since then, ozone losses have been documented around the globe. Each spring, for more than a decade, a hole in the ozone layer has appeared over Antarctica. In 1995, for the first time, a hole also appeared over the North Pole. Most scientists now think that ozone is being depleted worldwide.

TABLE 3.1

Selected Missions to Study Ozone

Mission	Launch	Scientific Objectives
Nimbus Series: Monitor of Ultraviolet Solar Energy (MUSE) Backscatter Ultraviolet Spectrometer (BUV) Solar Backscatter Ultra violet/Total Ozone Mapping Spectrometer (SBUV/TOMS)	1969-1978	Monitored solar energy controlling stratospheric ozone Mapped total global ozone and the vertical distribution of ozone in the atmosphere
Stratospheric Aerosol and Gas Experiments: (SAGE II)/Earth Radiation Budget Satellite (ERBS) (SAGE III)/Meteor-3M (Russia)	1984 1998	Map ozone at all latitudes
NOAA-9, 11, 14, L, N, (SBUV/2)	1984-2001	Map total ozone and the vertical distribution of ozone in the stratosphere
Shuttle Solar Backscatter Ultraviolet (SSBUV)	1989-2000	In-orbit calibration checks of satellite instruments
Total Ozone Mapping Spectrometer (TOMS) /Meteor-3 (Russia) /Earth Probe (U.S.) /ADEOS (Japan) /Meteor-3M (Russia)	1991-2000	Daily high-resolution global mapping of total ozone (TOMS was first flown on Nimbus-7 in 1978)
Upper Atmosphere Research Satellite (UARS)	1991	Monitor solar energy, upper atmospheric chemistry and dynamics
ATLAS/STS Payloads	1992-1995	Monitor solar energy, upper atmospheric chemistry and dynamics
Earth Observing System (EOS)	1998-	Comprehensive investigations of ozone processes, particularly with data from the EOS-Chemistry mission beginning in 2002, and SAGE III starting in 1998

SOURCE: NASA Facts. NF-198. U.S. National Aeronautics and Space Administration, May 1998

1.3 million tons. By the late 1990s, production had dropped to an estimated 300,000 tons. However, there is a significant lag between the decline of emissions at the earth's surface and the point at which ozone levels in the stratosphere recover. Estimates suggest that, if all countries comply with the Montreal Protocol (see discussion later) which banned the production of CFCs as of 1996, the ozone shield should begin to heal itself within two years, and full recovery could be expected by the middle of the twenty-first century.

BLACK MARKET FREON. Under the Montreal Protocol, developing countries are permitted to produce CFCs until January 1, 2010. It is illegal, however, to import CFCs into the United States. The ban on CFCs—including Freon, which was widely used for air conditioning automobiles—created a black market for the product. Demand by the approximately 80 million American owners of older cars, which still use the refrigerant in their air conditioning systems, caused prices to spiral upward.

Several U.S. government agencies—the Environmental Protection Agency (EPA), the Customs Service, the Commerce Department, the Internal Revenue Service, and the Justice Department—began an intensive anti-smuggling effort. In 2000 the EPA reported that, since the 1996 phaseout of CFCs began, approximately two million pounds of CFCs and other ozone-depleting substances had been seized and impounded, and more than 90 individuals and businesses had been charged with smuggling. Early arrests for CFC smuggling were made primarily in Miami; authorities then began to notice these illegal imports

TABLE 3.2

Ozone-Destroying Chemicals

Name	Use	When U.S. production ends*
CFCs (chlorofluorocarbons)	solvents, aerosol sprays (most spray can uses banned in 1970s) foaming agents in plastic manufacture	January 1, 1996
Halons	fire extinguishers	January 1, 1994
Carbon tetrachloride	solvents, chemical manufacture; carbon tetrachloride causes cancer in animals	January 1, 1996
Methyl chloroform (1,1,1-trichloroethene)	very widely-used solvent; in many workplace and consumer solvents, including auto repair and maintenance products	January 1, 1996
HCFCs (hydro CFCs)	CFC substitutes, chemicals slightly different from CFCs	January 1, 2003**

*The 1990 Clean Air Act includes a schedule for ending United States production of ozone-destroying chemicals and provisions for speeding up the phase out schedule if that is necessary. The dates in this table are "speeded-up" dates, proposed by EPA in early 1993

**Production of the HCFC with the most severe ozone destroying effects will end by January 1, 2003. Production of the rest of the HCFCs will end by January 1, 2030.

SOURCE: *The Plain English Guide to the Clean Air Act.* U.S. Environmental Protection Agency, 1993

creeping into other port cities, such as Houston and Baltimore. They also made their way into Europe, which banned CFCs one year before the United States.

The London-based Environmental Investigation Agency, in *A Crime Against Nature: The Worldwide Illegal Trade in Ozone-Depleting Substances* (1998), report-

TABLE 3.3

Estimated Number of CFC-12 Units in Operation in the U.S., 1999–2005

	Average Charge per Unit (pounds)	Average Annual Loss per Unit (Percent of Charge)	Estimated Number of CFC-12 Units in Operation						
			1999	2000	2001	2002	2003	2004	2005
Mobile Air Conditioners (MACs) source: (EPA 1998)	2.9	12%	55,000,000	44,000,000	35,000,000	27,000,000	20,000,000	13,000,000	7,000,000
Chillers source: EPA & ARI estimates	1,400	8%	7,500	6,000	4,500	3,000	1,500	0	0
Commercial Refrigeration (Supermarkets) source: EPA estimates	1,500	25%	6,600	6,000	5,100	4,200	3,200	2,200	1,200
Commercial Refrigeration (Cold Storage Warehouses) source: EPA estimates*	0.0022	15%	2,692	2,353	2,015	1,677	1,338	1,000	662
Refrigerated Transport source: EPA estimates	16.6	18%	185,000	148,000	120,000	92,000	63,000	32,000	2,000
Industrial Process Refrigeration source: EPA estimates	2,200	5%	6,200	5,500	4,800	4,100	3,500	3,000	2,500
Refrigerated Appliances source: EPA estimates	0.38	0.9%	102,000,000	94,000,000	87,000,000	80,000,000	72,000,000	66,000,000	60,000,000

*NOTE: For cold storage warehouses, the average charge per unit is reported in pounds per cubic feet and the estimated number of units in operation is reported in millions of cubic feet of refrigerated space.

SOURCE: *Report on the Supply and Demand of CFC-12 in the U.S. 1999*. U.S. Environmental Protection Agency, 1999

ed that the illegal trade in CFCs had moved from Russia, where progress had been made in controlling exports, to China. Successful smuggling of CFCs from China to the United States and Europe typically involves false labeling, counterfeit paperwork, and fake export corporations.

Although CFCs are no longer used in new applications, existing users can continue using them provided they are maintained under strict regulation, such as being replenished and "reclaimed" by authorized technicians. There are other exceptions to the ban, including medical inhalers, which commonly use CFCs as propellants. Refrigerated appliances account for, by far, the largest number of units using CFC, followed by mobile air conditioners. (See Table 3.3.) The EPA, in its *Report on the Supply and Demand of CFC-12 in the United States 1999* (1999), estimated that the number of units using CFC-12, the predominant form of CFC, will fall steadily through the year 2010.

Also, demand for CFC-12 is expected to decline as substitutes increase and aging equipment is replaced. The EPA report estimated demand for CFC-12 in 1999 at 23 million pounds. Mobile air conditioning accounted for 82 percent of annual demand, followed by chillers and refrigeration in cold-storage warehouses (4 percent each), industrial processes and commercial refrigeration in supermarkets (3 percent each), refrigerated transport (2 percent), and refrigerated appliances (1 percent). By 2002 U.S. consumers will demand approximately 12 million pounds of CFC-12, and by 2005, 3 million pounds. (See Table 3.4.)

The EPA's Significant New Alternative Policy (SNAP) program evaluates alternatives to ozone-depleting substances and determines their acceptability for use. Submissions for evaluation include those that could be used in a variety of industrial applications, including refrigeration and air conditioning, foam blowing, and fire suppression and protection.

Another Villain: Halon

While CFCs account for approximately 75 percent of ozone loss, other gases account for the rest. One of those gases is halon, which contains bromine, and is an even more damaging ozone destroyer than chlorine. Like chlorofluorocarbons, halon is inert at ground level but long-lived in the atmosphere. Halon is used primarily for fighting fires. Civilian and military fire-fighting training accounts for much of the halon emission. In 1992, 87 nations convened in Copenhagen, Denmark, and agreed to halt the manufacture of new halon as of January 1994. (Developing nations were given a 10-year grace period to phase out manufacture of CFCs and halon.)

Recycled halon and inventories produced before January 1, 1994, are the only supplies now available. It is legal under the Montreal Protocol and the U.S. Clean Air Act to import recycled halon, but each shipment requires approval from the EPA. Certain uses, such as fire protection, are classified as "critical use" and are permitted as long as supplies remain. The EPA maintains a list of acceptable substitutes for halon, although by 2000, none had been found with the exact properties of halon.

Methyl Chloroform

Another ozone destroyer is methyl chloroform, a man-made chemical used as an industrial solvent and cleaning agent. Scientists have detected the presence of methyl chloroform in the atmosphere since 1978 and have found it lingers in the lower atmosphere for about 4.6 years. By contrast, CFCs remain in the atmosphere for 50 to 150 years. Many scientists believe this is encouraging evidence that short-lived man-made chemicals can eventually be eliminated from the atmosphere.

Under the Montreal Protocol, methyl chloroform is to be phased out by 2005. Dr. A. R. Ravishankara, a chemist with the National Oceanic and Atmospheric Administration (NOAA), believes that the decline of methyl chloroform levels noted in recent years is proof that the Montreal Protocol is working. "If the other compounds that have been slated for reduction are being phased out, we should see the ozone hole disappear by 2050," said Dr. Ravishankara.

WHY DO HUMANS NEED THE OZONE LAYER?

The ozone layer acts as a protective shield against ultraviolet radiation. As ozone diminishes in the upper atmosphere, the earth receives more ultraviolet radiation. Increased radiation, especially of the frequency known as ultraviolet-B (UV-B), the most damaging wavelength, promotes skin cancers and cataracts, suppresses the human immune system, and produces wrinkled, leather-like skin. It also reduces crop yields and fish populations, damages some materials such as plastics, and creates smog.

Threats to Human Health

The skin is the largest organ in the human body. It covers and protects the organs inside the body. Globally, the incidence of skin cancer is rising. There are two types of skin cancer: non-melanoma and melanoma.

The American Cancer Society reported in 2000 that approximately 1.3 million cases of non-melanoma skin cancer are diagnosed in the United States each year and predicted that 1,900 people will die from non-melanoma skin cancer in 2000. The incidence of cancer is closely tied to cumulative exposure to ultraviolet radiation. Each 1 percent drop in ozone is projected to result in a 4 to 6 percent increase in these types of skin cancer.

Melanoma skin cancer is less common but far more deadly, accounting for just 4 percent of skin cancer cases but a whopping 79 percent of skin cancer deaths. Melanoma is more likely to metastasize (spread to other parts of the body, particularly major organs such as the lungs, liver, and brain). The American Cancer Society predicted that, in 2000, 47,700 Americans would be diagnosed with melanoma and 7,700 would die from the dis-

TABLE 3.4

Estimated Demand for CFC-12 in the U.S., 1999–2005

	Estimated Demand for CFC-12 (million pounds)						
	1999	2000	2001	2002	2003	2004	2005
Mobile Air Conditioners (MACs)	18.9	15.1	12.0	9.3	6.9	4.5	2.4
Chillers	0.84	0.67	0.50	0.34	0.17	0.00	0.00
Commercial Refrigeration (Supermarkets)	2.5	2.3	1.9	1.6	1.2	0.8	0.5
Adjusted Commercial refrigeration (Supermarkets)**	0.8	0.7	0.6	0.5	0.4	0.3	0.1
Commercial Refrigeration (Cold Storage Warehouses)	0.9	0.8	0.7	0.6	0.4	0.3	0.2
Refrigerated Transport	0.55	0.44	0.36	0.27	0.19	0.10	0.01
Industrial Process Refrigeration	0.68	0.61	0.53	0.45	0.39	0.33	0.28
Refrigerated Appliances	0.34	0.32	0.29	0.27	0.24	0.22	0.20
Total Demand	**23**	**19**	**15**	**12**	**9**	**6**	**3**

** NOTE: Two-thirds of the demand for R-12 in supermarkets is assumed to be handled internally through refrigerant management plans. Therefore, because most national and regional supermarket chains have refrigerant management plans (RMPs), they do not demand any R-12 from the open market. On the other hand, smaller grocery stores (which do not have the capital or resources needed to operate an RMP) are the only portion of this end-use that may need to purchase R-12 from the open market. Therefore, this analysis assumes that one-third of the units in operation are located in smaller grocery and convenience stores.

SOURCE: *Report on the Supply and Demand of CFC-12 in the U.S. 1999.* U.S. Environmental Protection Agency, 1999

ease. The EPA estimates that ozone depletion will result in an additional 31,000 to 126,000 cases and 7,000 to 30,000 fatalities among white Americans born before 2075. Melanoma appears to be associated with acute radiation exposure, such as severe sunburns, which are more likely to occur when the ozone hole is larger.

The EPA estimates that, among Americans born before 2075, depletion of the ozone layer could be responsible for 555,000 to 2.8 million additional cases of cataracts. Cataracts cause clouding of the eye's lens, which results in blurred vision and—if left untreated—blindness. It is also expected that victims will be stricken at younger and younger ages.

Some medical researchers theorize that ultraviolet radiation also depresses the human immune system, lowering the body's resistance to tumors and infectious diseases.

Damage to Plant and Animal Ecosystems

Terrestrial and aquatic ecosystems are also affected by the depletion of the ozone layer. Ultraviolet radiation alters photosynthesis, plant yield, and growth in plant species. Phytoplankton, one-celled organisms found in the ocean, are the backbone of the marine food web. Studies have found that a 25 percent reduction in ozone would decrease their productivity by about 35 percent. Fish species that live solely on phytoplankton would likely disappear. As ecosystems are altered, important fish species will become vulnerable. These fish are a vital component in feeding the increasing human population. In addition, scientists have identified the rise in ultraviolet radiation caused by the thinning of the ozone layer as the culprit behind the decline in the number of frogs and other amphibians.

Deterioration of Materials

Increased ultraviolet radiation also affects synthetic materials. Plastics are especially vulnerable, tending to weaken, become brittle and discolored, and break.

A LANDMARK IN INTERNATIONAL DIPLOMACY: THE MONTREAL PROTOCOL

CFCs and halons were widely used in thousands of products; they represented a significant share of the international chemical industry, with billions of dollars in investment and hundreds of thousands of jobs. Ozone depletion was a global problem that necessitated international cooperation, but nations mistrusted one another's motives. As with the issues of global warming and pollution, developing countries resented being asked to sacrifice their economic development for a problem they felt the industrialized nations had created. To complicate matters, gaps in scientific proof led to disagreements over whether a problem actually existed.

In 1985, as the first international response to the ozone threat, 20 nations signed an agreement in Vienna, Austria, calling for data gathering and cooperation and a political commitment to take action at a later date. In 1987 negotiators meeting in Montreal, Canada, finalized a landmark in international environmental diplomacy: the Montreal Protocol on Substances That Deplete the Ozone Layer. The agreement called for industrial countries to cut CFC emissions in half by 1998 and to reduce halon emissions to 1986 levels by 1992. Developing countries were granted deferrals to compensate for their low levels of production. Industrial countries agreed to reimburse developing countries that complied with the protocol for "all agreed incremental costs," which means all additional costs above any they would have expected to incur had they developed their infrastructure in the absence of the accord. And, very importantly, the protocol also called for further amending as new data became available.

The Montreal Protocol was hailed as a historic event—the most ambitious attempt ever to combat environmental degradation on a global scale. It was signed on the spot by 24 nations and the European Community, and has since been ratified by 175 countries. It ushered in a new era of environmental diplomacy.

Some historians view the signing of the accord as a defining moment, the point at which the definition of international security was expanded to include environmental issues as well as military matters. In addition, an important precedent was established—that science and policymakers had a new relationship. Many observers thought that the decision to take precautionary action in the absence of complete proof of a link between CFCs and ozone depletion was an act of foresight that would now be possible with other issues.

Two major problems have emerged in implementing the Montreal Protocol. The economic and political chaos in the former USSR and in some parts of Eastern Europe has slowed progress in eliminating CFCs. The nine countries that existed in this part of the world when the accord was first signed have since divided into 27, only 21 of which have signed the protocol. The second stumbling block is the growing black market in smuggled CFCs that threatens to undermine the phase-out in industrial countries. Russia, China, and India are believed to be major sources of this trade.

THE RECORD TO DATE

In the industrial world, many countries did more than was required by the protocol. As a result, when the official CFC phaseout date arrived, most industrial nations were ready, some phasing out ozone-depleting substances before they were required to. By the end of 1994, the European Union no longer permitted CFCs. In addition to banning Freon in 1997, the Clean Air Act required the United States to end the use of methyl bromide by 2001—nine years ahead of protocol requirements.

Title VI of the 1990 Clean Air Act Amendments (PL 101-549) is the United States's primary response to ozone depletion. As part of the act, the U.S. Congress approved a provision requiring the president to speed up the schedule for chemical phaseout if new evidence warranted it. In fact, new data did become available, including worrisome evidence that showed ozone depletion was occurring over the Northern Hemisphere.

New Evidence

Although the ozone agreement was a major achievement in international negotiations, the world community could not enjoy the success for long. In 1991 new scientific information revealed that ozone depletion was occurring twice as fast as expected. This news spurred a call to again revise the treaty.

In 1992, spurred by further evidence that the protective shield was more depleted than feared, representatives from 87 nations met in Copenhagen to advance the deadline for halting production of CFCs to January 1996, and for halons to January 1994. The delegates also set a deadline for eliminating hydrochlorofluorocarbons (HCFCs), which are temporary substitutes for the more potent CFCs. Their phase-out was to begin in 2004 and end no later than 2030.

FIGURE 3.3

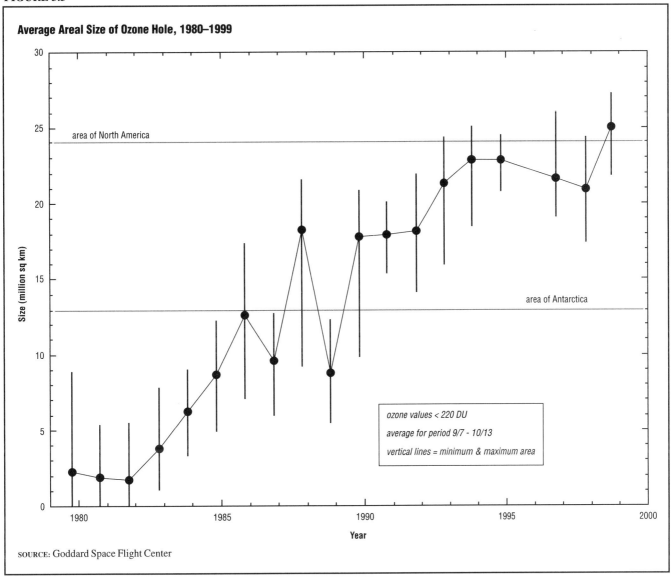

Average Areal Size of Ozone Hole, 1980–1999

ozone values < 220 DU

average for period 9/7 - 10/13

vertical lines = minimum & maximum area

SOURCE: Goddard Space Flight Center

In 1993 satellite measurements indicated a 10 to 20 percent reduction in ozone levels over parts of Canada, Scandinavia, Russia, and Europe from 1992 levels when the ozone hole extended more than 9.4 million square miles. In 1995 the hole covered an area twice the size of Europe and was growing at the unprecedented rate of 1 percent per day.

In 1996 researchers announced that, for the first time, the quantity of ozone-destroying chemicals in the atmosphere was declining. This meant that ozone destruction could peak by the end of the decade, as scientists previously predicted. The new measurements, taken in the United States, Canada, Australia, Antarctica, Samoa, and Hawaii by the NOAA, showed an outright reduction in chlorine concentrations. Scientists predicted that it would take several more years for the air masses at ground level to migrate to the stratosphere and affect the ozone. They cautioned that the decline in the quantity of ozone-destroying chemicals in the atmosphere would have to

continue for a decade or more before the ozone layer would start to recover.

NASA's Goddard Space Flight Center has measured the size of the ozone hole over Antarctica every year since 1980. These annual measurements are always taken during the same five-week period, from early October to mid-November. Figure 3.3 shows the average area of ozone depletion (in Dobson units, or DUs) during the period from 1980 to 1999 in both North America and Antarctica. Average ozone loss was higher in Antarctica than in North America.

In 2000 the British Broadcasting Corporation (BBC) reported the results of the most recent study of ozone levels over the Arctic. In a cooperative effort between NASA's Sage III Ozone Loss and Validation Experiment and the Third European Stratospheric Experiment on Ozone sponsored by the European union, a range of satellites, aircraft, balloons, and ground-based instruments collected data from above the polar region between January and March 2000.

The international group of researchers found cumulative ozone losses of more than 60 percent about 11 miles above the area. The scientists noted that these were among the largest chemical losses observed at that altitude in the past decade. The European Union stated that the report did not point to a hole in the ozone layer, such as the one over the Antarctic, but to a diminishment of ozone content in the stratosphere. Scientists expressed concern, however, since this was clear evidence that ozone depletion was not a problem confined to the Southern Hemisphere.

LOOKING TO NEW TECHNOLOGIES

To phase out CFCs and halons, substitute chemicals and technologies began to be developed. However, new substitutes cost more and still have to be tested for safety and performance. Should the new chemical replacements not perform well in safety, health, and marketplace tests, political leaders face tough decisions: either shut down major parts of modern society's technology, or accept a slower pace of substitution with the accompanying harm to the environment.

Under pressure from environmentalists, industrial engineers seek out new technologies, including semiconductors that cool down when charged with electricity, refrigeration that uses plain water as a refrigerant, and sound energy, or what is called "thermoacoustics." Also, some experts proposed returning to the refrigerant gases that were used before the invention of CFCs. These include sulfur dioxide, ammonia, and various other hydrocarbon compounds, which are often toxic.

CHAPTER 4
WASTE DISPOSAL

About 500 B.C.E. ("before the common era"), Athens in Greece issued the first known edict against throwing garbage into the streets and organized the first municipal dumps by mandating that scavengers transport wastes to no less than one mile from the walls of the city. This method was not practiced in medieval Europe (c. 500–1485). Parisians in France and Londoners in England continued to toss trash and sewage out their windows until the 1800s. The west end of London and the west side of Paris became fashionable in the late seventeenth and the eighteenth centuries because the prevailing winds blew west to east, carrying the smell of rotting garbage with them.

Industrialization brought with it a greater need for collection and disposal of refuse. Garbage was transported beyond the city limits and dumped in piles in the countryside. As cities grew, the noxious odors and rat infestations at the dumps became intolerable. Freestanding piles gave way to pits, but that solution soon became unsatisfactory.

In 1874 the first systematic incineration (burning) of municipal waste was tested in England. Burning reduced waste volume by 70 to 90 percent, but the expense of building incinerators and the reduced air quality caused many cities to abandon the method. Waste burial remained the most widely practiced form of disposal.

EARLY EFFORTS TO MAKE USE OF TRASH

Most Americans produced little trash until the twentieth century. Food scraps were boiled into soups or fed to animals, which returned the favor with eggs or meat. Durable items were passed on to the next generation or to people more in need. Objects that were of no further use to adults became toys for children. Broken items were repaired or dismantled for reuse. Many Americans possessed the skills required for repairing. Things that could no longer be used were burned for fuel, especially in the homes of the poor. Even middle-class Americans traded rags to peddlers in exchange for buttons or tea kettles. These "ragmen" worked the streets, begging for or buying for pennies items such as bones, paper, old iron, rags, and bottles. They then sold the "junk" to dealers who marketed it to manufacturers.

Spending time to prolong the useful lives of items and to use scraps saved money. In 1919, in *Save and Have, A Book of 'Saving Graces' for American Homes* (New York), the University Society discussed habits of thrift for American housewives. The Society recommended keeping cake fresh by storing it with an apple, and "turning" worn sheets by tearing them down the middle and sewing up the good sides, extending their life. Other suggestions included collecting grease to make soap, reusing flour sacks for dish towels or clothing, using jars for drinking glasses, and keeping a can on the stove containing grease that could be used over and over.

Besides giving away clothes, mending and remaking them, and using them as rags for work, women reworked textiles into useful household furnishings such as quilts, rugs, and upholstery. Rags were also important materials collected for recycling in factories: Paper mills used rags to make paper, and a growing paper industry made it profitable for thrifty housewives to save rags.

This trade in used goods provided crucial resources for early industrialization, but these early systems of recycling began to pass into history around the turn of the twentieth century. Sanitary reformers and municipal trash collection did away with scavenging. Technology made available cheap and new alternatives. People made fewer things, and they bought more than previous generations had. They saved and repaired less and threw out more.

However, old-fashioned reuse and recycling did not cease overnight. During the first decades of the twentieth century, most people still threw away relatively little as

FIGURE 4.1

Philadelphia sanitation workers dump their horse drawn garbage cart in the street during a strike in 1938. The way Americans deal with their garbage changed tremendously during the twentieth century. (*Corbis Corporation. Reproduced by permission.*)

old ways still survived in the midst of the new. Nonetheless, as the century progressed, middle-class people learned to throw things away, attracted by convenience and wanting to avoid any association with scavenging and poverty. Success often meant that one did not have to use second-hand things. As municipalities became responsible for collecting and disposing of refuse, Americans found it easier to throw things out.

DOING SOMETHING ABOUT GARBAGE

Throughout the nineteenth century many cities passed anti-dumping ordinances, but many people ignored them. Some landowners and merchants resented ordinances, which they considered infringements of their rights. As cities grew so did garbage, becoming not only a public eyesore but a threat to public health.

City leaders began to recognize that they had to do something about the garbage. By the turn of the twentieth century, most major cities had set up garbage collection systems. (See Figure 4.1.) Many cities introduced incinerators to burn some of the garbage. By the time the First World War began in 1914 about 300 incinerators were operating in the United States and Canada.

Cities located downstream from other cities that were pouring their garbage into rivers sued the upstream cities because the water was polluted. As a result, more and more cities stopped dumping their garbage into rivers and began to build landfills or garbage dumps to get rid of their waste. Many coastal cities began to take their refuse out into the ocean and dump it, although much of the garbage they poured into the sea washed back to dirty the beaches.

Some health officials and reformers knew that pollution was unhealthy and could lead to sickness, but most people were not concerned about the long-term effects of garbage and pollution on the environment. As the twentieth century progressed, more and more Americans became concerned that garbage and pollution were harming the environment.

IN A CONSUMER SOCIETY—A GARBAGE GLUT

The United States is facing a problem with its ever-growing mountains of garbage. America generates more garbage than any other nation on Earth, twice as much per person as in Europe. As with most environmental issues, waste disposal has grown to crisis proportions. The cost of handling garbage is the fourth biggest item—after education, police, and fire protection—in many city budgets. Some 80 percent of the nation's solid waste is being dumped in landfills, but sites are rapidly filling up and many are leaking toxic substances into the nation's water supply.

The more that is learned about garbage, the more apparent it is that trucking garbage to landfills does not necessarily eliminate it. As a result, municipal governments worldwide are struggling to find the best methods for managing waste. Growing populations, rising incomes, and changing consumption patterns have combined to complicate the waste management problem. Garbage generation expands as a city grows in size and, as consumers earn more money, their demand for consumer goods increases. This includes "convenience foods" with their packaging that immediately becomes waste. Other convenience items, such as disposable diapers, also add to the mountains of waste.

Before the days of densely populated urban areas, waste disposal was eased by the apparent ability of the surrounding land and water to absorb that waste. At the beginning of the twentieth century, farm communities created little waste. By the end of that century, farm communities produced waste in almost similar proportion to urban areas, not to mention the runoff created by insecticides and fertilizers used in food-producing fields.

One-Time Use

The widespread human appetite for all materials has defined this century in much the same way that stone, bronze, and iron characterized previous eras.
—United States Geological Survey, *Mineral Commodity Summaries*, 1998

The scale of materials use by Americans, Europeans, Japanese, and other industrialized countries dwarfs that of a century ago. The stock of materials used by the dawn of 2000 drew from all 92 naturally occurring elements in the periodic table compared with just 20 as the century began. The United States Geological Survey (USGS) estimates that, in the United States alone, consumption of metal, glass, wood, cement, and chemicals has grown 18-fold since 1900 and that the nation uses one-third of all materials used throughout the world.

Much of the waste disposal problem exists because of a growing population and because most consumer goods are designed for short-term use. This contrasts sharply with the practices of earlier eras when materials were reused or transformed into other uses. As Susan Strasser notes, in *A Social History of Trash* (New York: Henry Holt and Co., 1999), more and more things today are being made and sold with the understanding that they will soon become worthless or obsolete.

The volume of waste increases with income—poor neighborhoods generate lower amounts of solid wastes per capita than richer neighborhoods. An inventory of what Americans throw away would reveal valuable metals, paper representing millions of acres of trees, and plastics incorporating highly refined petrochemicals.

During the late twentieth century, however, interest in recycling grew and most states began making recycling an important part of waste collection and disposal. Nonetheless, the nation's garbage dumps began filling up and new ones had to be constructed. Most states began sending some of their garbage to other states that would accept it, or even to other countries. By the beginning of the twenty-first century, some states and countries would no longer accept other people's garbage.

The Sorting Process

Nothing is inherently trash. Trash is produced by a human behavior called sorting. Items in people's lives eventually require a decision—to keep or to throw away. Some things go here; some things go there.

The sorting process varies from person to person, from place to place, and changes over time. What is considered rubbish changes from decade to decade. Some societies value saving things more than others: Nomadic people, who must travel light, save less. Age matters as well. The youth of the late twentieth and early twenty-first century have more readily adopted the notions of convenience and disposability than their parents and grandparents, and are less likely to conserve.

Sorting is also an issue of class. Trash making creates social differences based on economic status. The wealthy can afford to be wasteful and discarding things may often be a way of demonstrating wealth and power.

At the turn of the millenium people in developed nations discarded things for reasons unheard of in developing nations or in earlier times—because they no longer wanted them. Disposing of out-of-style clothes and outmoded equipment reflected a worship of newness that was not widespread before the twentieth century. Dump-sters filled with "perfectly good stuff" that was simply not new anymore—stuff the owner had tired of.

Economic growth during the twentieth century was fueled—in part—by a growing consumer culture that demanded a continual supply of new products, disposables, and individually packaged consumer items. This, combined with "planned obsolescence" on the part of manufacturers, produced increasing volumes of garbage. Colored plastic trash bags represent the contemporary attitude about trash, far from the homemade soup, darned underwear, and flour-sack dresses of an earlier time.

HOW MUCH GARBAGE?

As the United States became richer, the nation produced more garbage and pollution. The Environmental Protection Agency (EPA), the government agency whose job it is to protect the environment, estimates that the average American man, woman, or child throws away about three-fourths of a ton of garbage a year. Since 1960 America's population has grown 34 percent, but the amount of garbage produced has increased 80 percent.

The EPA estimates that somewhere between 12 and 13 billion tons of garbage are produced annually. Of this figure about 0.7 billion tons are hazardous waste and 11 to 12 billion tons are non-hazardous waste.

Most of the waste that people see is produced by ordinary households throwing out their uneaten food, yesterday's newspapers, packaging materials, lawn clippings, and branches from bushes and trees. This is the type of garbage that the EPA calls municipal solid waste (MSW). Garbage produced by industry and agriculture is classified as non-household waste.

Municipal Solid Waste

According to the EPA in *Municipal Solid Waste Generation, Recycling, and Disposal in the United States: Facts and Figures for 1998* (2000), in 1998, Americans produced 220.2 million tons of MSW, up from 205.2 million tons in 1990 and 88.1 million tons in 1960. Per capita generation held at about 4.46 pounds per person per day, up from 2.7 pounds per day in 1960. (See Figure 4.2.) Municipal solid waste represented only about 2 percent of all solid waste generated in 1998. The rest was non-household waste.

NON-HOUSEHOLD WASTES

According to the EPA more than half (57 percent, or seven billion tons) of all 1998 wastes resulted from industrial wastes (waste produced by factories and manufacturers). Special wastes (mining, oil and gas production, electric utilities, cement kilns, and agriculture) comprised 39 percent of solid waste in 1998.

FIGURE 4.2

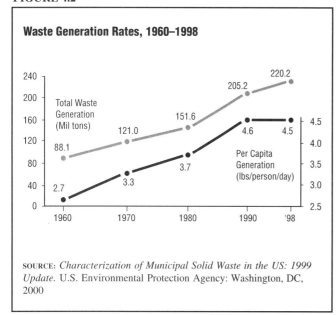

Waste Generation Rates, 1960–1998

SOURCE: *Characterization of Municipal Solid Waste in the US: 1999 Update.* U.S. Environmental Protection Agency: Washington, DC, 2000

FIGURE 4.3

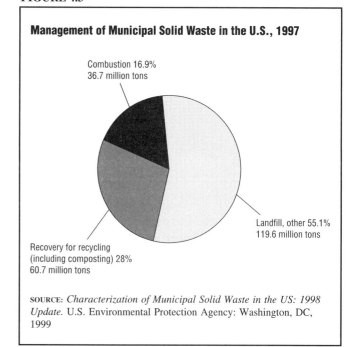

Management of Municipal Solid Waste in the U.S., 1997

SOURCE: *Characterization of Municipal Solid Waste in the US: 1998 Update.* U.S. Environmental Protection Agency: Washington, DC, 1999

Manufacturing produces huge amounts of waste. The paper industry, which uses many chemicals to produce paper, accounted for the largest proportion of manufacturing wastes—35 percent of the total. The iron and steel (20 percent) and chemical industries (14 percent) produced most of the rest. Many of the big manufacturing plants have sites on their own property where they can dispose of waste or treat it so it will not become dangerous. Still others ship it to private disposal sites for dumping or for treatment. Smaller manufacturers might use private waste disposal companies or even the city garbage company.

Mining also produces much waste, most of it rock and tailings. Normally, miners have to move rock to retrieve the ore or minerals. Tailings are left over after miners have sifted through the rocks and dirt for the ore or minerals. Chemicals used to remove minerals from ore become waste after they have done their job. Sometimes these chemical wastes are liquid, and sometimes solid. Either way, they must be disposed of appropriately so as not to pollute the environment.

Almost all (96–98 percent) of the waste from gas and oil drilling is water. Water is either pumped out of the ground before the oil is found, or it is found mixed with oil. This water is often salt water and it must be separated from the oil and gas before those natural products can be turned into refined products for use in automobiles or home heating. Other waste comes from mud and rock extracted by the drilling process. Most oil and gas companies dispose of their own waste.

When an electric company burns coal to heat water to make electricity, about 90 percent of the coal is burned up but it leaves about 10 percent in the form of ash. This waste must then be discarded somewhere.

HOW DO HUMANS DISPOSE OF GARBAGE?

There are four general ways to get rid of garbage:

• Garbage can be dumped into a landfill or "garbage dump."

• Trash can be burned in incinerators.

• Refuse can be recovered and recycled.

• Garbage can be composted.

The EPA's *Characterization of Municipal Solid Waste in the United States: 1998 Update* reported that more than 55 percent of MSW goes to landfills, while 28 percent is recovered for recycling and almost 17 percent is incinerated. (See Figure 4.3.)

Manufacturing uses all four methods to get rid of waste. Almost all of the garbage produced by cities and construction wastes ends up in landfills. Refuse that is incinerated produces residue that is then processed using one of the three methods.

LANDFILLS

Landfills are areas set aside specifically for garbage dumping. There are several types of landfills. In the most common type, garbage is dumped into a large pit and ultimately buried with earth. One is called "surface impoundment," a large pond in which liquid wastes can be stored and then treated so they can be safely disposed of. Almost all wastes from oil and gas production, mining, and agriculture end up in surface impoundments. Another is "land application" where waste is taken to a designated area and spread over the surface of the land. Garbage may also be

FIGURE 4.4

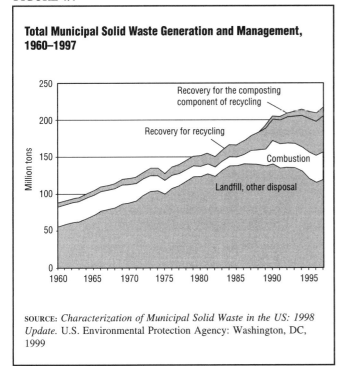

Total Municipal Solid Waste Generation and Management, 1960–1997

SOURCE: *Characterization of Municipal Solid Waste in the US: 1998 Update.* U.S. Environmental Protection Agency: Washington, DC, 1999

TABLE 4.1

Number of Municipal Solid Waste Landfills and Incinerators, Average Tip Fees, and Capacity, by State, 1999

	Landfills			Incinerators		
State	**Number**	**Average Tip Fee ($/ton)**	**Remaining Capacity (years)**	**Number**	**Average Tip Fee ($/ton)**	**Daily Capacity (tons/day)**
Alabama	29	33	10+	1	40	700
Alaska	239	70	n/a	4	80	210
Arizona	47	22	n/a	0	–	–
Arkansas	23	24	30	1	15	30
California	184	39	28	3	n/a	4,940
Colorado	115	n/a	n/a	1	n/a	n/a
Connecticut	1	n/a	n/a	6	n/a	7,360
Delaware	3	59	30	0	–	–
Dist. of Columbia	0	–	–	0	–	–
Florida	57	43	n/a	13	55	19,200
Georgia	70	27	23	1	59	500
Hawaii	10	47	3-50	1	81	2,000
Idaho	29	n/a	n/a	0	–	–
Illinois	58	28	17	1	n/a	1,400[1]
Indiana	37	n/a	12	2	n/a	2,175
Iowa	61	32	25	1	47	n/a
Kansas	54	23	n/a	1	50	15
Kentucky	35	n/a	n/a	0	–	–
Louisiana	23	25	n/a	0	–	–
Maine	8	55	7-10	4	55	2,850
Maryland	22	58	10+	4	n/a[2]	3,650
Massachusetts	43	n/a	1.8	9	n/a	8,600
Michigan	53	n/a	17	4	n/a	n/a
Minnesota	22	45	7.5	9	50	4,680
Mississippi	19	18	15	1	30	150
Missouri	26	27	5-10	2	n/a	n/a
Montana	33	13	1,546	0	–	–
Nebraska	24	27	n/a	0	–	–
Nevada	26	23	50+	0	–	–
New Hampshire	4	60	10	2	55	700
New Jersey	12	60	n/a[3]	5	51	6,490
New Mexico	77	25	50	0	–	–
New York	28	n/a	n/a[4]	10	n/a	12,000[5]
North Carolina	39	31	n/a	1	34	380
North Dakota	14	25	40+	0	–	–
Ohio	49	30	20	0	–	–
Oklahoma	41	18	20+	1	44	1,125
Oregon	29	25	40+	2	67	600
Pennsylvania	49	n/a	12	5	n/a	8,950
Rhode Island	4	40	10	0	–	–
South Carolina	19	32	n/a[6]	1	60	700
South Dakota	15	32	n/a	0	–	–
Tennessee	51	30	5-10	2	38	1,250
Texas	184	25	31	9	n/a	n/a
Utah	36	n/a	2%	1	n/a	420[7]
Vermont	5	70	12.5[8]	0	–	–
Virginia	65	n/a	n/a	6	n/a	n/a
Washington	21	n/a	34	5	n/a	n/a
West Virginia	19	45	n/a	0	–	–
Wisconsin	46	30	6+	2	46	350
Wyoming	58	n/a	n/a	1	n/a	20
Total	**2,216**			**122**		

[1] Average ton/day capacity based on total of 422,000 tons received in 1999 at state's only incinerator. Of total received, 261,000 tons processed—118,000 tons recycled and 43,000 tons landfilled;

[2] Tipping fee range in state is $35 to $67.50/ton;

[3] 30 million tons of capacity remaining with 660,000 tons being added;

[4] 50 million tons of capacity remaining with 62 million tons being added;

[5] Average ton/day capacity based on total of 3.6 million tons/year of capacity;

[6] 76.3 million tons of capacity remaining with about 1.2 million tons being added;

dumped onto a waste pile on the ground where it is stored and may eventually be treated.

Landfills are popular because, when compared with the cost of alternative disposal methods, dumping waste in the ground is a relatively cheap solution to an immediate problem. About 55 percent of all of America's solid waste ends up in landfills. (See Figure 4.3.) Because landfills are becoming filled and use of other disposal methods is increasing, that percentage is beginning to decline. (See Figure 4.4.)

Biocycle magazine's twelfth annual *State of Garbage in America* survey reported that, in 1999, 2,216 landfills operated across the United States, down from 2,314 in 1998 and approximately 8,000 in 1988. Among the states that reported its data, their remaining landfill capacity ranged from 1.8 years in Massachusetts to 1,546 years in Montana. (See Table 4.1.)

The growing amount of waste has led to a depletion of landfill capacity. According to the EPA and the now-defunct U.S. Office of Technology Assessment (OTA), 75 to 80 percent of the nation's active landfills will close over the next 10 to 15 years. Cities that exhaust their landfills are forced to find sites elsewhere, usually in more remote areas, which increases cost for transportation and landfill fees. Most of the states reporting less than 10 years of landfill capacity were located in the eastern half of the United States.

New landfills are becoming harder to find. Many local officials blame the lack of new facilities on public opposition, the "not in my backyard" syndrome rather than a lack

TABLE 4.1

Number of Municipal Solid Waste Landfills and Incinerators, Average Tip Fees, and Capacity, by State, 1999 [CONTINUED]

[7] Average ton/day capacity based on total of 125,000 tons/year of capacity;

[8] 12.5 years remaining capacity if landfill at permitted rate (6.1 years if all MSW went only to Vermont landfills—currently 30% goes out of state).

SOURCE: *The State of Garbage in America.* Biocycle, vol. 41, no. 4, p. 34, April 2000

---

of available space. Landfills, like prisons and interstate highways, have never been welcome in neighborhoods.

Many states began exporting their garbage to other states that would accept waste from out of state. States in the Northeast were most likely to export garbage and the least likely to import it. (See Table 4.2.) Some local governments are shipping waste abroad, often to underdeveloped countries. Because this activity is unregulated, there are no statistics on the amount exported, but it is thought to be a growing practice.

Some states have tried to enact bans on importing garbage into their states, but the Supreme Court, in *Chemical Waste Management v. Hunt* (112 S. Ct. 2009, 1992), ruled that such shipments are protected by the constitutional right to conduct commerce across state borders. (States accepting garbage charge fees for garbage dumping.) Newer, state-of-the-art landfills are now being built with multiple liners to prevent leaks and equipment to treat emissions. (See Figure 4.5.) Experts point out that many of these landfills will have to accept waste from a wide region to be financially viable. According to industry executives, it may cost as much as $400,000 an acre to build these state-of-the-art landfills.

Perhaps the most ominous example of the nation's garbage situation sits on New York's Staten Island. (See Figure 4.6.) Listed in the *Guinness Book of World Records* as the largest anywhere, the 50-year-old Fresh Kills landfill is about as big as the largest of the Egyptian pyramids in height and volume. The landfill operated 24 hours a day, six days a week, receiving about 10,000 tons of waste per day. By its scheduled closing on December 31, 2001, it would be more than 500 feet tall and rival the Great Wall of China as the largest human-made structure in the world. Fresh Kills, which opened in 1948 on a swampy lowland, would become the tallest "mountain" on the Atlantic coast between Florida and Maine. The city proposed turning it into a grass-covered park.

Cleaning Up

Experts agree that even the most advanced landfills will eventually leak, releasing hazardous materials into surface or underground water. Methane, a flammable gas, is produced when organic matter decomposes in the absence of oxygen.

TABLE 4.2

Municipal Solid Waste Generation, Waste Imports and Exports, by State, 1999

State	Generation (unless noted) (tons/yr)	Imported (tons/yr)	Exported (tons/yr.)
Alabama	5,700,000[1]	210,000	n/a
Alaska	675,000[2]	0	20,000
Arizona	5,187,000	422,400	<1,000
Arkansas	1,643,000[2]	n/a	n/a
California	59,700,000[1]	12,700	791,500
Colorado	6,455,000[3]	n/a	n/a
Connecticut	3,157,000[2]	234,000	267,100
Delaware	823,000[1]	0	n/a
Dist. of Columbia	220,000[4]	n/a	800,000
Florida	28,585,000[2, 5, 6]	n/a	n/a
Georgia	11,420,000[7]	453,900	n/a
Hawaii	1,873,000[3]	0	0
Idaho	794,000[2]	0	65,500[2]
Illinois	13,515,000[6, 8]	15,978,000	n/a
Indiana	12,000,000[1, 2]	2,185,000	220,000
Iowa	3,500,000[6]	425,000	100,000
Kansas	3,000,000	500,000-700,000	17,000
Kentucky	4,077,000[3, 7]	475,400	n/a
Louisiana	4,600,000	n/a	n/a
Maine	1,635,000[6, 9]	n/a	n/a
Maryland	6,300,000[2]	46,300	1,304,500
Massachusetts	7,936,000[2]	55,500[2]	945,700[2]
Michigan	19,500,000[7]	2,116,600	n/a
Minnesota	5,298,000[2]	n/a	450,000
Mississippi	2,264,000[2]	799,900	0
Missouri	8,013,000[2, 3]	143,400	1,551,400
Montana	1,082,000[2, 6]	30,000	n/a
Nebraska	2,200,000[1, 2]	125,000	20,000
Nevada	3,389,000[2]	513,000	0
New Hampshire	1,284,000[2]	742,000[9]	93,700[9]
New Jersey	7,800,000[2]	600,000	1,600,000
New Mexico	2,732,000[2, 6]	n/a	n/a
New York	29,900,000[3, 9]	300,000[2]	4,600,000[9]
North Carolina	13,000,000[1]	91,000	1,200,000
North Dakota	498,000[6]	54,800	5,500
Ohio	12,428,000[10]	n/a	n/a
Oklahoma	3,545,000[3, 10]	n/a	n/a
Oregon	4,302,000[2, 6]	1,248,500	19,900
Pennsylvania	10,337,000[6]	7,974,500	300,000
Rhode Island	421,000[11]	0	0
South Carolina	9,409,000[1]	862,900	57,400
South Dakota	514,000	n/a	n/a
Tennessee	9,213,000[2]	297,100[2]	64,000[2]
Texas	36,401,000[2, 12]	35,700[2]	395,000[2]
Utah	2,188,000[2]	11,400	n/a
Vermont	598,000[2, 13]	n/a	n/a
Virginia	8,136,000[2]	n/a	n/a
Washington	6,212,000[2]	307,900	986,800
West Virginia	1,800,000[14]	250,000	215,000
Wisconsin	3,800,000[2, 15]	1,200,000[2]	n/a
Wyoming	530,000[2]	0	0
Total	**389,939,000**		

Note: Municipal solid waste is residential, commercial, institutional streams. Additional fractions of waste stream are noted if included in number provided.

[1] Includes industrial, agricultural waste and construction and demolition (C&D) debris;

[2] 1998 data;

[3] Includes industrial waste and C&D debris;

[4] All waste is exported. Number provided is only residential waste collected by District of Columbia;

[5] Total amount of MSW collected;

[6] Includes C&D debris;

[7] Total amount disposed-not generated;

[8] Generation calculated from county solid waste management plans dated 1988 to 1996. That is used as basis for calculating state recycling rate;

TABLE 4.2

Municipal Solid Waste Generation, Waste Imports and Exports, by State, 1999 [CONTINUED]

⁹ 1997data;
¹⁰ 1996 data;
¹¹ Includes industrial waste and small amount of C&D debris;
¹² Includes 5% industrial waste and 18% C&D debris;
¹³ Includes C&D if mixed in with MSW;
¹⁴ Includes agricultural waste, C&D debris;
¹⁵ Includes only MSW disposed in landfills from Wisconsin generators.

SOURCE: *The State of Garbage in America.* Biocycle, vol. 41, no. 4, p. 33, April 2000

If not properly vented or controlled, it can cause explosions and underground fires that smolder for years. Increasingly, this gas is being recovered through pipes inserted into landfills and distributed or used to generate energy.

Regulations passed under the Resource Conservation and Recovery Act (RCRA; PL 95-510), which took effect in 1993, required landfill operators to do several things to lessen the chance of pollution. The most important standard requires all landfills to monitor groundwater for contaminants—a procedure complied with by only one-fourth of all landfills.

The rules also require plastic liners for dump sites, and all debris must be covered with soil to prevent odors and trash from being blown away. Methane gas must be monitored, and the owner is responsible for cleanup of any contamination. These rules must be observed for a 30-year period after the landfill is closed to prevent pollution of the environment. Many landfills have closed and many others likely will, at least in part because of non-compliance with the rules.

According to the EPA, less than one-third of the nation's toxic waste dumps in 2000 met requirements under disposal laws for monitoring underground water supplies near their sites. The virtual disappearance of affordable environment-impairment liability insurance forced many dumps to shut down. The government also planned to close all dumps that failed to meet requirements; however, it was slow to act because it lacked the resources to prosecute violators.

What Is in a Landfill?

The EPA reported that containers and packaging (32.9 percent) made up the largest amount of MSW by weight, followed by non-durable goods (27.4 percent), durable goods (16.6 percent), yard waste (12.6 percent), food waste (10 percent), and other (1.5 percent). Small amounts of food products, textiles, plastics, petroleum byproducts, and agricultural chemicals make up the remainder. (See Figure 4.7.)

In 1998 most waste was paper (38.2 percent), followed by yard waste (12.6 percent), plastics (10.2 percent), food waste (10 percent), metals (7.6 percent), rubber, leather, and textiles (7 percent), glass (5.7 percent), wood (5.4 percent), and other materials (3.3 percent). (See Figure 4.8.)

Landfills of the Twenty-first Century

Landfills are, and will continue to be, the cornerstone of the nation's waste services system. However, a number of changes will occur in site design and function. Sites

FIGURE 4.5

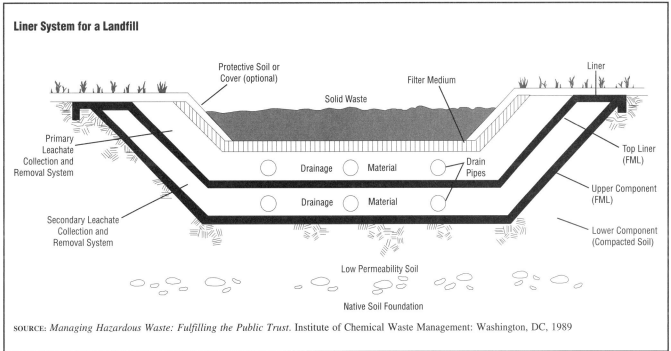

Liner System for a Landfill

SOURCE: *Managing Hazardous Waste: Fulfilling the Public Trust.* Institute of Chemical Waste Management: Washington, DC, 1989

FIGURE 4.6

Fresh Kills Municipal Landfill

LEGEND

Fresh Kills Municipal Landfill

Census Tacl Boundary

Road

River/Creek

Water

SOURCE: Hazardous Substances and Public Health, vol. 8, no. 2, Winter 1998. Agency for Toxic Substances and Disease Registry

will become more standardized, especially in the areas of liners and in the collection of landfill gas, which is expected to stimulate the development of new landfill gas-to-energy plants.

The number of landfills is expected to decrease due to the stricter standards and the need for operators to provide assurance that they can fund closure, cleanup, and security in the event of contamination. In many cases, it will be difficult to economically justify small-scale sites. The result will be fewer, but larger and more regional operations. Most waste will move away from its point of generation, resulting in increased interdependence among communities and states in waste disposal. More waste will cross state lines.

The volume of waste traditionally handled at landfills will also decrease. Landfills will provide diverse services—burial of waste, bioremediation, recycling facilities, leachate collection (contaminants picked up through the leaching of soil), and gas recovery. To make landfills more pleasing to neighborhoods, operators will establish larger buffer zones and more green space and will show more sensitivity to land-use compatibility and landscaping.

ALTERNATIVES TO LANDFILLS

Many states and the EPA endorse a system that prioritizes waste management methods as follows: source reduction (including composting), recycling, incineration, and landfilling.

Source Reduction

Many experts believe that the primary solution to the world's mounting garbage problem is "source reduction," or waste prevention. The less waste people create, the less there is to be thrown away. Source reduction involves minimizing the amount or toxicity of materials in products and/or reducing the amount of wastes produced during manufacturing. Manufacturing, packaging, or processing goods in certain ways "up front" will generate less refuse to be discarded.

In 1996 the EPA performed the first-ever comprehensive nationwide analysis of source reduction. Its report,

the *National Source Reduction Characterization Report* (1999), found that a total of 23 million tons—11 percent of that year's MSW generated—were source reduced. (See Table 4.3.) Yard trimmings and food scraps accounted for, by far, the greatest reductions. (Note: Some categories, such as paper boxes, experienced an increase known as "source expansion.") The decrease in containers and packaging was largely due to *materials substitution*. Manufacturers are replacing glass with plastic and

FIGURE 4.7

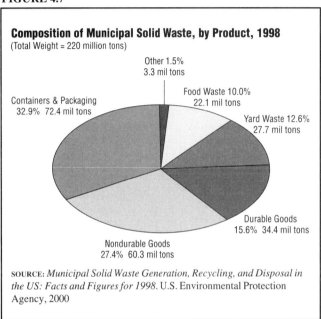

Composition of Municipal Solid Waste, by Product, 1998
(Total Weight = 220 million tons)

Other 1.5%
3.3 mil tons

Food Waste 10.0%
22.1 mil tons

Yard Waste 12.6%
27.7 mil tons

Durable Goods
15.6% 34.4 mil tons

Nondurable Goods
27.4% 60.3 mil tons

Containers & Packaging
32.9% 72.4 mil tons

SOURCE: *Municipal Solid Waste Generation, Recycling, and Disposal in the US: Facts and Figures for 1998*. U.S. Environmental Protection Agency, 2000

FIGURE 4.8

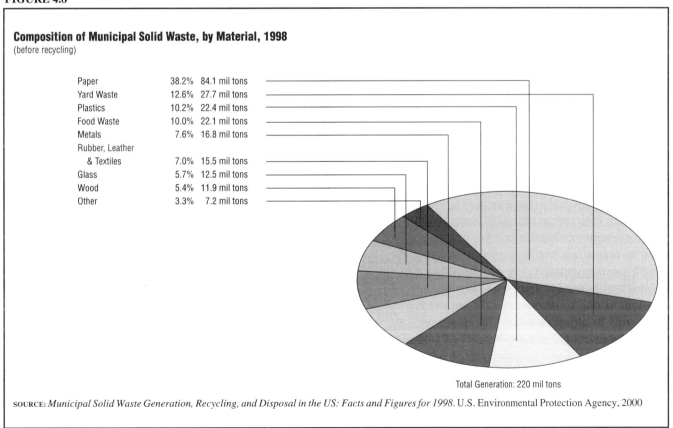

Composition of Municipal Solid Waste, by Material, 1998
(before recycling)

Material	%	mil tons
Paper	38.2%	84.1 mil tons
Yard Waste	12.6%	27.7 mil tons
Plastics	10.2%	22.4 mil tons
Food Waste	10.0%	22.1 mil tons
Metals	7.6%	16.8 mil tons
Rubber, Leather & Textiles	7.0%	15.5 mil tons
Glass	5.7%	12.5 mil tons
Wood	5.4%	11.9 mil tons
Other	3.3%	7.2 mil tons

Total Generation: 220 mil tons

SOURCE: *Municipal Solid Waste Generation, Recycling, and Disposal in the US: Facts and Figures for 1998*. U.S. Environmental Protection Agency, 2000

TABLE 4.3

Source Reduction/Expansion for Functional Categories of Products and Materials, 1996

Product	Source Reduction/(Expansion)* (Based on consumer spending and change in waste generation rate)
Durable Goods	
Miscellaneous Durables	2,145
Furniture/Furnishings	388
Major Appliances	237
Tires	188
Batteries, Lead Acid	(96)
Small Appliances	(258)
Carpets/Rugs	(426)
Subtotal	2,179**
Nondurable Goods	
Publications	4,581
Office Paper	616
Tissue Paper/Towels	380
Miscellaneous Nondurables	351
Other Nonpackaging Paper	289
Towels, Sheets, Pillowcases	56
Trash Bags	25
Disposable Diapers	15
Third Class Mail	(174)
Plates and Cups	(284)
Clothing/Footwear	(788)
Other Commercial Printing	(1,497)
Subtotal	3,571**
Containers & Packaging	
Wood Packaging	2,806
Beverage Containers	1,785
Food Containers	878
Bags and Sacks	497
Wrapping	(34)
Miscellaneous Packaging	244
Paper Boxes	(2,174)
Subtotal	4,002
Other MSW	
Yard Trimmings	11,731
Food Scraps	1,711
Miscellaneous Inorganics	92
Subtotal	13,534**
Grand Total	**23,286****

* Parentheses denote negative numbers, or source expansion. Positive numbers indicate source reduction.

** Discrepancies in calculations may occur due to rounding.

SOURCE: *National Source Reduction Characterization Report*. U.S. Environmental Protection Agency, 1999

FIGURE 4.9

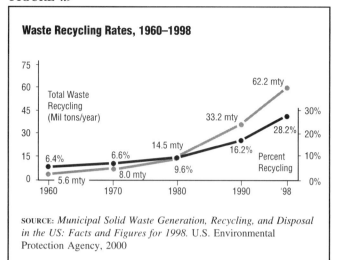

Waste Recycling Rates, 1960–1998

SOURCE: *Municipal Solid Waste Generation, Recycling, and Disposal in the US: Facts and Figures for 1998*. U.S. Environmental Protection Agency, 2000

other packaging materials that are, on average, lighter than glass.

COMPOSTING. Composting is a form of source reduction that involves the mixing of vegetable and organic refuse in order to speed the natural decomposition into fiber and micronutrients. For example, a compost pile in one's backyard recycles food scraps and lawn clippings that are deposited and mixed periodically. The decomposed product can then be used as fertilizer for the garden. Organic waste gives off energy in the form of heat when microorganisms metabolize the waste, causing it to lose between 40 and 75 percent of its original volume. After decontamination and refinement processes have been completed, the finished product is often used in landscaping, land reclamation, landfill cover, farming, and for nurseries. About one-third of trash is organic material—food and yard waste.

Composting is a particularly promising method of disposal of household wastes in developing countries and is also very advanced in Europe. Yard waste is a prime candidate for composting due to its high moisture content. Because composting contains moisture and micronutrients, slows soil erosion, and improves water retention, it is an alternative to the use of environmentally dangerous chemical fertilizers. More than half the states in the nation have bans on dumping yard waste, and several others are considering similar laws. The tonnage of yard trimmings dumped in landfills has declined, most likely due to these bans that cause more people to use backyard composting and mulching lawnmowers.

Recycling

Recycling urban solid waste offers many economic advantages. It conserves the energy used in incineration, reduces environmental degradation and chemicals that pollute water resources, generates jobs and small-scale enterprise, reduces dependence on foreign imports of metals, and conserves water. Some analysts claim that more than half the consumer waste can be economically recycled.

Every state has some type of recycling program. The oldest recycling law is the Oregon Recycling Opportunity Act, passed in 1983 and put into effect in 1986. A growing number of states require that many items sold must be made from recycled products. At least nine states require that recycled paper be used to make newspapers; many require that recycled materials be used in making telephone directories, trash bags, glass, and plastic contain-

FIGURE 4.10

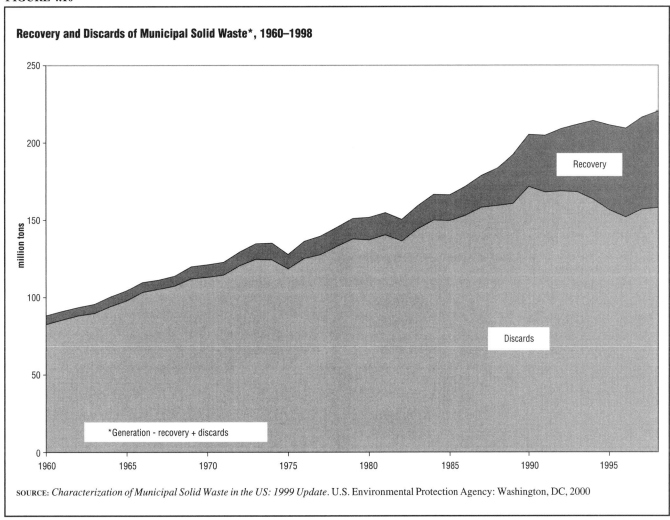

Recovery and Discards of Municipal Solid Waste*, 1960–1998

SOURCE: *Characterization of Municipal Solid Waste in the US: 1999 Update.* U.S. Environmental Protection Agency: Washington, DC, 2000

ers. Most states have goals to recycle from 25 to 70 percent of municipal solid waste. Rhode Island (70 percent) and New Jersey (60 percent) have the highest recycling goals of all the states; Maryland had the lowest (20 percent). More than 30 states bar some recyclable materials from being thrown into landfills. These include car and boat batteries, grass cuttings, tires, used oil, glass, plastic containers, and newspapers. Almost all books and pamphlets printed by the U.S. Government Printing Office are printed on recycled paper.

The EPA reported in *Municipal Solid Waste Generation, Recycling, and Disposal in the United States: Facts and Figures for 1998* (2000) that Americans recycled in record numbers in 1998. Total recycling was 28 percent of municipal solid waste produced, up from 16.2 percent in 1990 and only 6.4 percent in 1960. (See Figure 4.9.) Among the states, Minnesota recycled the greatest amount of its waste (46 percent), followed by New Jersey (43 percent), Wisconsin, Florida, and Tennessee (40 percent each), and Washington (39 percent). This was an all-time high since 1960. (See Figure 4.10 for the trend in recovery and discards from 1960 to 1998.)

In 1998, 56 percent of material recovered was paper and paperboard, followed by yard trimmings (20 percent), metals (10 percent), other materials (7 percent), glass (5 percent), and plastics (2 percent). (See Figure 4.11.)

Incineration

Some observers think incinerators are the best alternative to landfills. The EPA reported that, in 1997, approximately 17 percent of waste in the United States was burned. (See Figure 4.3.) When an incinerator burns waste, it reduces the amount of garbage. About 75 percent of the weight of the garbage burns off. Some incinerators not only burn garbage; by using the heat from the burning garbage to make energy, they can also be waste-to-energy (WTE) facilities. Waste-to-energy incinerators are preferred over older incinerator models because they use an improved combustion process, have better pollution-control technology, and produce energy from trash. Incinerators vary greatly in size. Smaller ones may burn only about 25 tons of garbage a day, while larger ones may burn up to 4,000 tons a day. They normally operate at a capacity of 85 percent over the course of a year.

Incinerators are very expensive to build. The country's largest incinerator, in Detroit, Michigan, cost $438 million. This huge incinerator produces enough steam to heat half of Detroit's central business district and enough electricity to supply 40,000 homes. However, most experts agree that energy recovery from MSW has the potential for making only a limited contribution to the nation's overall energy production. While the current contribution of waste-derived energy production is less than one-half of 1 percent of the nation's total energy supply, the DOE has set a goal for waste-derived energy at 2 percent of the total supply by 2010.

During the operation of a typical incinerator, trucks dump waste into a pit; the waste is moved to the furnace by a crane; and the furnace burns the waste at a very high temperature, heating a boiler that produces steam for generating electricity and heat. Ash collects at the bottom of the furnace where it is later removed and dumped in a landfill. (See Figure 4.12.)

Most experts believe that incineration can never serve as a primary method of garbage disposal because it (1) produces residue that must then be transferred to a landfill and (2) produces poisonous gases, primarily dioxin

FIGURE 4.11

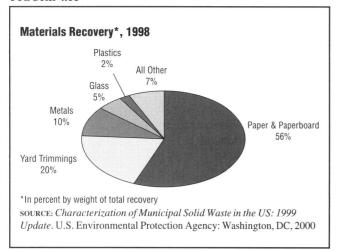

Materials Recovery*, 1998

Plastics 2%
All Other 7%
Glass 5%
Metals 10%
Yard Trimmings 20%
Paper & Paperboard 56%

*In percent by weight of total recovery
SOURCE: *Characterization of Municipal Solid Waste in the US: 1999 Update*. U.S. Environmental Protection Agency: Washington, DC, 2000

FIGURE 4.12

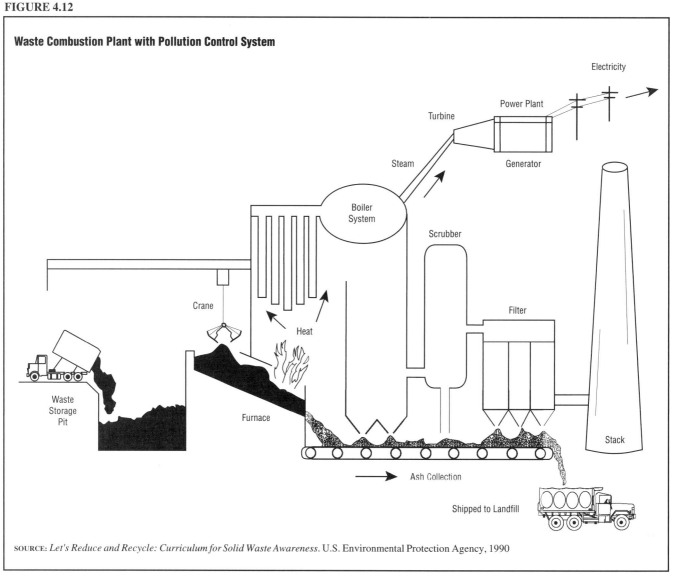

Waste Combustion Plant with Pollution Control System

SOURCE: *Let's Reduce and Recycle: Curriculum for Solid Waste Awareness*. U.S. Environmental Protection Agency, 1990

TABLE 4.4

Projections of Materials Generated* in the Municipal Waste Stream, 2000 and 2005
(In thousands of tons and percent of total generation)

Materials	Thousands of tons 2000	Thousands of tons 2005	% of total 2000	% of total 2005
Paper and Paperboard	87.7	94.8	39.3%	39.6%
Glass	11.9	11.2	5.3%	4.7%
Metals	17.6	18.7	7.9%	7.8%
Plastics	23.4	26.7	10.5%	11.2%
Wood	14.0	15.8	6.3%	6.6%
Others	19.7	22.2	8.8%	9.3%
Total Materials in Products	**174.3**	**189.4**	**78.1%**	**79.1%**
Other Wastes				
Food Wastes	22.5	23.5	10.1%	9.8%
Yard Trimmings	23.0	23.0	10.3%	9.6%
Miscellaneous Inorganic Wastes	3.4	3.6	1.5%	1.5%
Total Other Wastes	**48.9**	**50.1**	**21.9%**	**20.9%**
Total MSW General	**223.2**	**239.5**	**100.0%**	**100.0%**

*Generation before materials recovery or combustion.
Details may not add to totals due to rounding.

SOURCE: *Characterization of Municipal Solid Waste in the US: 1998 Update.* U.S. Environmental Protection Agency: Washington, DC, 1999

FIGURE 4.13

Historical and Projected Generation of Municipal Solid Waste, 1960–2005

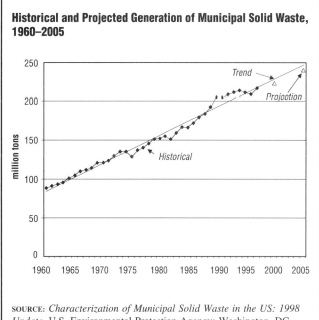

SOURCE: *Characterization of Municipal Solid Waste in the US: 1998 Update.* U.S. Environmental Protection Agency: Washington, DC, 1999

and mercury, which are increasingly being found to be dangerous. It may, however, be useful to augment landfill and recycling. WTE plants have the same problems. Mercury is largely impossible to screen with pollution control devices such as scrubbers (an air pollution device that uses a spray of water or reactant to trap pollutants). In the process of burning paints, fluorescent lights, batteries, or electronics, mercury is released as a gaseous vapor that is poisonous to humans and to the environment. Most of the first incinerators built have been retired because they failed to meet subsequent air quality standards. Some analysts are not satisfied that the emissions problems have been solved, especially the problems of burning materials containing chlorine. Chlorine molecules, when burned, create dioxin, a known carcinogen.

Regulators are also concerned about the acid gases and heavy metals released from WTE plants. Scrubbers reduce but do not eliminate these emissions. Even when the toxic elements are largely removed from emissions, the resulting ash is still toxic and, when put in landfills, can leach into the groundwater. Thus, toxic compounds in incinerator ash are simply removed from one environmental medium to enter another. Toxic compounds still end up in the soil. By law, toxic residue created by burning waste in incinerators must be treated as hazardous waste and must not be dumped in ordinary landfills.

PROJECTIONS FOR MUNICIPAL SOLID WASTE GENERATION

The EPA expects Americans will produce 223 million tons of garbage a year in the year 2000 and 240 million tons in 2005. Generation of waste made of paper and paperboard, metals, plastics, wood, and other materials such as rubber and textiles, is projected to increase. Only glass is expected to decline. (See Table 4.4.) Food waste is projected to increase at the same rate as the population but the dumping of yard trimmings per capita has been decreasing due to legislation regulating their disposal. Overall, municipal solid waste generation is projected to increase at a rate of 1 percent per year through 2000, then increase 1.4 percent annually from 2000 through 2005. (See Figure 4.13.)

PACKAGING

Nearly one dollar out of every 12 dollars Americans spend for food and beverages pays for packaging. According to the United States Department of Agriculture, packaging is the second largest portion of the cost of marketing food (advertising is the largest portion). The increasing numbers of women in the work force and changes in family structure have resulted in greater demand for convenience products—carry-out meals and frozen and vacuum-packed foods. One way to reduce waste is to trim the amount of packaging.

Soft-drink consumption has also risen, increasing waste in the form of cans and plastic containers. Aluminum, the most abundant metal manufactured on the earth, was first refined into a valuable product in the 1820s. Its use has continually escalated and beverage cans are the largest single use of aluminum. They are also a source of much waste.

FIGURE 4.14

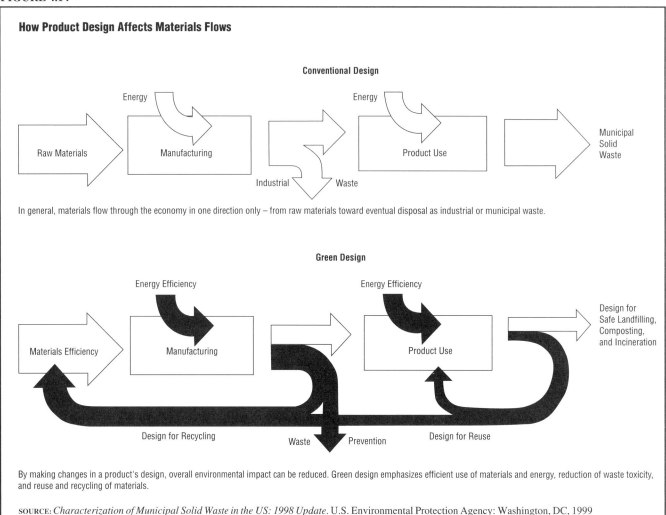

How Product Design Affects Materials Flows

Conventional Design

In general, materials flow through the economy in one direction only – from raw materials toward eventual disposal as industrial or municipal waste.

Green Design

By making changes in a product's design, overall environmental impact can be reduced. Green design emphasizes efficient use of materials and energy, reduction of waste toxicity, and reuse and recycling of materials.

SOURCE: *Characterization of Municipal Solid Waste in the US: 1998 Update.* U.S. Environmental Protection Agency: Washington, DC, 1999

The advent of low-priced petrochemicals in the early twentieth century ushered in the age of plastics. Several times more plastics—a family of more than 46 types—are now produced in the United States than aluminum and all other nonferrous metals combined. Most of these plastics are non-biodegradable and, once discarded, remain relatively intact for many years.

Many Americans claim they would pay more for a product with environmental benefits. Marketing which considers these consumer preferences is known as "green marketing." "Green design" can make products more environmentally safe. (See Figure 4.14.)

LAWS GOVERNING THE DISPOSAL OF GARBAGE

The Resource Conservation and Recovery Act (RCRA; PL 94-580), the major federal law on waste disposal, was passed in 1976. This law primarily covers hazardous waste, which is only a small part (approximately 6 percent) of all garbage. State and local governments are mainly responsible for passing laws concerning non-haz-

ardous waste, although the federal government will supply money and guidance to local governments so they can better manage their garbage systems.

Other federal laws cover other areas of waste disposal. For example, the Clean Water Act (PL 95-217) regulates wastewater disposal, the Safe Drinking Water Act (PL 93-523) controls underground injections (when wastewater is dumped into deep wells), and the Clean Air Act (PL 95-95) governs air pollution.

THE STATE OF GARBAGE IN 2000—STABILITY

Fueled by legislative mandates, landfill bans on certain items, funding for state recycling programs, and the 1994 federal deadline for the closing of substandard landfills, the early 1990s saw significant jumps in recycling rates, big declines in the number of landfills, and increases in composting of yard trimmings. In the late 1990s, however, the rates of change have become more gradual, reflecting both saturation and maturation of programs and stabilization of landfill closures. Furthermore, nothing by the year 2000 was likely to change slowdown. No major federal or state initiatives were anticipated that would significantly alter waste management trends.

HAZARDOUS WASTE

What Is Hazardous Waste?

Some solid, liquid, or gas waste can be harmful to human beings or the environment when not handled properly. Waste of this nature is commonly called hazardous waste. This waste material may be the byproduct of manufacturing or the accidental spill or release of a chemical product that is toxic, corrosive, ignitable, explosive, or radioactive. Because of these potential dangers, hazardous waste requires special care when being stored, transported, or disposed of.

Hazardous waste is the inevitable by-product of industrialization. Manufacturers use many chemicals to create their products. Hazardous waste is generated by big industries like automobile and computer manufacturers and by small businesses like neighborhood photo shops or cleaners. Although people can reduce quantities of hazardous waste through careful management, it is not possible to entirely eliminate hazardous residues because of the continual demand for goods.

The EPA estimates that between 2 and 4 percent of all waste is hazardous (Subtitle C) waste. Every year, Americans produce more than a ton of hazardous waste for every man, woman, and child in the country. Ninety percent of all hazardous waste in the United States is produced by about 14,000 large waste generators (manufacturing facilities that producing more than 1,000 kilograms of hazardous waste a month each). The chemical industry is by far the largest producer, followed by petroleum refiners and the metal-processing industry. The remaining 10 percent comes from more than 100,000 small-quantity generators—businesses that produce less than 1,000 kilograms of hazardous waste per month. These include photo labs, service stations, dry cleaners, body shops, printers, laboratories, and private homes.

Industrial wastes are usually a combination of compounds, one or more of which may be hazardous; for example, used pickling solution from a metal processor can also contain residual acids and metal salts. A mixture of waste produced regularly as a result of the industrial process is called a waste stream, and it generally consists of diluted rather than full-strength compounds. Often the hazardous components are diluted in a mixture of dirt, oil, or water.

Contamination of the air, water, and soil with hazardous wastes can frequently lead to serious health problems. The EPA estimates that roughly 1,000 cases of cancer annually, as well as degenerative diseases, mental retardation, birth defects, and chromosomal changes, can

be linked to public exposure to hazardous waste. While most scientists agree that exposure to high levels of hazardous waste is dangerous, there is less agreement on the danger of exposure to low levels.

Methods of Dealing with Hazardous Waste

In North America 96 percent of all hazardous waste is treated and disposed of at the site where it is produced. Four percent is treated and disposed of by commercial waste service companies. A variety of techniques exist for safely managing hazardous wastes:

- Reduction—This approach reduces the waste stream at the outset. Waste generators change their manufacturing and materials in order to produce less waste. For example, a food packaging plant might replace solvent-based adhesives used to seal packages with water-based adhesives.

- Recycling—Some waste materials become raw material for another process, or can be recovered, reused, or sold.

- Treatment—A variety of chemical, biological, and thermal processes can be applied to neutralize or destroy toxic compounds. For example, microorganisms or chemicals can remove hydrocarbons from contaminated water.

- Incineration—Hazardous waste can also be burned. Unfortunately incineration has a flaw—as waste is burned, hot gases spew into the atmosphere, carrying toxic materials not consumed by the flames. In 1999 the Clinton Administration imposed a ban on new hazardous waste incinerators.

- Land disposal—Some hazardous wastes are buried in landfills. State and federal regulations require the pretreatment of most hazardous wastes before they can be disposed of in landfills. These treated materials can only be placed in specially designed land disposal facilities.

BURYING THE WASTE—LAND DISPOSAL AND CONTAMINATION. Groundwater is a major source of drinking water for many parts of the nation. If not properly constructed, land disposal facilities for hazardous waste may leak contaminants into the underlying groundwater. The Resource Conservation and Recovery Act (RCRA; PL 94-580) imposed control over such disposal facilities to minimize their adverse environmental impacts. The EPA, in order to implement the act, requires that owners/operators of hazardous waste sites install wells to monitor the groundwater under their facilities. (See Figure 4.15.)

SHIPPING ELSEWHERE. Many states have refused to accept toxic trash from states that have not developed their own disposal programs. Their position has been undermined, however, by the U.S. Supreme Court's determina-

FIGURE 4.15

Cross Section of a Minimal Groundwater Monitoring System

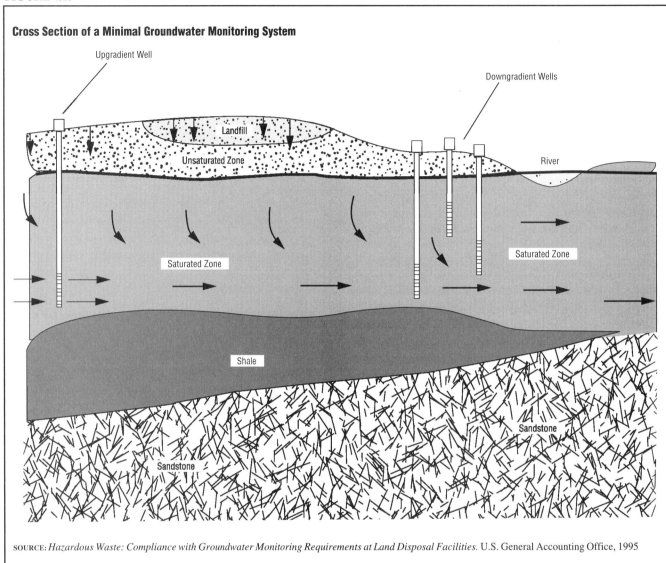

SOURCE: *Hazardous Waste: Compliance with Groundwater Monitoring Requirements at Land Disposal Facilities.* U.S. General Accounting Office, 1995

tion that waste is a commodity in interstate commerce and could be subject to federal, not state, regulation.

Because of the problems in finding disposal sites, the United States is sending larger and larger amounts of toxic waste out of the country. Mexico and Central and South America have become preferred spots for disposing of sludge and incinerator ash. Unfortunately, toxic waste is sometimes mislabeled nontoxic by the time it arrives in South American countries. Until recently, Africa had been a favorite location for dumping toxic waste. In 1988, however, most African countries signed agreements that restricted importation of dangerous materials.

THE NATIONAL PRIORITIES LIST—THE SUPERFUND

The Comprehensive Environmental Response, Compensation, and Liability Act of 1980 (CERCLA; PL 96-510) established the Superfund to pay for cleaning up abandoned disposal sites. The Superfund—initially a $1.6 billion, five-year program—was intended to clean up leaking dumps that jeopardized groundwater and posed public health risks. During the act's original mandate, only six sites were cleaned up, and when it expired in 1985, many observers viewed the program as a billion-dollar fiasco rampant with scandal and mismanagement. The negative publicity surrounding the program increased public awareness of the magnitude of the cleanup job required in America to reduce the risk to public health. (See Figure 4.16.) Consequently, in 1986 and 1990, the Superfund was reauthorized.

A Huge Project

CERCLA requires the government to maintain a National Priorities List (NPL) of sites that pose the highest potential threat to human health and the environment.

FIGURE 4.16

Types of Environmental and Public Health Risks Addressed at Superfund Sites

Contaminated Air

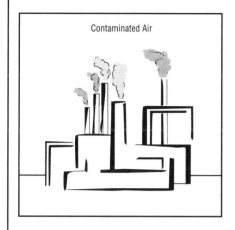

Direct Contact With Hazardous Waste

Contaminated Drinking Water

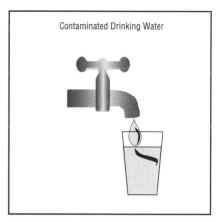

Ecological Damage

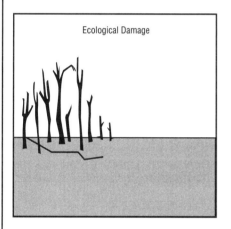

Fire and Explosion Hazard

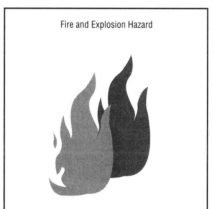

Exposure Through Food Chain

Contaminated Groundwater

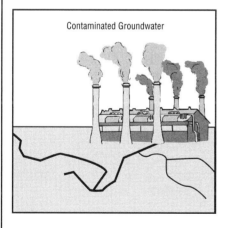

Contaminated Soil

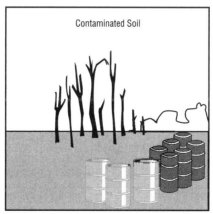

Contaminated Surface Water

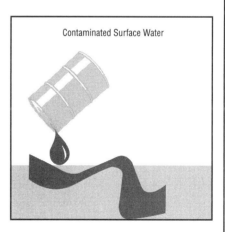

SOURCE: *Superfund Program Management.* U.S. General Accounting Office, 1992

(See Table 4.5.) As of May 2000 the NPL included 1,227 sites. Most were general sites, such as landfills used by manufacturers and municipal landfills. About one-fifth of the sites slated for cleanup by the Superfund were municipal landfills. Some were federal sites, including those for the disposal of nuclear materials from bombs or U.S. Air Force bases that did not properly dispose of fuels and other dangerous materials.

Many sites are still years away from being cleaned up. (See Figure 4.17.) As of 1999, 34 percent were classified as cleanup-completed, 30 percent had remedies

TABLE 4.5

EPA's 16 Superfund Site Types

Contaminant-based Categories

Asbestos
Dioxin
Metals
Metals/organic compounds
Organic compounds
Polychlorinated biphenyls (PCBs)
Pesticides
Radioactive/mixed waste
Solvents

Former Use Categories

Battery recycling/lead
Metal-plating
Industrial landfill
Mining waste
Municipal landfill
Munitions
Wood-preserving

SOURCE: *Superfund: Problems with the Completeness and Consistency of Site Cleanup Plans.* U.S. General Accounting Office, 1992

FIGURE 4.17

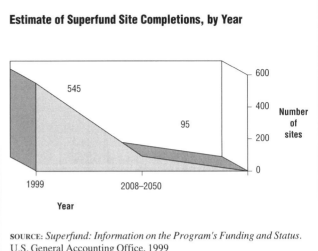

Estimate of Superfund Site Completions, by Year

SOURCE: *Superfund: Information on the Program's Funding and Status.* U.S. General Accounting Office, 1999

pending, while 14 percent had been deleted because they had been satisfactorily completed.

In addition to the sites that have been officially listed on the NPL, another 1,789 sites have been preliminarily judged as potentially eligible for the NPL. (See Table 4.6.) Most of those sites have already contaminated ground or drinking water. Metals are the principal contaminant at these sites. Additional contaminants include pesticides, volatile organic compounds (VOCs), dioxins, and polychlorinated biphenyls (PCBs).

When the Superfund was created, the program was expected to deal with a limited number of sites over a relatively short time. It eventually became clear that the number of sites needing attention was much larger than originally believed and that the program could run several more decades. Since 1980 almost 37,000 sites have been reported to the EPA's inventory. The number of sites reported has declined steadily since 1985. The percentage of sites that the EPA believes warrant further consideration after initial investigation has remained relatively constant since 1984, with about 40 percent of the recommended sites needing further action. The average site cleanup takes approximately 12 years to complete.

What is the Cost and Who Pays the Bill for Superfund?

According to the U.S. General Accounting Office (GAO), in *Superfund: Information on the Program's Funding and Status* (1999), as of 1999, Superfund cleanups had cost more than $14 billion. The GAO estimated that another $9.9 billion would be needed to complete cleanup at the remaining sites. (See Table 4.7.)

While federal funds are available to finance cleanup, the EPA is authorized to compel parties responsible for creating hazardous pollution, such as waste generators or haulers or site owners or operators, to clean up the sites. If these parties cannot be found or if a settlement cannot be reached, the EPA can conduct a cleanup, using funds from a trust fund established by CERCLA to perform such cleanups. This trust fund is financed primarily by a tax on crude oil and certain chemicals and by environmental taxes on corporations. After completing a cleanup, the EPA can take action against the responsible parties to recover costs and replenish the fund. The average cost of cleanup is about $30 million per site, large enough to make it worthwhile for parties to pursue legal means to spread the costs among large numbers of responsible parties. Many cleanups involve dozens of parties.

Disputes have arisen between industries and cities over who is responsible for a cleanup, and numerous lawsuits have been filed by industries against cities over responsibility for what could be a huge expense. Many businesses and municipalities may be unable to assume such expense. The EPA reports that the government currently collects only one-fifth of the cleanup costs that could be recovered from polluters under the Superfund law. According to the EPA, in many cases, the polluters have disappeared or are unable to pay. In other cases, the agency lacks the manpower or evidence to proceed with lawsuits.

BROWNFIELDS

Many former industrial sites have become eyesores of urban scenery. These trash-strewn plots, concentrated mostly in the Northeast and Midwest, are called brownfields and number nearly 450,000 sites. The EPA defines brownfields as abandoned, idled, or underused industrial or commercial sites where expansion or redevelopment is complicated by real or perceived environmental contamination.

TABLE 4.6

Potentially Eligible Superfund Sites Contributing to Specific Adverse Conditions

Conditions resulting from contamination at 1,789 potentially eligible sites	Number of potentially eligible sites with condition	Percentage of potentially eligible sites with condition	Percentage of potentially eligible sites for which presence of condition is uncertain
Workers/visitors may have direct contact with contaminants	981	55	21
Trespassers may come into direct contact with contaminants	969	54	20
Fences/barriers/signs are erected to keep residents or others out of contaminated areas	618	35	19
Residents/community have concerns about contamination or potential health effects caused by this site	548	31	35
Fish could be unsafe to eat	486	27	29
Institutional restrictions[a] are necessary because of site's contamination	410	23	46
Residents/others should avoid exposure to contaminated dust on some days	355	20	23
Sources of drinking water permanently changed[b]	215	12	20
Obnoxious odors are present	194	11	24
Residents advised not to use wells	150	8	20
Fish, plants, or animals are sick/dying	143	8	33
Residents, workers, etc. use water (for bathing, landscaping, etc.) that fails to meet water quality standards	102	6	29
Recreation (e.g., fishing, swimming) is stopped or restricted	85	5	23
Residents advised to use filtered water	75	4	21
Residents advised to use bottled water	72	4	20
Residents advised not to let children play/dig in their yards	55	3	20
Crops are irrigated with contaminated water	52	3	29
Livestock drink contaminated water	44	3	28

[a]Institutional restrictions include limitations on uses of a property such as deed restrictions that limit a property to industrial use or legal limits placed on the depth of a well at a site.
[b]For example, by connecting residents to municipal water supplies in place of well water.

SOURCE: *Hazardous Waste: Unaddressed Risks at Many Potential Superfund Sites.* U.S. General Accounting Office, 1999

Many of the properties are polluted. They are shunned by developers, often stalling efforts to revive poor, inner-city neighborhoods. Until 1995, developers and buyers had avoided some 38,000 sites listed as possible targets under the Superfund Law, which says that anyone involved in the management of a property can be held liable for the entire cost of cleanup. Many of those sites had, in fact, been passed over by the EPA as not contaminated enough for Superfund action. Nonetheless, many of those properties were deemed untouchable by the real estate industry. A 1995 survey of the American Bankers Association showed that 83 percent of smaller banks had refused to make loans to projects because of concerns about environmental liability.

To help the reclamation effort the EPA, in 1995, removed 25,000 of the least-polluted sites from the list. The sites will require some type of cleanup but will not be subjected to the tougher Superfund standards. In addition to restoring the environment, the purpose of recla-mation programs is to encourage the reuse of abandoned sites, revitalize cities, create jobs, and generate municipal tax revenues. Redevelopment of polluted sites is becoming a thriving business. Experts estimate that about one-third of real estate sales involve sifting data bases of environmental agencies for records of toxic spills before a real estate transaction can take place. Sensing a new business possibility, several insurance companies have created divisions offering policies that protect developers of polluted real estate against unforeseen cleanup costs or lawsuits.

In 1997 President Clinton signed the Taxpayer Relief Act (PL 105-34 and PL 105-32), both of which included new tax incentives to spur the cleanup and redevelopment of brownfields. The acts enable taxpayers to consider any qualified environmental remediation expenditure as tax deductions in the year paid rather than having to be capitalized over time. The incentives were expected to return approximately 14,000 brownfields to productive use.

TABLE 4.7

Estimated Annual Costs to complete Cleanup of 85% of the National Priorities List Superfund Sites, by 2008

Dollars in millions

Fiscal year	Percent of total costs	Annual costs, assuming completion of 85 percent of Operable Units by 2008							
		Contractor cleanup costs	Site-specific support costs	Subtotal for cleanup and site-specific support	Non-site-specific support costs	Subtotal for cleanup and support	O&M	Total costs
2000	9.4	$366.2	$137.7	$503.9	$310.9	$814.8	$12.0	$826.8
2001	9.4	366.2	137.7	503.9	310.9	814.8	24.0	838.8
2002	9.4	366.2	137.7	503.9	310.9	814.8	36.0	850.8
2003	9.4	366.2	137.7	503.9	310.9	814.8	48.0	862.8
2004	9.4	366.2	137.7	503.9	310.9	814.8	60.0	874.8
2005	9.4	366.2	137.7	503.9	310.9	814.8	72.0	886.8
2006	9.4	366.2	137.7	503.9	310.9	814.8	84.0	898.8
2007	9.4	366.2	137.7	503.9	310.9	814.8	96.0	910.8
2008	9.4	366.2	137.7	503.9	310.9	814.8	108.0	922.8
2009	7.7	300.0	112.8	412.7	254.7	667.4	118.0	785.4
2010	7.7	300.0	112.8	412.7	254.7	667.4	116.0	783.4
2011	a	a	a	a	a	a	104.0	104.0
2012	a	a	a	a	a	a	92.0	92.0
2013	a	a	a	a	a	a	80.0	80.0
2014	a	a	a	a	a	a	68.0	68.0
2015	a	a	a	a	a	a	56.0	56.0
2016	a	a	a	a	a	a	44.0	44.0
2017	a	a	a	a	a	a	32.0	32.0
2018	a	a	a	a	a	a	20.0	20.0
2019	a	a	a	a	a	a	10.0	10.0
Total	100	$3,895.6	$1,464.7	$5,360.3	$3,307.4	$8,667.7	$1,280.0	$9,947.7

Note: Totals may not add because of rounding.
aCleanups completed except for O&M.

SOURCE: *Superfund: Information on the Program's Funding and Status.* U.S. General Accounting Office, 1999

The brownfields program has been popular in a Congress that appropriated increasing amounts for it—$36.8 million in fiscal year (FY) 1997, $85 million in FY 1998, $91.3 million for FY 1999, and $91.7 million for FY 2000.

NUCLEAR WASTE

There is currently no agreed-upon safe way to dispose of nuclear waste. None of the current options guarantee to protect the biosphere from radiation, which survives for many thousands of years. Because of the scientific and political difficulties with geologic burial and other methods, above-ground "temporary" storage, despite its dangers, may remain the only option well into the twenty-first century.

Low Level Radioactive Waste

Low-level radioactive waste decays in 10 to 100 years. Until the 1960s, the United States dumped low-level wastes into the ocean. The first commercial site to house such waste was opened in 1962, and by 1971 six sites were licensed for disposal. The volume of low-level waste increased during the initial years (1963–80) of commercially generated waste disposal, until passage of the Low Level Waste Policy Act of 1980 (PL 96-573). Since then, volume has decreased.

By 1979 only three commercial low-level waste sites were still operating—Hanford, Washington; Beatty, Nevada; and Barnwell, South Carolina. In response to the threatened closing of the South Carolina site, Congress passed the Low Level Radioactive Waste Policy Act of 1980 (PL 96-573), calling for the establishment of a national system of such facilities. Every state would be responsible for finding a low-level disposal site by 1986 for wastes generated within its borders. It also gave states the right to bar imports of low-level wastes if they were engaged in regional compacts for waste disposal. The disposal of high-level wastes, however, remains a federal responsibility.

COMPACTS. The 1980 law encouraged states to organize themselves into compacts to develop new low-level waste facilities. As of 1999, 10 compacts serving 44 states had been approved by Congress. (See Table 4.8.) Compacts and unaffiliated states have confronted significant barriers to developing disposal sites, however, including: public health and environmental concerns, antinuclear sentiment, substantial financial requirements, political issues, and "not in my backyard" campaigns by some citizen activists.

No compact or state had successfully developed a new disposal facility for low-level wastes by 1999. California had planned a facility, but the land could not be obtained. Texas received federal approval for a site at Sierra Blanca, which would receive waste from Maine and Vermont. The plan still requires the approval of the Texas Natural Resource Conservation Commission, however. Certain conditions have led some states to remain uncommitted to disposal development and to consider other options. The reopening of the Barnwell, South Carolina, facility in 1995 eased some of the pressure on the states. The emergence in 1995 of new private sector nuclear waste handlers—Envirocare of Utah, Inc.—has increased interest in the possibility of privately operated waste disposal facilities. Collectively the Barnwell, Hanford, and Envirocare facilities provided disposal capacity for almost all types of low-level wastes in 2000.

Spent Fuel and High-Level Radioactive Waste

The most dangerous radioactive waste is irradiated uranium from commercial nuclear power plants. Spent fuel, the used uranium fuel removed from a nuclear reactor, is far from being completely "spent." It contains highly penetrating and toxic radioactivity and requires isolation from living things for thousands of years. It still contains significant amounts of uranium, as well as plutonium created during the nuclear fission process. Spent fuel is a serious problem for nuclear power plants that will be decommissioned before a long-term, high-level waste disposal repository is available. (See Figure 4.18.) Unless a temporary site becomes available, decommissioned plants have the following options:

- Leave the fuel in place.

- Use on-site storage casks. This is not an option for hot fuel (fuel that is less than five years out of the core).

- Ship the spent fuel abroad for reprocessing. France, which is heavily dependent on nuclear power, developed the technology to reprocess spent fuel, something not available in the United States. In 1993 the British government also opened a nuclear fuel reprocessing plant that reprocesses spent fuel from nuclear power generators around the world. Nuclear watch-groups and some Americans fear that shipping spent fuel abroad will undermine efforts to halt the spread of nuclear arms because the process of transporting such materials increases the possibility for theft or accident.

- Continue to operate the unit.

- Ship the fuel to a monitored retrievable storage facility, if there is one available.

The Vestiges of Nuclear Disarmament

Nuclear disarmament resulted in the dismantling of much of the United States' nuclear arsenal and the resulting need to store tons of plutonium. The federal government has proceeded to take apart as many as 15,000 warheads with intentions of eventually storing them at

TABLE 4.8

Status of Low-Level Radioactive Waste Disposal Compacts and Unaffiliated States
Dollars in millions

State compacts (Host state and state members)	Status of disposal siting efforts	Development costs
Appalachian compact (Pennsylvania, Delaware, Maryland, West Virginia)	Halted.	$37.0
Central compact (Nebraska, Arkansas, Kansas, Louisiana, Oklahoma)	License application denied by Nebraska. Nebraska to withdraw from compact.	95.6
Central Midwest compact (Illinois, Kentucky)	Halted.	95.8
Midwest compact (No host state, Indiana, Iowa, Minnesota, Missouri, Ohio, Wisconsin)	Halted.	Not available
Northeast compact (Dual hosts: Connecticut, New Jersey)	Connecticut: halted disposal facility siting, considering storage for 100 years or longer.	15.2
	New Jersey: halted siting effort.	9.7
Northwest compact (Washington, Alaska, Hawaii, Idaho, Montana, Oregon, Utah, Wyoming)	Uses existing Richland disposal facility located on DOE's Hanford site.	Not applicable
Rocky Mountain compact (No host state, Colorado, Nevada, New Mexico)	Contracted with Northwest compact to use the Richland facility.	Not applicable
Southeast compact (North Carolina, Alabama, Florida, Georgia, Mississippi, Tennessee, Virginia	North Carolina halted licensing process for disposal facility, shut down its siting agency, and, on July 26, 1999, enacted legislation withdrawing from the compact.	112.0
Southwestern compact (California, Arizona, North Dakota, South Dakota)	Halted.	92.6
Texas compact (Texas, Maine, Vermont)	Halted, initial license application for original site denied by state's licensing authority.	52.0
Unaffiliated states		
District of Columbia	No plans to site a facility.	Not applicable
Massachusetts	Halted.	Not available
Michigan	No efforts under way.	12.6
New Hampshire	No plans to site a facility.	Not applicable
New York	Halted.	62.7
Puerto Rico	No plans to site a facility.	Not applicable
Rhode Island	No plans to site a facility.	Not applicable
South Carolina	Host state for Barnwell facility.	Not applicable
Totals		**$585.2**

SOURCE: *Low-Level Radioactive Wastes: States Are Not Developing Disposal Facilities.* U.S. General Accounting Office, 1999

one of two former nuclear weapons-making plants—Pantex, near Amarillo, Texas, and Savannah River, South Carolina. DOE officials predict that dismantling will continue through the year 2003. The government must then decontaminate buildings used at those facilities, dispose of millions of gallons of boiling radioactive water, and decontaminate hundreds of square miles of desert at the Nevada nuclear test site.

FIGURE 4.18

Locations of Spent Nuclear Fuel and High-Level Radioactive Waste Ultimately Destined for Geologic Disposal

■ Pearl Harbor

Symbols do not reflect precise locations

●	Commercial Reactors	▲	Non-DOE Research Reactors
✖	Shut-down Reactors with Spent Nuclear Fuel Onsite	#	Non-DOE Shut-down Research Reactors with Spent Nuclear Fuel Onsite
∅	Shut-down Reactors without Spent Nuclear Fuel Onsite	■	Navy Reactor Fuel
✳	Commercial Spent Nuclear Fuel Storage Facility	▼	DOE-Owned Spent Nuclear Fuel and High-Level Radioactive Waste

SOURCE: OCRWM Bulletin, Summer/Fall 1994

Radioactive metals will be disposed of in one of two ways—immobilization in glass or ceramic containers, or burning as a mixed oxide fuel (MOX). Immobilization is the preferred method among nuclear watchdogs who fear increased nuclear waste through burning. MOX fuel could be used in existing reactors, a process used in Europe but never tried in the United States. The Clinton Administration prefers the dual-track approach, using both options.

A Serious Leak of Radioactive Waste—Hanford

In 1997 scientists discovered that about 900,000 gallons of radioactive waste had leaked into the soil from 68 of the 149 tanks at the nuclear weapons plant in Hanford, Washington. Eventually all the tanks are expected to leak. The leak contaminated underground water moving toward the nearby Columbia River. Managers at the plant maintained that the leaks were insignificant because radioactive materials would be trapped by the area above the water table (the "vadose zone"). Furthermore officials had been saying for decades that no waste from the tanks would reach the groundwater in the next 10,000 years.

Nonetheless, the groundwater under more than 85 square miles of the site is already contaminated. Washington's governor Gary Locke called it a "Chernobyl waiting to happen." A threatened lawsuit by the State of Washington against the U.S. DOE over the leaks at the Hanford site resulted in an agreement to clean up the two indoor pools near the Columbia River by 2007.

FEDERAL NUCLEAR WASTE REPOSITORIES

In the United States the federal government is focusing on two locations as eventual long-term nuclear waste repositories: the Waste Isolation Pilot Plant (WIPP) in southeastern New Mexico for transuranic (defense) waste, and Nevada's Yucca Mountain for civilian waste.

The Waste Isolation Pilot Plant (WIPP)

The Waste Isolation Pilot Plant (WIPP) became the world's first deep depository for nuclear waste when it received its first shipment of waste on March 26, 1999. This large facility is located near Carlsbad, New Mexico. (See Figure 4.19.) WIPP is 655 meters below the earth's surface in the salt beds of the Salado Formation and is intended to house up to 6.25 million cubic feet of transuranic waste for more than 10,000 years.

Under congressional mandate the WIPP facility will not accept commercial or high-level waste; only transuranic waste will be accepted. More than 99 percent of transuranic waste is temporarily stored in drums at nuclear defense sites in California, Colorado, Idaho, Illinois, Nevada, New Mexico, Ohio, Tennessee, South Carolina, and Washington. So far, the Energy Department has authorized the plant to receive waste from the Los Alamos National Laboratory in California, from the Idaho National Engineering and Environmental Laboratory, and from Rocky Flats, a former nuclear trigger factory near Denver. The waste is tracked by satellite and is moved only at night when traffic is lighter. It can be transported only in good weather and must be routed around major cities. Figure 4.20 shows transport routes, and Figure 4.21 shows a tractor-trailer container transporting radioactive waste to the WIPP.

By the beginning of the twentieth century about 61 million Americans lived within 50 miles of a military nuclear waste storage site. By the time the WIPP has been in operation for 10 years, the number should drop to 4 million. By 2035, barring court challenges, almost 40,000 truckloads of nuclear waste will be trucked across the country to the WIPP.

Yucca Mountain

The centerpiece of the federal government's plan to dispose of highly radioactive waste is a proposed facility at Yucca Mountain in Nevada. The Nuclear Waste Policy Act of 1982 requires the secretary of energy to investigate the site and, if it is suitable, recommend to the president that the site be established. The investigation of Yucca Mountain has taken a long time. The DOE's 1998 objective was to begin disposing of waste in the repository in 2010, 12 years later than originally expected.

It is not clear whether the Yucca Mountain site is suitable. (See Figures 4.22 and 4.23.) It is located near volcanic and earthquake activity and rights to the land are being contested by local American Indian tribes. Scientists have also discovered areas of "perched water" above the water table that could be affected by the facility.

FIGURE 4.19

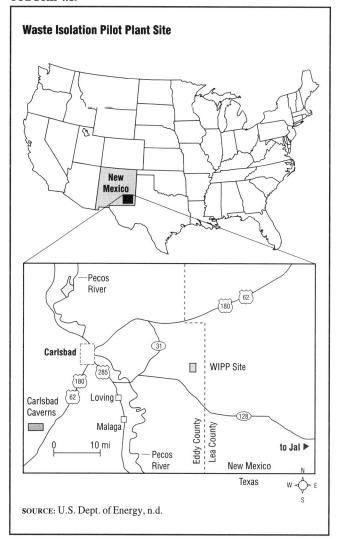

Waste Isolation Pilot Plant Site

SOURCE: U.S. Dept. of Energy, n.d.

Although these areas may not ultimately prove to be a problem, researchers must determine their size and number in order to know for certain. In addition, some western states feel they have long been targeted for hazardous facilities. If the site is found to be acceptable, the president approves it, and a recommendation goes to the Congress, the state of Nevada is expected to file a notice of disapproval.

STANDARDS FOR CONTAINMENT. In order for the Yucca Mountain Repository to be built the DOE must satisfactorily demonstrate to the Nuclear Regulatory Commission (NRC) that the combination of the site and the repository design complies with the standards set forth by the EPA. The EPA's standard is based on a new approach of using numerical probabilities to establish requirements for containing radioactivity within the repository. Their quantitative terms are:

- Cumulative releases of radioactivity from a repository must have a likelihood of less than one chance in ten of exceeding limits established in the standard, and a

FIGURE 4.20

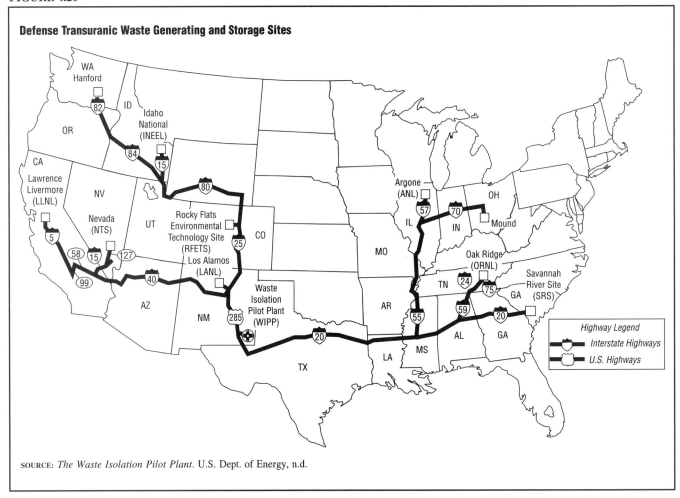

Defense Transuranic Waste Generating and Storage Sites

SOURCE: *The Waste Isolation Pilot Plant.* U.S. Dept. of Energy, n.d.

likelihood of less than one chance in 1,000 of exceeding ten times the limits, for a period of 10,000 years.

- Exposures of radiation to individual members of the public for 1,000 years must not exceed specified limits.

- Limits are placed on the concentration of radioactivity for 1,000 years after disposal from the repository to a nearby source of groundwater that 1) currently supplies drinking water for thousands of persons and 2) is irreplaceable.

- Prescribed technical or institutional procedures or steps must provide confidence that the containment requirements are likely to be met.

Crisis in the Industry

The long delay in providing disposal sites for nuclear wastes, coupled with the accelerated pace at which nuclear plants are being retired, has created a crisis in the industry. Several aging plants are being maintained—at a cost of $20 million a year for each reactor—simply because there is no place to send the waste once the plants are decommissioned. Under the Nuclear Waste Policy Act of 1982 the DOE was scheduled to begin picking up waste on January 31, 1998. The utilities have been paying one-tenth of one cent per kilowatt-hour produced by the reactors to finance a repository.

Although the 1987 waste amendment designated Yucca Mountain as the site, little progress has been made in approving the project. In 1996 the U.S. Court of Appeals, in *Indiana Michigan Power Co. v. Department of Energy* (88F3d1272 [D.C. Ctr., 1996]), ruled that the nuclear waste act created an obligation for the DOE to start disposing of utilities' waste no later than January 31, 1998. Because the DOE missed the deadline, more than 20 utilities have sued.

In February 1999 the DOE announced that, because it was unable to receive nuclear waste for permanent storage, it would take ownership of the waste and pay temporary storage costs with money the utilities have paid to develop the permanent repository. The waste will stay where it currently is being stored, and the DOE will pay the storage costs.

Both the Senate and the House passed legislation to build a temporary repository in Nevada. The Clinton Administration, however, vetoed the temporary site,

FIGURE 4.21

Using specially designed containment canisters, a truck hauls radioactive waste to the Waste Isolation Pilot Plant in New Mexico, where it can be safely stored. (*U.S. Department of Energy.*)

claiming that it has not been proven safe and would deflect funds and engineering talent needed to build the permanent facility. Even without the expense of temporary storage, the nuclear waste fund (from the 1 cent per 10 kilowatt-hours of nuclear power generated by the utilities) is many billions short of what Yucca Mountain is expected to cost.

FIGURE 4.22

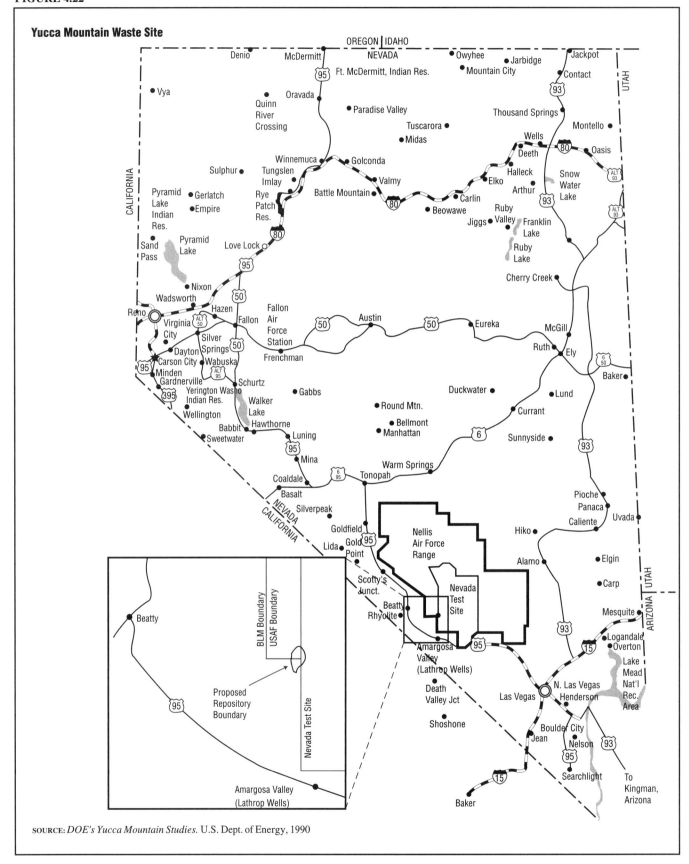

Yucca Mountain Waste Site

SOURCE: *DOE's Yucca Mountain Studies.* U.S. Dept. of Energy, 1990

FIGURE 4.23

An aerial view of Yucca Mountain, Nevada. Yucca Mountain is the proposed site for a major, long-term, nuclear waste storage facility. (*U.S. Department of Energy, Office of Civilian Radioactive Waste Management.*)

CHAPTER 5

AIR QUALITY

THE GLOBAL HEALTH THREAT

It is possible for Americans to avoid city garbage dumps and even dirty water. If they stay away from landfills they can avoid seeing the mountains of garbage building up; if their drinking water is clean and they do not visit the rivers, the lakes, or the seashore, they can avoid seeing polluted waters.

But all Americans must breathe air and some of that air is dangerous. Air pollution can make people sick and damage the environment. Trees, lakes, and animals can be harmed by air pollution. Air pollutants have made the ozone layer around Earth thinner, leading to increased skin cancer and cataracts (a disease of the eye that causes blindness) in humans. Air pollution dirties and eats away the stone in buildings, monuments, and statues. Air pollution also causes haze that decreases visibility.

THE AIR PEOPLE BREATHE

Fossil fuels and chemicals have played a major role in society's pursuit of economic growth and higher standards of living. But the burning of fossil fuels and the use of chemicals alter Earth's chemistry and can threaten the food, water, and air supplies life depends on.

Smog

Polluted air is everywhere. In the United States the list of cities darkened by smog has grown every year since 1970. Smog, a word made up from "smoke" and "fog," is probably the most well known form of air pollution.

Smog is made up mainly of ground-level ozone, a gas that forms naturally in the upper atmosphere. Ozone can be good or bad, depending on where it is located. (See Chapter 3.) When ozone is high in the atmosphere, it shields Earth from the ultra-violet light radiated by the sun and protects human health and the environment. When the ozone is at ground level, however, it becomes the most harmful element in smog. Ground-level ozone is produced by the mixing of pollutants from many sources, including smokestacks, car exhausts, and paints. When a car burns gasoline or a painter paints a house, dangerous fumes rise into the air.

Wind often blows smog-forming pollutants away from their sources. The reaction that creates smog occurs while pollutants are being blown through the air by the wind. This is why smog is often more serious miles away from where the pollutants were created. The smog-forming pollutants are brought together in the sky and, if it is hot and sunny, smog easily forms. Weather and location determine where smog goes and how bad it will become.

Unlike other pollutants, ozone is not emitted directly into the air. It, too, forms on sunny, hot days by complex chemical reactions among other pollutants. Without sunlight, ozone breaks down quickly. When temperature inversions occur (the warm air stays near the ground instead of rising) and winds are calm, such as during the summer, smog may hang over a huge area for days at a time. As traffic and other pollution sources add more pollutants to the air, the smog gets worse.

Most people associate dirty air with cities and the areas around them. There is good reason for this since some of the worst smog in the country occurs in cities such as Los Angeles, California—a city known for its air quality problems.

In a major industrial nation such as the United States, however, smog is not limited just to cities. The Great Smoky Mountains, located in western North Carolina and eastern Tennessee, are seeing more air pollution. Harmful emissions from various coal-burning facilities, located outside the mountain range, as well as pollution from motor vehicles, are damaging the mountain's environment.

Breathing dirty air is unhealthy and the damage polluted air does to buildings and statues indicates how

TABLE 5.1

Number of Areas Exhibiting National Air Quality Standards Nonattainment Status, By Pollutant, 1999

Pollutant	Original # areas	1999 # areas	1999 Population (in 1000s)
CO	43	20	33,230
Pb	12	8	1,116
NO$_2$	1	0	0
O$_3$	101	32	92,505
PM$_{10}$	85	77	29,880
SO$_2$	51	31	4,371

SOURCE: *National Air Quality and Emissions Trends Report 1998.* U.S. Environmental Protection Agency, 2000

TABLE 5.2

Air Pollutants, Health Risks, and Contributing Sources

Pollutants	Health risks	Contributing sources
Ozone[1] (O$_3$)	Asthma, reduced respiratory function, eye irritation	Cars, refineries, dry cleaners
Particulate matter (PM-l0)	Bronchitis, cancer, lung damage	Dust, pesticides
Carbon monoxide (CO)	Blood oxygen carrying capacity reduction, cardiovascular and nervous system impairments	Cars, power plants, wood stoves
Sulphur dioxide (SO$_2$)	Respiratory tract impairment, destruction of lung tissue	Power plants, paper mills
Lead (Pb)	Retardation and brain damage, esp. children	Cars, nonferrous smelters. battery plants
Nitrogen dioxide (NO$_2$)	Lung damage and respiratory illness	Power plants, cars, trucks

[1]Ozone refers to tropospheric ozone which is hazardous to human health.

SOURCE: *Healthy People 2000—Statistical Notes.* Centers for Disease Control and Prevention, 1995

potent it is. Most television and radio weather reports in major cities include ozone and "air quality" readings as part of the daily weather statistics.

Air pollution is not only an American problem, however. Mexico City, the capital city of Mexico, is generally considered to have the worst air in the world while the recent surge in economic development in China has led to extremely polluted air in many of its cities.

Throughout the world, poor air quality contributes to hundreds of thousands of deaths and diseases each year, not to mention dying forests and lakes and the corrosion of stone buildings and monuments. In the United States, in 1999, an estimated 151 million people lived in areas designated by the Environment Protection Agency (EPA) as nonattainment areas, that is, where the air has been declared unfit to breathe because of high levels of six criteria pollutants—carbon monoxide, lead, nitrogen dioxide, ozone, particulate matter, and sulfur dioxide. (See Table 5.1.) As of September 1999 a total of 121 areas in the United States failed to meet pollution standards compared to 158 areas in 1997.

Air pollution has three principal sources created by humans—energy use, vehicular emissions, and industrial production. All of these generally increase with economic growth unless pollution control measures are established. Air pollution, particularly that which causes acid rain, devastates forests, crops, and waterways and works its way into the water cycle and food chains.

HEALTH EFFECTS

The quality of the environment plays a major role in public health. However, while focusing on the environment is an obvious and often successful approach to improving public health, it is a complex problem. Among the factors that must be considered are the levels of pollutants in the air, the levels of individual exposure to these pollutants, individual susceptibility to toxic substances, and exposure times related to ill effects from certain sub-

stances. Attributing health effects to specific pollutants is also complicated by the impact of non-environmental causes of the same health effects (for example, smoking, heredity, or diet).

Air pollution is related to a number of respiratory diseases including bronchitis, pulmonary emphysema, lung cancer, bronchial asthma, eye irritation, weakened immune system, and premature lung tissue aging. In addition lead contamination causes neurological and kidney disease and can be responsible for impaired fetal and mental development. The American Lung Association estimates the annual health costs of exposure to the most serious air pollutants at $40 to $50 billion.

Emissions of toxic materials into the air cause an estimated 2,000 cancer deaths, alone, a year. Because research has generally included only one-third of the known carcinogens (and many more have yet to be discovered), this is believed to be an underestimate. Cancer rates have been known to be higher among persons who live near certain types of factories. Researchers believe that as many as 50,000 to 60,000 deaths a year from different diseases and disorders are caused by particle (soot) pollution. They report that some types of particles can be fatal, even when under the legal limit. Much legislation being considered at the beginning of the twenty-first century addressed small particle pollution.

WHAT ARE THE AIR POLLUTANTS?

The Clean Air Act of 1970 (PL 91-604) and the Clean Air Act Amendments of 1990 (PL 101-549) address the six pollutants associated with the National Ambient Air Quality Standards (NAAQS). These pollutants—ozone, particulate matter, carbon monoxide, sulfur dioxide, nitrogen dioxide, and lead—are called criteria pollutants and are identified as serious threats to human health. (See Table 5.2.)

The EPA has documented air pollution trends in the United States annually since 1973. Its *National Air Quality and Emissions Trends Report, 1998* (2000) reports two kinds of trends for criteria pollutants. Emissions are calculated estimates of the total tonnage of these pollutants released into the air annually. Air quality concentrations measure pollutant concentrations in the air at monitoring stations. From 1989 to 1998 air quality concentrations improved for all six criteria pollutants. (See Table 5.3.) Emissions of all criteria pollutants also improved, with the exception of nitrogen dioxide.

Ozone

Ozone (O_3) is the principal component of urban smog. Ozone is a gas formed when energy from sunlight causes hydrocarbons (from industry and automobile emissions) to react with nitrogen oxides (from power plants and automobiles). Ozone has become a persistent problem in many parts of the world. Even the smallest amounts of ozone can cause breathing difficulty. Ground-level ozone is the most complex, pervasive, and difficult to control of the six criteria pollutants. Between

TABLE 5.3

Percent Decrease in National Air Quality Pollutant Concentrations, 1989–1998

Carbon Monoxide	39
Lead	56
Nitrogen Dioxide	14
Ozone*	4
Particulate Matter (PM$_{10}$)	25
Sulfur Dioxide	39

*based on 1-hour level.

SOURCE: *National Air Quality and Emissions Trends Report 1998*. U.S. Environmental Protection Agency, 2000

FIGURE 5.1

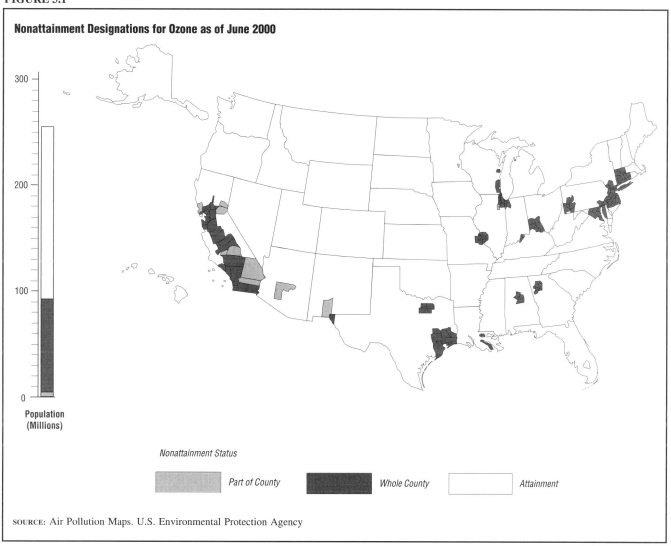

Nonattainment Designations for Ozone as of June 2000

Population (Millions)

Nonattainment Status

Part of County   Whole County   Attainment

SOURCE: Air Pollution Maps. U.S. Environmental Protection Agency

FIGURE 5.2

FIGURE 5.3

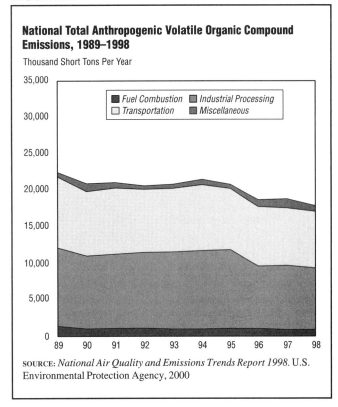

National Total Anthropogenic Volatile Organic Compound Emissions, 1989–1998

Thousand Short Tons Per Year

SOURCE: *National Air Quality and Emissions Trends Report 1998.* U.S. Environmental Protection Agency, 2000

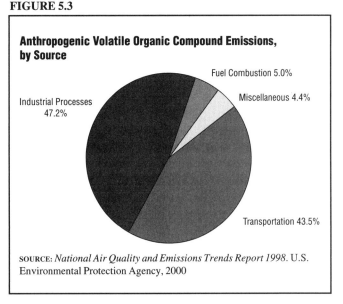

Anthropogenic Volatile Organic Compound Emissions, by Source

SOURCE: *National Air Quality and Emissions Trends Report 1998.* U.S. Environmental Protection Agency, 2000

1989 and 1998 ozone concentrations in the United States declined 4 percent, although concentrations increased slightly (4 to 5 percent) over 1997 readings. Emissions declined 20 percent since 1989 and 5 percent since 1997. (See Chapter 3.)

After southern California, already under special mandate to reduce emissions because of its high ozone levels, the nation's worst smog problems are in the Chicago, Houston, and New York metropolitan areas. (See Figure 5.1.) Eleven Atlantic coast states (Maine, New Hampshire, Vermont, New York, Pennsylvania, Massachusetts, Rhode Island, Connecticut, New Jersey, Maryland, and Delaware) and the District of Columbia came under regulatory order to control smog due to their high levels of ozone. New York is scheduled to set up pollution meters, signs that provide pollution readings for the public in an effort to raise public awareness. Texas will likely fall under stricter standards if air quality levels do not improve.

New regulations require extensive changes to dozens of utility generating stations that use oil, coal, and natural gas. Because of the city's efforts to reduce ozone pollution, measurements of smog in the Los Angeles, California, area are the lowest in a generation. The Phoenix, Arizona, suburb of Glendale has initiated an innovative means of reducing pollution. City employees give up driving their cars at least one day a week. In exchange the city donates to the employees the use of a bicycle taken from the city's unclaimed stolen bicycles and persons who carpool get valet parking or shaded parking spaces.

The American Lung Association, in *State of the Air: 2000* (May 2000), assessed the quality of air in U.S. communities giving them grades ranging from "A" through "F" based on how often their air quality exceeds the "unhealthful" limits of the EPA's air quality index. The report focuses on ground level ozone. The 2000 report found that 132 million Americans live in areas that received an "F." That is about 72 percent of the nation's population who live in counties where there are ozone monitors. Of the 678 counties examined almost half (333) received an "F." Living in the counties that received a failing grade are an estimated 16 million Americans over the age of 65 years, more than 7 million asthmatics, 29 million children under the age of 14, and 7 million adults with chronic bronchitis. The agency found that, although cities in southern California generally scored poorly, several "sunbelt" states such as Texas, Tennessee, and Georgia also scored low. Ozone levels in Houston, Texas, exceeded levels in Los Angeles, historically the highest in the nation.

A CONTRIBUTOR TO OZONE—VOLATILE ORGANIC COMPOUNDS. Volatile organic compounds (VOCs), pollutants that have health and environmental effects similar to smog, are released from burning fuels such as coal, natural gas, gasoline, and wood. Cars are a major source of VOCs. From 1940 through 1970 VOC emissions increased about 77 percent, mainly because of the increase in car and truck traffic and industrial production. Since 1970 national VOC emissions have decreased as a result of emission controls placed on cars and trucks and less open burning of solid waste. VOC emissions dropped 20 percent from 1989 to 1998. (See Figure 5.2.) Most of the decline has occurred in the transportation area and the burning of fuel, while VOC pollution from industry and solid-waste disposal has changed little. Industrial processes and transportation still account for almost all VOC emissions. (See Figure 5.3.)

FIGURE 5.4

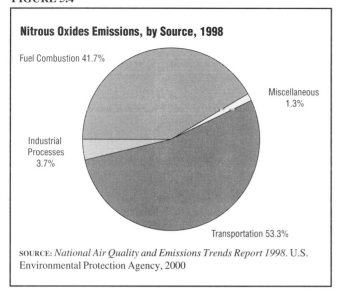

Nitrous Oxides Emissions, by Source, 1998

Fuel Combustion 41.7%

Miscellaneous 1.3%

Industrial Processes 3.7%

Transportation 53.3%

SOURCE: *National Air Quality and Emissions Trends Report 1998*. U.S. Environmental Protection Agency, 2000

FIGURE 5.5

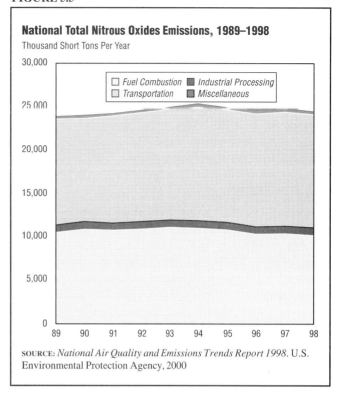

National Total Nitrous Oxides Emissions, 1989–1998

Thousand Short Tons Per Year

Legend: Fuel Combustion, Industrial Processing, Transportation, Miscellaneous

SOURCE: *National Air Quality and Emissions Trends Report 1998*. U.S. Environmental Protection Agency, 2000

FIGURE 5.6

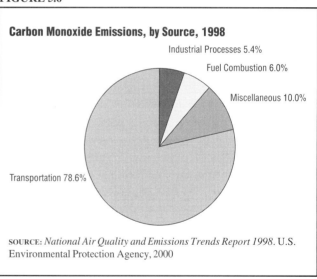

Carbon Monoxide Emissions, by Source, 1998

Industrial Processes 5.4%

Fuel Combustion 6.0%

Miscellaneous 10.0%

Transportation 78.6%

SOURCE: *National Air Quality and Emissions Trends Report 1998*. U.S. Environmental Protection Agency, 2000

Nitrogen Dioxide

Nitrogen dioxide (NO_2) comes from burning fuels such as gasoline, natural gas, coal, and oil. Transportation—cars, trucks, and other vehicles—is the major source of nitrogen dioxide (53 percent), followed by fuel burning (42 percent). (See Figure 5.4.) Nitrogen dioxide is a major part of smog and causes the same health and environmental effects as smog. Nitrogen dioxide is also in acid rain that damages trees and lakes and eats away at stone buildings and statues. (See Chapter 7.)

From 1940 to 1970 nitrogen oxide emissions increased because of more burning of natural gas and the increase in gasoline consumption in the growing number of cars and trucks. Overall nitrogen dioxide air concentrations fell 14 percent from 1989 to 1998, although emissions showed a 2 percent increase over that time. (See Figure 5.5.) All U.S. counties attained the EPA standards for nitrogen dioxide in 1999.

Carbon Monoxide

Like most criteria pollutants carbon monoxide (CO) comes from burning gasoline, natural gas, wood, coal, and oil, and is created primarily in the transportation industry. (See Figure 5.6.) Carbon monoxide is a dangerous gas that reduces the ability of blood to bring oxygen to the body cells. More than a half-century ago, in 1940, cars and trucks created about 28 percent of the carbon monoxide emissions, while homes burning coal and oil made up about 50 percent. From 1940 through 1970 emission from cars and trucks nearly tripled, by 1970, it accounted for 71 percent of all carbon monoxide, and a dozen years later—in 1982—it produced about 80 percent of the total carbon monoxide emission.

Total air concentration of carbon monoxide fell 39 percent from 1989 to 1998. (See Figure 5.7.) The total amount of carbon monoxide emissions dropped 16 percent during that time period, due primarily to an estimated 26 percent drop in highway vehicle emissions. Nevertheless, there were still a number of areas of the country that failed to meet air quality standards in 2000. (See Figure 5.8.)

Particulate Matter

Particulate matter (PM_{10}) is the general term for solid or liquid particles—dust, smoke, and soot—that come from burning fuels in industrial plants, from farmland, and from unpaved roads. It includes particles so small that an electron microscope is required for identification. Particulate matter can irritate the nostrils, throat, and lungs. When particulate matter hangs in the air, it creates a haze.

From 1940 to 1971 particulate matter generally increased. Pollution control laws, however, led to a drop in particulate matter, most of which occurred during the 1970s. Between 1988 and 1998 overall air concentrations of particulates declined 25 percent; particulate emissions dropped 19 percent. However, there were still areas where particulate matter levels did not meet air quality standards in 2000. (See Figure 5.9.)

Sulfur Dioxide

Sulfur dioxide (SO_2) is formed when fuel containing sulfur—mainly coal and oil—is burned. Some industrial processes and metal smelting also cause sulfur dioxide to form. Inhaling sulfur dioxide can lead to serious breathing problems. Together, sulfur dioxide and nitrogen dioxide are the major components of acid rain that can damage surface and groundwater and organisms living there. From 1940 to 1970 sulfur dioxide emissions increased as a result of the growing use of fossil fuels, especially coal, by industry and electrical utility plants. Since 1970, total sulfur dioxide emissions have dropped because of cleaner fuels with lower sulfur content and the greater use of pollution devices such as scrubbers that clean factory emissions.

Most sulfur emission still comes from fuel combustion. (See Figure 5.10.) From 1989 to 1998 air concentrations decreased 39 percent and emissions decreased 16 percent. (See Figure 5.11.) Seven sulfur dioxide nonattainment areas still remained in the United States as of June 2000. (See Figure 5.12.)

Lead

The Centers for Disease Control and Prevention (CDC) labels lead (Pb) poisoning as the most common and most devastating environmental disease affecting young children. Lead can damage the brain and nervous system. Until recently the main source of lead pollution was leaded gasoline. Smelters and battery plants followed by highway vehicles are the leading sources of lead emissions. From 1989 to 1998 overall air concentrations of lead fell 56 percent while emissions decreased 27 percent over that time, primarily because highway vehicles no longer use leaded gasoline. There was a 96 percent decline in lead concentrations from 1979 through 1998. (See Figure 5.13.) However, some areas in the United States still failed to meet lead standards as of June 2000. (See Figure 5.14.)

The primary cause of the most severe cases of lead poisoning in children is exposure to lead-based paint, once widely used in housing in the United States. The Office of Housing and Urban Development (HUD) estimates that 57 million, or about three-fourths, of the 77

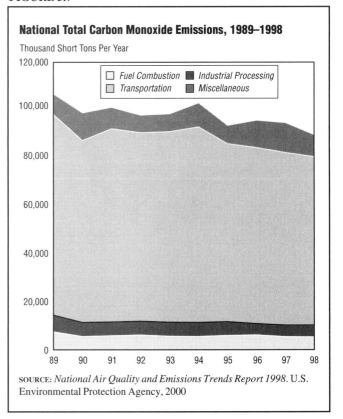

FIGURE 5.7

National Total Carbon Monoxide Emissions, 1989–1998

Thousand Short Tons Per Year

Legend: Fuel Combustion, Industrial Processing, Transportation, Miscellaneous

SOURCE: *National Air Quality and Emissions Trends Report 1998.* U.S. Environmental Protection Agency, 2000

million privately owned and occupied homes built before 1980 contain lead-based paint. Almost 10 million of these homes are occupied by families with children under seven years of age—the age group most vulnerable to lead poisoning. In 1997 the EPA and HUD issued a regulation requiring the disclosure of known lead-based paint hazards when homes are sold or rented. In addition many states have passed laws requiring disclosure of lead in real estate transactions. (See Figure 5.15 and Chapter 8.)

AIR TOXICS

Hazardous air pollutants (HAPs), also referred to as air toxics, are pollutants that may cause severe health effects or ecosystem damage. Approximately 3.7 million tons of air toxics are released into the air each year. The Clean Air Act (CAA) lists 188 substances as HAPs and targets them for regulation in section 112 (b) (1). Examples of such toxins are benzene, dioxins, arsenic, beryllium, mercury, and vinyl chloride. (See Table 5.4.) Air toxins are emitted from many sources including industrial and mobile (motor-vehicle and non-road equipment) sources. The air toxics program complements the NAAQS program.

The Sector Facility Indexing Project—"Right to Know"

In the mid-1990s the Clinton Administration sought to expand "right to know" initiatives—environmental

FIGURE 5.8

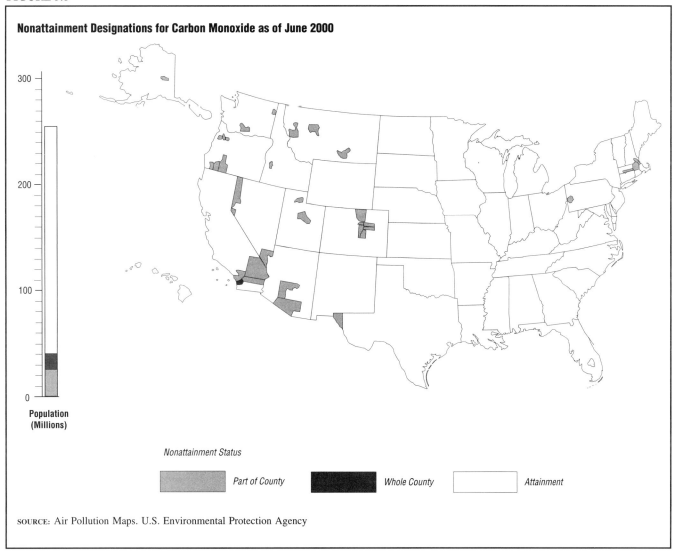

Nonattainment Designations for Carbon Monoxide as of June 2000

Population
(Millions)

Nonattainment Status

Part of County   Whole County   Attainment

SOURCE: Air Pollution Maps. U.S. Environmental Protection Agency

programs designed to inhibit pollution not with legislation and regulation but by exposing polluters to pressure from a well-informed public. The Sector Facility Indexing Project (SFIP) provides data on the Internet about companies in five industries—automobile, steel, metals, oil refining, and papermaking. The environmental profiles include inspections, noncompliance, enforcement, pollution releases and spills, as well as demographics on these companies such as how many people live within three miles of a plant.

THE TOXICS RELEASE INVENTORY. The Toxics Release Inventory (TRI) is a publicly available database (www.epa.gov/tri/tri98/) that reports information on toxic chemical releases by various industries. Under the Emergency Planning and Community Right-to-Know Act of 1986, facilities in metal mining, coal mining, electrical utilities, hazardous waste treatment and disposal facilities, chemicals distributors, petroleum facilities, federal facilities, and solvent recovery services are required to report their releases of nearly 650 toxic chemicals. In 1998,

23,000 facilities reported a total of 7.3 billion pounds of releases to air, land, water, and underground injection.

THE AUTOMOBILE'S CONTRIBUTION TO AIR POLLUTION

Automobiles dominate the transportation sector's share of energy-related carbon emissions. The transportation sector accounts for more than 65 percent of U.S. petroleum consumption and more than 78 percent of carbon monoxide emissions.

Concerns about energy security and air pollution from transportation have encouraged the development of alternative fuel technologies to reduce dependence on traditional gasoline and diesel fuels. Standards established in the 1970s improved efficiency markedly during the 1980s but have since leveled off, leading to a decline in fuel economy. In the absence of tighter standards emissions from road travel are expected to double by 2020 because of increased usage. With industrialization occurring in many

FIGURE 5.9

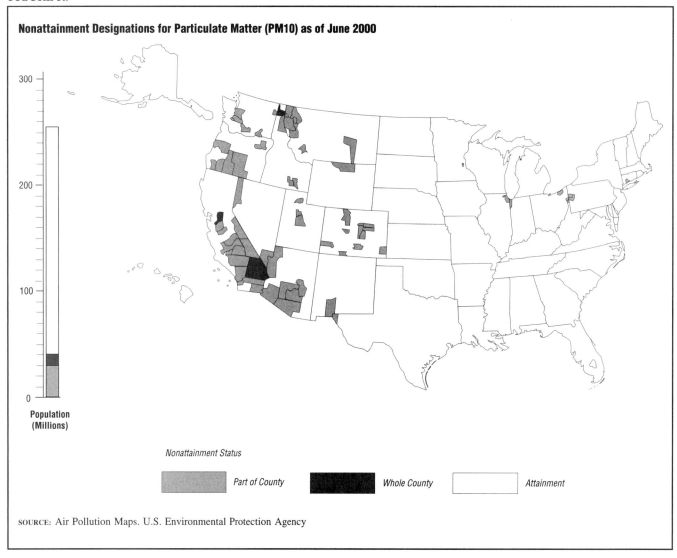

Nonattainment Designations for Particulate Matter (PM10) as of June 2000

Population (Millions)

Nonattainment Status

Part of County   Whole County   Attainment

SOURCE: Air Pollution Maps. U.S. Environmental Protection Agency

FIGURE 5.10

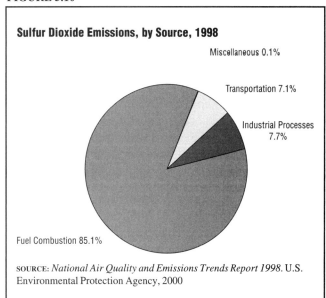

Sulfur Dioxide Emissions, by Source, 1998

Miscellaneous 0.1%

Transportation 7.1%

Industrial Processes 7.7%

Fuel Combustion 85.1%

SOURCE: *National Air Quality and Emissions Trends Report 1998.* U.S. Environmental Protection Agency, 2000

developing countries, the increase in global automobile use and the emissions that accompany them is inevitable.

American states that do not meet CAA standards must do something to bring emissions into compliance with national standards. Because of California's extreme air pollution problems, the CAA Amendments of 1990 (PL 101-549, see below) allowed the state to set stricter emission standards than those required by the amendments, which it did. These included strict new laws on automobile pollution. The remaining 49 states were given the option of choosing the standards of California or the federal CAAA.

States have the freedom to cut their emissions in whatever manner they choose. Some states are phasing in tougher tests for auto emissions. In major metropolitan areas, particularly in the Northeast, owners of cars and light trucks would be required to take their vehicles to centralized, high-technology inspection stations. The EPA estimates that approximately three-fourths of the vehicles will pass the inspections on the first try. For

FIGURE 5.11

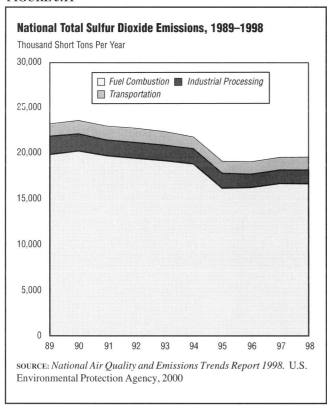

National Total Sulfur Dioxide Emissions, 1989–1998

Thousand Short Tons Per Year

Legend: □ Fuel Combustion ■ Industrial Processing ▨ Transportation

SOURCE: *National Air Quality and Emissions Trends Report 1998.* U.S. Environmental Protection Agency, 2000

those that do not pass repairs are expected to cost between $38 and $120; repairs for old "clunkers" will most likely cost more. Public protest over inspections such as these caused several states to rescind vehicle checks—at least temporarily. Those states must find other ways to cut their emissions. If they cannot bring their emissions down to comply with EPA standards they may have to reinstate vehicle inspections. Politicians are finding such measures very unpopular with their constituents.

One of the major failings in reducing auto-induced smog is that efforts have focused on reducing tailpipe emissions instead of eliminating their formation in the first place. Car-makers have shown that they can adapt to tighter emission standards by introducing lighter engines, fuel-injection, catalytic converters, and other technological improvements. Some experts believe, however, that efforts could be better spent by drastically reducing emissions and promoting alternative transportation such as mass transit systems, carpools, and bicycles, and alternative energy.

Alternative Approaches

One option under consideration to reduce emissions is the introduction of vehicles powered by alternative fuels such as methanol, ethanol, natural gas, hydrogen, and electricity. Although these fuels offer advantages their use may substitute one problem for another. For example, methanol reduces ozone formation but increases formaldehyde, a human carcinogen, and is twice as toxic as gasoline if it comes in contact with the skin. Engines require twice as much methanol as gasoline to travel a similar distance. Natural gas reduces hydrocarbons and carbon monoxide but increases nitrogen oxides. In the long run electricity and hydrogen, clean burning fuels, seem the most promising of the alternative fuels for vehicles. However, electricity generated from non-polluting, renewable sources is not on the immediate horizon and hydrogen produced by solar cells is still too costly. Many experts believe the most feasible solution in the near future will be a hybrid vehicle using a combination of gasoline and one of the other fuel sources, most likely a fuel cell or hydrogen.

REFORMULATED GASOLINE. Many oil companies researched alternative fuels to cut emissions even further in anticipation of CAA standards that went into effect in 1995. The Act required those areas with the worst polluted air to sell reformulated gasoline (RFG) beginning January 1995. Other areas could voluntarily choose to participate.

Carbon monoxide emissions result from incomplete combustion of fuel. Adding oxygen (*oxygenation*) to the fuel makes combustion more complete, especially in older cars, and carbon dioxide rather than carbon monoxide, results. The most frequently used oxygenates are methyl tertiary butyl ether ethanol (MTBE), which accounts for about 98 percent of RFG, and ethanol, which accounts for the remaining 2 percent. RFG results in a lower-octane fuel and could possibly cause an increase in price (from 4 to 5 cents per gallon to as much as 25 cents per gallon).

In cities with the highest ozone levels, fuels with added oxygen are already required. Denver, Colorado, known for its "brown cloud" of pollution, enacted the nation's first oxygenated fuels program—an entire winter period during which all fuels sold at gas stations were required to have a 3 percent oxygen content. However, customer complaints over price have caused many areas that had voluntarily opted to sell RFG to back out.

MTBE is becoming a source of great concern. It is very soluble in water and therefore tends to migrate into water supplies. Testing by the United States Geological Survey found MTBE in approximately 20 percent of the water in RFG areas compared to 2 percent in non-RFG areas. EPA administrator Carol Browner called for reductions in the use of MTBE, and some legislators are taking steps to ban its use. Consequently, the use of ethanol is expected to rise.

CONVERTERS. Tailpipe catalytic converters are one of the most successful technologies in the history of smog control, eliminating 96 to 98 percent of the carbon monoxide and hydrocarbons. Striving to eliminate pollutants entirely, researchers are now perfecting new "pre-heating" converters. Converters typically cannot function properly until the car has warmed to a specific temperature causing emissions to be greatest during the first few miles. The new converters would function as soon as a car is started.

FIGURE 5.12

Nonattainment Designations for Sulfur Dioxide as of June 2000

Population (Millions)

Nonattainment Status

Part of County   Whole County   Attainment

SOURCE: Air Pollution Maps. U.S. Environmental Protection Agency

In 1998 the EPA expressed concerns, however, that catalytic converters rearrange nitrogen-oxygen compounds to form nitrogen oxide, the most common of the global warming gases. Nitrogen oxide levels may increase, at least in part, from the growth in the number of vehicle miles traveled by cars that have catalytic converters.

Government Regulation—Corporate Average Fuel Economy (CAFE) Standards

In 1973 the Organization of Petroleum Exporting Countries (OPEC) imposed an oil embargo that provided a painful reminder to America of how dependent it had become on foreign sources of fuel. Although the United States makes up only 5 percent of the world's population, it consumes approximately one-quarter of the world's supply of oil, much of which is imported from the Middle East. The 1973 oil embargo prompted Congress to pass the 1975 Automobile Fuel Efficiency Act (PL 96-426) which set the initial Corporate Average Fuel Efficiency standards (commonly called the CAFE standards).

CAFE standards required each domestic automaker to increase the average mileage of the new cars sold to 27.5 miles per gallon (mpg) by 1985. Under CAFE rules car manufacturers could still sell the big, less efficient cars with powerful eight-cylinder engines, but to meet the average fuel efficiency rates they also had to sell smaller, more efficient cars. Automakers that failed to meet each year's CAFE standards were required to pay fines. Those that managed to surpass the rates earned credits that they could use in years when they fell below CAFE requirements.

Faced with CAFE standards, automobile companies became more inventive and managed to keep their cars relatively large and roomy with such innovations as electronic fuel injection and front-wheel drive. Ford's prestigious Lincoln Town Car managed to achieve better mileage in 1985 than its small Pinto did in 1974.

Some legislators have proposed further increases in CAFE standards but, as of 2000, no action had been successful. Opponents to raising CAFE standards believe that the congressional fuel economy campaign would saddle American motorists with car features they would not like and would not buy. They have also pointed out that the only way to raise fuel efficiency levels is to manufacture much smaller cars and trucks and to limit the number of larger vehicles. They claim smaller cars would raise the numbers of highway deaths and injuries, limit consumer choice of larger and family-sized vehicles, and place thousands of auto-related jobs at risk.

In the absence of any further mandate to increase CAFE standards since 1985, mileage remained steady briefly and then declined. (See Figure 5.16.) The average fuel economy for 1999 light vehicles was 23.8 miles per gallon (mpg), 28.1 for passenger cars, and 20.3 mpg for light trucks—the lowest since 1980. The required efficiency standard has not improved further due to America's preference for larger, less-efficient cars, minivans, and sport-utility vehicles (SUVs). States also raised speed limits which lowers fuel efficiency.

Industry officials claim that increasing fuel efficiency is not cost-effective. With each mile added in efficiency, the costs to obtain that improvement increases to the point that it is no longer cost-effective. This is the same objection that is made, in general, about cleaning up many environmental hazards—that the first and most drastic improvements are the least expensive and thereafter clean-up becomes more costly.

In contrast, for 2005, the European Commission has proposed an ambitious target of 47 mpg for gasoline-driven cars (compared to the current average of 29 mpg) and 52 mpg for diesel-powered cars. It must be noted that, while European countries do not generally legislate fuel efficiency, the cost of gasoline is more than twice that in the United States. That serves as a powerful incentive to European drivers to buy fuel-efficient vehicles.

Although some manufacturers in the United States claim to be able to produce more efficient cars, they also contend that American consumers would not buy them at this time. They believe consumers seek, instead, features that raise both the size of a car and its price. In essence, manufacturers claim, there is little market for such a vehicle. The EPA reports in *Light-Duty Automotive Technology and Fuel Economy Trends Through 1999* (1999) that "the trend has clearly been to apply new technologies to increase average new vehicle weight, power, and performance while maintaining fuel economy," not improving fuel economy. This is reflected in heavier average vehicle weight (up 20 percent since 1986), rising horsepower (up 58 percent since 1986), and lower 0 to 60 mile-per-hour acceleration time (19 percent faster that in 1986), but a fuel economy decrease of 7 percent. Manufacturers continue research in AFVs such as electric cars, which they hope will eventually satisfy American tastes and pocketbooks and yet be fuel-efficient, as well. In 1999 the EPA

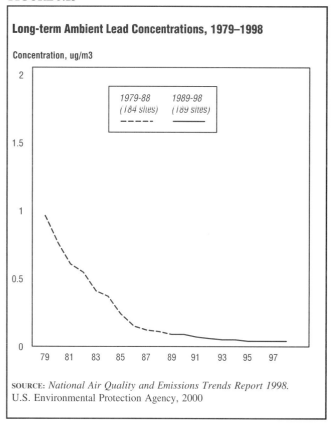

SOURCE: *National Air Quality and Emissions Trends Report 1998.* U.S. Environmental Protection Agency, 2000

entered into a partnership with Saturn Corporation to promote cleaner production practices and reduce pollution by Saturn vehicles.

The Impact of Sport-Utility Vehicles

Big SUVs and pickups are the fastest-growing segment of the auto industry, accounting for 46 percent of the U.S market and producing most of the profits for the major auto companies. (See Figure 5.17.) These types of vehicles fall under less stringent emissions standards than automobiles because they originated as modifications of light-truck bodies and are classified as trucks. Automakers and buyers of trucks and SUVs oppose tightening restrictions on emissions of those vehicles, although critics contend that new SUVs are more like cars than trucks in design.

In 1999 the EPA proposed new regulations tightening emissions standards on cars, mini-vans, SUVs under 8,500 pounds, and small pickup trucks. The regulations would, however, not affect the rules for big SUVs and pickups, allowing these bigger vehicles to emit up to as much as five times more nitrogen oxides than cars. Under rules still enforced by July 2000, these vehicles can emit three times more nitrogen oxides, the main cause of smog.

Denver, Colorado, has experienced two winters without violating federal standards for carbon monoxide, particulate matter, or ozone. Experts attribute the improvement to automakers' compliance with federal

FIGURE 5.14

Nonattainment Designations for Lead as of June 2000

Population (Millions)

Nonattainment Status

Part of County   Whole County   Attainment

SOURCE: Air Pollution Maps. U.S. Environmental Protection Agency

pollution standards. However, the "brown cloud" still hangs over the city on windless days, and scientists warn of a possible return to former pollution levels since about half the vehicles in the city are SUVs.

PARTNERSHIP FOR A NEW GENERATION OF VEHICLES (PNGV). In September 1993 President Bill Clinton, along with the Big Three automakers (General Motors, Ford, and Chrysler), announced the formation of a government and industry research program called the Partnership for a New Generation of Vehicles (PGNV) aimed at developing an environmentally friendly "supercar." The car would more than triple the fuel efficiency of today's mid-size cars without sacrificing cost, safety, and performance. Prototypes are expected to be ready by 2004. Development is to focus on hybrid electric drive, direct-injection engines, fuel cells, and more lightweight materials. By tripling the fuel economy of a mid-size car to around 80 mpg, PNGV would offer environmental benefits both in reducing fuel consumption and in reducing carbon dioxide emissions.

Mandating of Alternative Fuel Vehicles (AFVs)

The Energy Policy Act of 1992 (EPACT; PL 102-486), passed in the wake of the 1991 Persian Gulf War, required that federal and state governments and fuel provider fleet owners increase the percentages of vehicles powered by alternative fuels. The fleet requirements affect those who own or control at least 50 vehicles in the United States and fleets of at least 20 vehicles that are centrally fueled (or capable of being centrally fueled) within a metropolitan area of 250,000 or more. (See Table 5.5.) In doing so many municipal governments and the U.S. Postal Service have put into operation fleets of natural gas vehicles such as garbage trucks, transit buses, and postal vans.

The Department of Energy (DOE) reported in *Alternatives to Traditional Transportation Fuels 1998* (2000) that, from 1992 to 2000, a total of 430,219 AFVs were put into use in the United States at an average annual growth rate of about 7 percent. Most of those vehicles used liquified petroleum gas (LPG) followed by those powered by compressed natural gas (CNG). (See Table 5.6.)

FIGURE 5.15

States With Lead Disclosure Laws

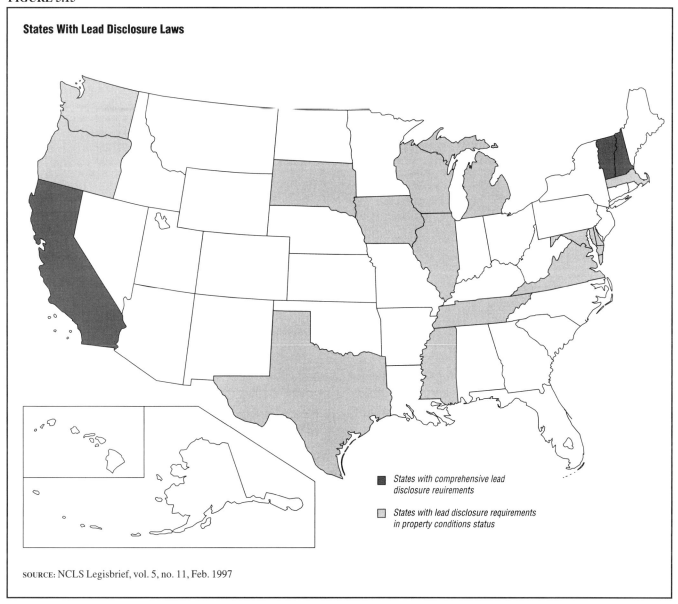

States with comprehensive lead disclosure reuirements

States with lead disclosure requirements in property conditions status

SOURCE: NCLS Legisbrief, vol. 5, no. 11, Feb. 1997

Alternative Fuels

The use of alternative, non-petroleum-based fuel is expanding worldwide and offers opportunities to reduce overall energy use and emissions.

GASOHOL. In 1991, to help reduce the nation's dependence on imported oil, the U.S. Congress enacted the National Defense Authorization Act (PL 101-510) which includes a provision directing federal agencies to purchase gasohol when it is available at prices equal to or lower than gasoline. Gasohol is a blend of 10 percent ethanol and 90 percent unleaded gasoline. Ethanol is derived from fermented agricultural products such as corn.

Executive Order 12759 of 1991 requires federal agencies operating more than 300 vehicles to reduce their gas consumption by 10 percent, another incentive to use gasohol. Despite these measures, however, use of gasohol has increased only slightly because of its cost

and frequent unavailability caused by the high cost of transporting and storing it. While there have been a number of proposals to drop gasohol altogether, senators and representatives from grain-growing states have used their influence to make sure that gasohol remains a part of the nation's fuel mix.

HYDROGEN. Hydrogen is the most simple naturally occuring element and can be found in materials such as water, natural gas, and coal. For decades advocates of hydrogen have promoted it as the fuel of the future—abundant, clean, and cheap. Hydrogen researchers from universities, laboratories, and private companies claim their industry has already produced vehicles that could be ready to market if problems of fuel supply and distribution could be solved. Other experts contend that economics and safety concerns will limit hydrogen's wider use for decades.

TABLE 5.4

List of Hazardous Urban Air Pollutants

VOCs	Metals (Inorganic Compounds)	Aldehydes (Carbonyl Compounds)	SVOCs & Other HAPs
acrylonitrile	arsenic compounds	acetaldehyde	2,3,7,8-tetrachlorodi benzo-p-dioxin (& congeners & TCDF congeners)
benzene	beryllium and compounds	formaldehyde	coke oven emissions
1,3-butadiene	cadmium compounds	acrolein	hexachlorobenzene
carbon tetrachloride	chromium compounds		hydrazine
chloroform	lead compounds		polycyclic organic matter (POM)
1,2 -dibromoethane (ethylene dibromide)	manganese compounds		polychlorinated biphenyls (PCBs)
1,3-dichloropropene	mercury compounds		quinoline
1,2-dichloropropane (propylene dichloride)	nickel compounds		
ethylene dichloride, EDC (1,2-dichlorethane)			
ethylene oxide			
methylene chloride (dichloromethane)			
1,1,2,2,-tetrachloroethane			
tetrachloroethylene (perchloroethylene, PCE)			
trichloroethylene, TCE			
vinyl chloride			

SOURCE: *National Air Quality and Emissions Trends Report 1998.* U.S. Environmental Protection Agency, 2000

FIGURE 5.16

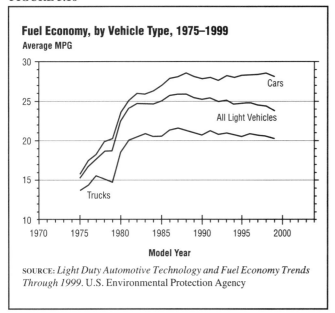

Fuel Economy, by Vehicle Type, 1975–1999
Average MPG

SOURCE: *Light Duty Automotive Technology and Fuel Economy Trends Through 1999.* U.S. Environmental Protection Agency

FIGURE 5.17

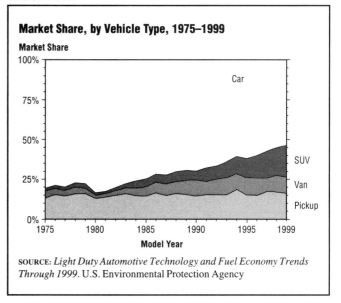

Market Share, by Vehicle Type, 1975–1999
Market Share

SOURCE: *Light Duty Automotive Technology and Fuel Economy Trends Through 1999.* U.S. Environmental Protection Agency

---

ALTERNATIVE FUEL AND THE MARKETPLACE. AFVs cannot become a viable transportation option unless a fuel supply is readily available. (See Table 5.7.) Ideally, the infrastructure for supplying alternative fuels would be developed simultaneously with the vehicles.

Many state policies and programs encourage the use of alternative fuels. California, for example, requires the sale of electric vehicles (EVs) by 2003. This has caused vehicle manufacturers to expedite vehicle research and development. In fact EVs are already selling there and some rental car agencies now offer EVs to customers at prices only slightly higher than gasoline-powered cars.

Chrysler Corporation stopped making natural gas-powered vehicles after the 1997 model year because it lost money on the vehicles. Chrysler sold only 4,000 natural gas-powered autos since it began production in 1992. General Motors, which had suspended sales of natural-gas vehicles in 1994, resumed sales again in 1997. Ford began selling some natural gas versions of its cars and trucks in 1995. Commercial fleets, not retail customers, are the main buyers of natural-gas vehicles.

Market success of alternative fuels and AFVs depends upon public acceptance. People are accustomed to using gasoline as their main transportation fuel and it is readily available. Perhaps as federal and state requirements for alternative fuels increase, so will its visibility to and acceptability by the general public.

Electric Cars—Promise and Reality

The electric car is not a new invention. Popular during the 1890s, the quiet, clean, and simple vehicle was expected to dominate the automotive market of the twentieth century. Instead, it quietly disappeared as automobile companies chose to invest billions of dollars in the internal combustion engine. It has taken a century but the electric car has returned.

Energy standards have created a clear market for alternative-energy cars. The combination of government mandates to reduce emissions and encouraging market opportunities has altered the automobile industry. By the year 2000 most car manufacturers either had an electric car on the roads or in testing.

The primary difficulty with electric vehicles (EVs) lies in inadequate battery power. The cars must be recharged often. These vehicles also use lead-acid or nickel-cadmium batteries and have a range of 70 to 100 miles on a single charge. The range is reduced by factors such as cold temperatures, the use of air conditioning, vehicle load, and steep terrain.

In addition EVs are expensive, although prices are coming down. General Motors' EV1 leases for between $480 and $640 per month. (See Figure 5.18.) This is less than the cost of luxury cars but more than the average mid-size American car. In California some car rental companies are offering EV rentals at rates only slightly more than gas-powered vehicles, and some of California's buses are electric-powered. In addition to General Motors' EV, Toyota produces the RAV-4-EV, Nissan makes the electric Altra, Ford the Ranger EV, and Daimler-Chrysler the EPIC.

Despite their high price, electric cars have many advantages. They are relatively noiseless and simple in design and operation. They cost less to refuel and service and have fewer parts to break down. Their owners are likely to spend less time on maintenance and, if they recharge at home, will rarely have to go to the service station. These time-savings have real value in the busy world of the twenty-first century. Over time the cost gap between cars that pollute and EVs that do not will narrow. With advances in battery development the gap could close entirely.

Automobile industry experts believe electric cars will assume a "second car" role for commuters and for short trips, much like the microwave oven has become not a replacement for, but an addition to, conventional ovens for cooking.

California led the development of EVs. In 1990 the California Air Resources Board (CARB), facing severe air pollution in Los Angeles and other cities, passed the toughest auto emissions standards in the world. Most

TABLE 5.5

Percentage of New Fleet Light Duty Purchases That Must be Alternative Fuel Vehicles

Year	Federal Government	State Government	Fuel Providers	Private/ Municipal†
1003	7,600*			–
1994	11,250*	–	–	–
1995	15,000*	–	–	–
1996	25	–	–	–
1997	33	10	30	–
1998	50	15	50	–
1999	75	25	70	–
2000	75	50	90	–
2001	75	75	90	–
2002	75	75	90	20
2003	75	75	90	30
2004	75	75	90	40
2005	75	75	90	50
2006	75	75	90	60
2007	75	75	90	70

*Actual number of vehicles

†Dependent upon DOE final rulemaking in 2000

SOURCE: *A Legislator's Guide to Alternative Fuel Policies and Programs.* National Conference of State Legislatures, 1997

notable was the requirement that 2 percent of cars sold in the state by the seven major carmakers in 1998 must be "zero-emission" and that proportion would rise to 10 percent by 2003. Auto-industry lobbyists protested and the 1998 mandate was lifted. But the big automakers were still required to achieve the 10 percent target by 2003. If the automakers actually achieved the goal, and this is a big if, approximately 800,000 zero-emission cars would be on California roads by 2010, up from 2,000 in use in the entire country in 1997.

In 1997 a federal judge ruled that the State of New York could order automobile manufacturers to sell thousands of electrically powered vehicles in New York in 1998, making it the only state to mandate the sale of EVs in 1998.

AIRPLANES

As air travel in affluent nations rose it caused a number of environmental problems. The average American flew 1,739 miles a year by the year 2000. Europeans, though they flew fewer miles, had the world's most crowded skies while the most rapid growth in flying was in Asia. Most air travel is done by a small portion of the world's people.

Flying carries an environmental price—it is the most energy-intensive form of transport. In much of the industrialized world air travel is replacing more energy-efficient rail or bus travel. Despite a rise in fuel efficiency of jet engines, jet fuel consumption has risen 65 percent since 1970.

Another problem with air travel is its impact on global warming. Airplanes spew nearly four million

TABLE 5.6

Estimated Number of Alternative Fuel Vehicles in Use in the U.S., by Fuel, 1992–2000

Fuel	1992	1993	1994	1995	1996	1997	1998	1999	2000	Average Annual Growth Rate (Percent)
Liquefied Petroleum Gases (LPG)[a]	221,000	269,000	264,000	259,000	263,000	263,000	266,000	268,000	270,000	2.5
Compressed Natural Gas (CNG)	23,191	32,714	41,227	50,218	60,144	68,571	78,782	89,633	101,991	20.3
Liquefied Natural Gas (LNG)	90	299	484	603	663	813	1,172	1,422	1,682	44.2
Methanol, 85 Percent (M85)[b]	4,850	10,263	15,484	18,319	20,265	21,040	19,648	19,497	18,725	18.4
Methanol, Neat (M100)	404	414	415	386	172	172	200	200	200	-8.4
Ethanol, 85 Percent (E85)[b,c]	172	441	605	1,527	4,536	9,130	12,788	22,359	30,017	90.6
Ethanol, 95 Percent (E95)[b]	38	27	33	136	361	347	14	14	14	-11.7
Electricity	1,607	1,690	2,224	2,860	3,280	4,453	5,243	6,417	7,590	21.4
Non-LPG Subtotal	**30,352**	**45,848**	**60,472**	**74,049**	**89,421**	**104,526**	**117,847**	**139,542**	**160,219**	**23.1**
Total	**251,352**	**314,848**	**324,472**	**333,049**	**352,421**	**367,526**	**383,847**	**407,542**	*430,219*	**6.9**

[a]Values are rounded to thousands. Accordingly, these estimates are not equal to the sum of Federal fleet data (for which exact counts are available) and non-Federal fleet estimates (rounded to thousands).

[b]The remaining portion of 85-percent methanol and both ethanol fuels is gasoline.

[c]In 1997, some vehicle manufacturers began including E85-fueling capability in certain model lines of vehicles. Those vehicles are capable of operating on E85, gasoline, or both. Alternative fuel vehicles (AFV's) in use include only those E85 vehicles believed to be intended for use as alternative-fuel vehicles (AFV's). These are primarily fleet-operated vehicles. All of the E85 vehicles are included in the data for "AFV's made available."

Note: Estimates for 1998-2000 have been revised since their original publication in October 1999. The revision reflects new data for the Federal sector only. Estimates for 1999 are preliminary and estimates for 2000 are based on plans or projections. Estimates for historical years may be revised in future reports if new information becomes available.

SOURCE: *Alternatives to Traditional Transportation Fuels 1998.* Energy Information Administration

TABLE 5.7

Alternative Fuel Refueling Sites Nationwide, by Fuel

Compressed Natural Gas	Liquefied Natural Gas	Ethanol E85	Methanol M85	Propane	Electricity
1,421	72	61	95	3,698	188

SOURCE: *A Legislator's Guide to Alternative Fuel Policies and Programs.* National Conference of State Legislatures, 1997

tons of nitrogen oxide into the air, much of it while cruising in the tropospheric zone five to seven miles above the earth where ozone is formed. The EPA estimates that air traffic accounts for about 3 percent of all global greenhouse warming. The Intergovernmental Panel on Climate Change (IPCC) notes that emissions deposited directly into the atmosphere do greater harm than those released at the earth's surface. (See Figure 5.19.)

In 1998 Pratt and Whitney announced plans to introduce a radical new engine design that would be cleaner, more efficient, and more reliable than conventional designs. The new engine, expected to enter service in 2002, would reduce emissions by 40 percent and exceed noise restrictions set to take effect in 2000. The engine is designed for use on single-aisle planes carrying 120 to 180 passengers, such as the Boeing 737 or the Airbus A320.

Although each generation of airplane engine gets cleaner and more fuel-efficient, there seems to be little that can be done about the increased amount of flying. There is, however, a movement toward doing something about other engines in the airline industry—those in trucks, cars, and carts that service airplane fleets. Electric utility companies, including the Edison Electric Institute and the Electric Power Research Institute, launched a program in 1993 to electrify airports. By converting terminal transport buses, food trucks, and baggage-handling carts to electricity, airports could reduce air pollution considerably.

FIGURE 5.18

The Saturn EV1 electric car is manufactured by General Motors Corporation. (*General Motors. AP/Wide World. Reproduced by permission.*)

REDUCING EMISSIONS—BUT NOT ENOUGH

The most widespread technological inventions to reduce emissions have been electrostatic precipitators (electrical cleaning systems) and filters designed to control emissions from power plants. These reduce particulate emissions from smokestacks by 99.5 percent but do nothing about gaseous emissions. The primary way to reduce sulfur dioxide has been the use of scrubbers (an air pollution device that uses a spray of water or reactant to trap pollutants) which remove 95 percent of the sulfur dioxide residue. The available technologies have limitations—they create scrubber ash, a hazardous waste, and do nothing to control carbon dioxide emissions.

With the aid of pollution control equipment and improvements in energy efficiency, many industrial countries have reduced emissions. Since the passage of the Clean Air Act there has been a 90 percent reduction in emissions of hydrocarbons and carbon monoxide from the average car in the United States and a 75 percent reduction in nitrogen oxides.

Other pollutants, however, have gone mostly unregulated, notably carbon dioxide, the inevitable byproduct of burning fossil fuels. The government has not tightened efficiency standards for years, and Americans are buying more sporty trucks and other gas-guzzlers and have been driving them more miles. One large unknown in predicting how much pollution cars will cause in the future is the question of how long this increase in driving will last. People cannot indefinitely expand the time they spend behind the wheel. And should fuel prices rise significantly, driving times could decline, reducing pollution rates.

California Leads the Way

In California, a state known for its noxious smog, pollution measurements show that cleanup efforts are steadily reducing air contamination. The stringent standards imposed in California have resulted in the introduction of scanners and sensors to detect and monitor emission levels at the source. These sensors are installed in each smokestack and the readings are transmitted directly to a regulatory agency. Products such as paints and solvents that contain reactive gases can be scanned with bar-code scanners, much like those used in the retail industry for pricing. The data collected can be transmitted electronically to air-quality agency computers.

The automobile industry is increasingly introducing "zero-emissions" vehicles and charging stations have

FIGURE 5.19

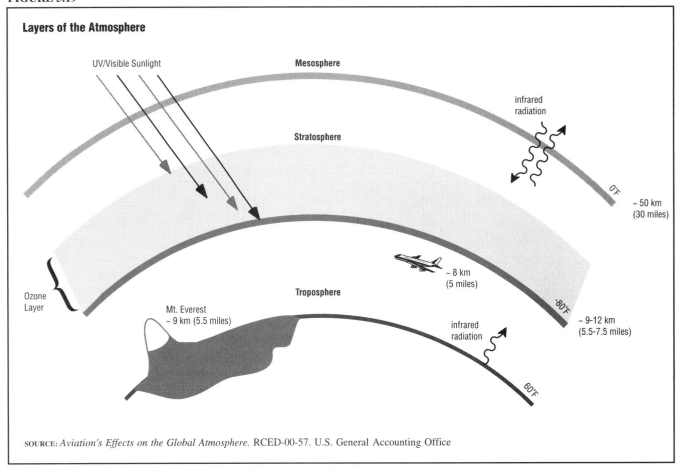

Layers of the Atmosphere

SOURCE: *Aviation's Effects on the Global Atmosphere*. RCED-00-57. U.S. General Accounting Office

sprung up at shopping malls. New cars made for sale in California are equipped with pollution controls that add $50 to $300 to their price but make them the cleanest internal combustion vehicles in the world. Newcomers from the other 49 states must pay a special $300 smog impact fee to register their cars.

In 1997 Los Angeles experienced the cleanest air in the four decades since official monitoring began. There was only one "Stage 1" smog alert compared to 122 in 1977 and 148 in 1970. (A "Stage 1" alert signifies the air is so polluted that people with respiratory ailments should stay indoors and others should refrain from vigorous outdoor exercise.) Since 1988 there has not been a "Stage 2" alert (ozone above 0.35 parts per million), when industries and driving are curtailed and school children are removed from playgrounds. Not since 1974 has there been a "Stage 3" alert (0.50 parts per million), when a general holiday is ordered. Nonetheless, California is still the most unhealthful region in the country. Linda Waade, executive director of the Coalition for Clean Air, a California public interest group, has expressed concern that governmental budget cuts now sweeping the United States could result in a loosening of the strict health standards that made California the national leader in environmental cleanup.

The Dollar Value of Air Pollution and Pollution Control

Putting a dollar value on the health costs of air pollution is very difficult because it requires judgments about the value of good health and human life. Health expert Thomas Crocker of the University of Wyoming estimated the cost to be as high as $40 billion annually in health care and lost productivity. The American Lung Association calculated the direct and indirect health-care costs associated with air pollution to be between $16 and $40 billion a year.

Professor Jane Hall, an environmental economist at California State University, concluded from studies of three decades of pollution control that pollution control efforts have shown no discernible negative economic effects on even the most heavily regulated sectors. She reported that the story of the past quarter-century has been that each successive step in pollution control has been declared undoable only until regulation pressured industry into coming up with cleaner technologies.

Like most environmental issues, pollution involves limits. There is only so much air to receive automobile emissions and so much land on which freeways can be built. Many transportation analysts think that interest in public transportation is long overdue. Several major cities, including San Francisco, Chicago, and New York, have had positive environmental results with mass transit systems.

The United States Is Missing Its Goals

Progress in cutting air pollution in the United States has been uneven. Although reductions in emissions of all six criteria pollutants have occurred, carbon dioxide emissions continue to rise. In 1990 carbon dioxide emissions were 1,340 million metric tons of carbon equivalent (MMTCE); by 1998 carbon dioxide emissions had reached 1,494 MMTCE. This is because the CAA of 1970 imposed controls on certain types of emissions but left others, such as carbon dioxide, unregulated.

Although a car built in 2000 produced only one-tenth the pollution per mile traveled as one built two decades earlier, fuel efficiency standards had not been tightened in years. Americans were buying less efficient vehicles and total miles traveled increase each year. The State Department stated that—fed by an economic boom that has stalled improvements in energy efficiency, affordable gas, and gas-guzzling vehicles—the United States will continue to increase emissions of carbon dioxide. Although data for 2000 would not be available until 2001, based on 1998 emissions, the goal of reducing carbon dioxide emissions in the United States to 1990 levels by 2000, as agreed at the 1992 United Nations Conference on the Environment in Rio de Janeiro, would not be reached. Other industrialized nations also expected to fall short of the treaty goal.

The 1997 Kyoto Global Warming Treaty

In December 1997 the United Nations convened a conference of 160 nations on global warming in Kyoto, Japan, in hopes of producing a new treaty on climate change that would place binding caps on industrial emissions. However, wide gulfs existed between rich and poor countries. There was disagreement even among rich countries, with the European community contending that the United States had not done enough to reach its goals.

The treaty, called the Kyoto Protocol to the United Nations Framework Convention on Climate Change (UNFCCC), binds industrialized nations to reducing their emissions of six greenhouse gases—carbon dioxide, methane, nitrous oxide, hydrofluorocarbons, sulfur dioxides, and perfluorocarbons—below 1990 levels by 2012, with each country having a different target. The United States must cut emissions by 7 percent, most European nations by 8 percent, and Japan by 6 percent. Reductions must begin by 2008 and be achieved by 2012. Developing nations are not required to make such pledges.

The United States had proposed a program of voluntary pledges by developing nations, but that section was deleted, as was a tough system of enforcement. Instead, each country decides for itself how to achieve its goal.

The treaty provides market-driven tools such as buying and selling credits for reducing emissions. It also set up a Clean Development Fund to help poorer nations with technology to reduce their emissions. Countries would decide on their own whether to sign and ratify the treaty.

Although it is the first time nations have made such sweeping pledges, many sources expect difficulty in getting ratification. In the United States, President Clinton signed the Protocol but the Senate did not ratify it. Business leaders believe the treaty goes too far, while environmentalists believe standards do not go far enough. Some experts doubt that any action emerging from Kyoto will be sufficient to prevent doubling of greenhouse gases. Representatives of the oil industry and business community contend the treaty will spell economic pain for the United States.

As of July 2000, 84 nations had signed the treaty and 22 had ratified it. The treaty becomes legally binding only when at least 55 countries have ratified. The UNFCCC was scheduled to meet again in September 2000 in Lyon, France, to finalize issues relating to the Protocol.

NO MORE CHEAP FIXES

Increasingly, a major problem with the effort to further reduce air pollution, and some other types of pollution as well, is that most of the relatively cheap fixes have already been made, and many economists argue the expensive ones may not be worth the price. The very premise of cleanup—that air pollution can be reduced to levels where it no longer poses any health risk at all—is questioned not just by industry but by observers as well.

Virtually all gains in the war on ozone have been achieved by reducing auto emissions. The costs for future air quality improvements may exceed, from some points of view, the value of any improvement, and the disparity may only get worse over time. For example, by 1994 the tailpipe pollution standard had reduced the exhaust of VOC pollutants (organic compounds released from burning fuel) by 98 percent. Getting the figure down to 99.5 percent will at least double the cost.

On the other hand some sources believe there are other technologically easy—if politically unpopular—steps that could be taken to improve air quality. Such steps could include forcing light trucks, minivans, and SUVs to meet the same smog standards as standard passenger cars.

Other possible improvements could come from changes in "grandfather" clauses, loopholes that exempt companies from compliance with laws because the companies existed prior to the law. As of the year 2000, compliance by industries operating under a grandfather clause has been on a voluntary basis, but voluntary cutbacks had not worked. Power plants rank first in grandfathered emissions. Other top industries affected by grandfather clauses include aluminum smelters, electric power plants, oil refineries, and carbon-black plants.

THE CLEAN AIR ACT—A HUGE SUCCESS

In 1970 the U.S. Congress passed the landmark Clean Air Act (CAA; PL 91-604), proclaiming that it would restore urban air quality. It was no coincidence that the law was passed during a 14-day Washington, D.C., smog alert. Although the CAA has had mixed results, and many goals remain to be met, most experts credit it with making great strides toward cleaning up the air. Since its adoption airborne lead has declined 90 percent, primarily due to the reduced sale of leaded gasoline, and most other measured emissions have also decreased. Los Angeles, for example, while still far from attaining air quality standards, has cleaner air than at any time since measurements were first taken in the 1940s. Nationwide, more than half the areas not meeting standards for ozone in 1990 did so by the year 2000, as did two-thirds of the areas not attaining the carbon monoxide standards in 1990.

The Clean Air Act Amendments of 1990

The overall goal of the Clean Air Act Amendments of 1990 (PL 101-549) was to reduce the pollutants in the air by 56 billion pounds a year—224 pounds for every man, woman, and child—when the law is fully phased in by the year 2005. Other aims were to cut acid rain in half by the year 2000, reduce smog and other pollutants, and protect the ozone layer by phasing out chlorofluorocarbons (CFCs) and related chemicals.

The CAA amendments of 1990 also encouraged states to pursue market-based approaches to improve air quality. One such program, the Accelerated Vehicle Retirement (AVR) program commonly known as Cash for Clunkers, provides economic incentives for the owners of highly polluting vehicles to retire their automobiles from use or repair them. The program gives pollution credits to private corporations for contributing funding to car dealers to entice car owners to trade in their old vehicles.

Resistance to CAA Gets Federal Concessions

During the mid- to late 1990s a number of states began balking at strict auto emissions testing that seemed necessary to comply with the Clean Air Act Amendments of 1990. Under the law, if a reduction does not come from auto emissions it would have to be made up by other sources, for example, smokestack industries. The states are free to implement whatever methods they choose to cut pollution, but most states with serious air quality problems had previously chosen, with EPA encouragement, the stricter car inspection programs. This meant many states were faced with testing that many consumers considered overly restrictive and expensive.

The EPA had counted on enforcing the program through sanctions specified by the 1990 CAA—cutting off highway money and other federal aid to the states. But some state legislatures seemed willing to forego this aid in what some sources considered an act of civil disobedience. Rather than provoking a confrontation with the states, the EPA chose to allow greater flexibility in auto emissions testing.

In *The Benefits and Costs of the Clean Air Act, 1970 to 1990* (1997), the first report mandated by the CAA on the monetary costs and benefits of controlling pollution, the EPA concluded that the economic value of clean air programs was 42 times greater than the total costs of air pollution control over the 20-year period. The study found that numerous positive economic consequences occurred in the U.S. economy because of CAA programs and regulations. The CAA affected industrial production, investment, productivity, consumption, employment, and economic growth. In fact, the study estimated that total agricultural benefits from the CAA were almost ten billion dollars. The EPA compared benefits to direct costs or expenditures. (See Table 5.8.) The total costs of the CAA were $523 billion for the 20-year period; total benefits equaled $22.2 trillion—a net benefit of approximately $21.7 trillion.

The National Conference of State Legislatures, in its *Two Decades of Clean Air: EPA Assesses Costs and Benefits* (1998), used data from the EPA analysis and found that the Act produced major reductions in pollution that caused illness and disease, smog, acid rain, haze, and damage to the environment. (See Figure 5.20.)

The second mandated review of the CAA, *The Benefits and Costs of the Clean Air Act Amendments of 1990* (2000), the most comprehensive and thorough review ever conducted, found similar results. Using a sophisticated array of computer models and the latest cost data, the EPA found that, by 2010, the Act will have prevented 23,000 Americans from dying prematurely and averted more than 1.7 million asthma attacks. The CAA will prevent 67,000 episodes of acute bronchitis, 91,000 occurrences of shortness of breath, 4.1 million lost work days, and 31 million days in which Americans would have had to restrict activity because of illness. Another 22,000 respiratory-related hospital admissions will be averted, as well as 42,000 admissions for heart disease and 4,800 emergency room visits.

The EPA estimated that the benefits of CAA programs will total about $110 billion in reduction of illness and premature death. By contrast the study found that the costs of achieving these benefits was only about $27 billion, a fraction of the value of the benefits. In addition the study reported there were other benefits that scientists

TABLE 5.8

Estimated Annual Clean Air Act Compliance Costs, 1973–1990

Year	Expenditures		Annualized Costs 1990 dollars at:		
	Current dollars	1990 dollars	3 percent	5 percent	7 percent
1973	7.2	19.6	11.0	11.0	11.1
1974	8.5	21.4	13.2	13.4	13.7
1975	10.6	24.4	13.3	13.6	14.0
1976	11.2	24.1	14.1	14.6	15.1
1977	11.9	24.1	15.3	15.9	16.6
1978	12.0	22.6	15.0	15.8	16.7
1979	14.4	24.8	17.3	18.3	19.3
1980	16.3	25.7	19.7	20.8	22.0
1981	17.0	24.4	19.6	20.9	22.3
1982	16.0	21.6	18.6	20.1	21.7
1983	15.5	20.1	19.1	20.7	22.5
1984	17.3	21.6	20.1	21.9	23.8
1985	19.1	22.9	22.5	24.4	26.5
1986	17.8	20.8	21.1	23.2	25.4
1987	18.2	20.6	22.1	24.2	26.6
1988	18.2	19.8	22.0	24.3	26.7
1989	19.0	79.8	22.9	25.3	27.8
1990	19.0	19.0	23.6	26.1	28.7

SOURCE: *Two Decades of Clean Air: EPA Assesses Costs and Benefits.* State Legislative Report, vol. 23, no. 3, Feb. 1998

and economists cannot quantify and express in dollar terms, such as controlling cancer-causing air toxics and benefits to crops and ecosystems by reducing pollutants.

At the same time, 30 years later, many cities are still not in compliance with the law. One reason efforts to clean the air have been only partly successful is that they have focused on specific measures to combat individual pollutants rather than addressing the underlying social and economic structures that create the problem, for example, the distance between many Americans' residences and their places of work.

Trading Pollution Credits

The CAA 1990 Amendments created pollution "credits," a free-market innovation that allowed companies that keep their emissions below standards to sell or trade their credits on the open market to other companies that do not keep their emissions below standard. This is often viewed essentially as permission to pollute. Companies can also choose to permanently retire their credits and thus reduce the potential of further pollution.

In 1996 New York's Governor, George Pataki, and state agencies accumulated millions of dollars in pollution credits and offered the credits free to companies willing to relocate to New York. Environmentalists voiced concerns over the policy, claiming that it subsidized polluters.

THE ANTI-REGULATORY REBELLION

The dissatisfaction with government regulation that developed in the 1980s grew even stronger in the 1990s.

FIGURE 5.20

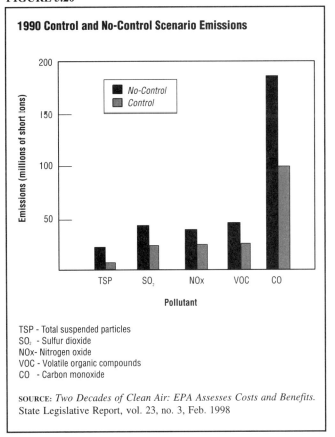

1990 Control and No-Control Scenario Emissions

TSP - Total suspended particles
SO₂ - Sulfur dioxide
NOx- Nitrogen oxide
VOC - Volatile organic compounds
CO - Carbon monoxide

SOURCE: *Two Decades of Clean Air: EPA Assesses Costs and Benefits.* State Legislative Report, vol. 23, no. 3, Feb. 1998

The Republican-led Congress took steps to put brakes on what they considered growing environmental regulation. As a result funding for environmental protection—including the Clean Air Act (PL 91-604), the Clean Air Amendments (PL 101-549), the Clean Water Act (PL 92-500), the Safe Drinking Water Act (PL 93-523), and other environmental statutes—was cut. Subsequent fiscal budgets continued those cuts by many millions of dollars.

Smog and Soot—Political Fallout and New Standards

The Clinton Administration struggled during the 1990s with two highly charged political issues—global warming and whether to tighten the Clean Air Act. Although the two issues are linked because they both involve the burning of fossil fuels, the American public seems less interested and compelled by global warming than by the smog they breathe and the way tiny particles can harm them and their loved ones. Many people have, on the one hand, applauded President Clinton's tightening of CAA standards while discounting the less immediate concern about eventual rising sea levels and scientific questions about subtle global warming.

The CAA requires the EPA to review public health standards at least every five years to ensure they reflect the best current science. In 1997—in response to what many consider compelling scientific evidence of the harm caused by ozone and fine particles to human health—the

EPA issued new, stricter air quality standards for ozone and particulate matter.

This was the first revision in ozone standards in 20 years and the first-ever standard for fine particulates. The provisions tightened the standard for ground-level ozone from the level of 0.12 parts per million (ppm) at the highest daily measurement to 0.08 ppm average over an eight-hour period. The new particulate matter standard included particles larger than 2.5 microns in diameter instead of the original standard of those larger than 10 microns. However, in May 1999 a three-judge federal appeals panel overturned the new standards. The EPA appealed but, in October 1999, the full U.S. Court of Appeals for the District of Columbia refused to overturn the decision. In January 2000 the American Lung Association petitioned the Supreme Court to review the decision. The Court was expected to decide the case in 2000.

Despite improvements in ozone levels, few large urban areas in the United States comply with ozone standards. Under the proposed standard, many smaller population centers would also become noncompliant, and states would be forced to take further pollution control measures to comply. Some critics contend that the standards are either unattainable or not worth the cost. The Global Climate Coalition, a business and trade organization, claims that the proposal would damage the U.S. economy while doing little to help the environment.

STATES ARE DIVIDED. Because of high ozone levels in the Northeast, many northeastern states already have stricter standards for ozone pollution than the federal government requires. These states believe that midwestern states should also be held to a stricter standard, particularly since pollutants are transported by westerly winds from the Midwest to the East Coast and even into Canada. The Ozone Transport Assessment Group, a coalition of 37 states east of the Rocky Mountains studying pollution, reported that up to half of Connecticut's pollution can be attributed to smog produced in midwestern states and carried by winds.

In 1997 the EPA proposed rules that would require 22 states east of the Mississippi River to sharply cut their emissions. The EPA urged the states to achieve those cuts by focusing on emissions from electric utilities and by trading pollution credits with other states. States would be free to meet emissions targets in any way they choose, but were expected to focus on utilities—the largest single source of ozone-causing pollution. The proposal would await a year-long period of public comment.

EFFECT OF UTILITY DEREGULATION ON AIR POLLUTION

Regulated for decades as "natural monopolies" electric utilities faced a radical shift by the year 2000 toward increased competition. As in the airline, trucking, natural gas, and telecommunications industries, more efficient technology and increasing demand for lower rates led regulators to consider some form of utility company deregulation. Thus, the Federal Energy Policy Act of 1992 (PL 102-486) gave other electricity generators access to the market.

Industrial customers eager to buy cheaper electricity from new sources or to build their own sources of power no longer viewed electricity as a non-negotiable cost of doing business but saw it as a commodity they could either provide for themselves or shop for. Under deregulation plans utilities would be free to market electricity anywhere and, by the beginning of the twenty-first century, more than half the state legislatures considered allowing utilities to compete for customers.

Environmental groups and some utilities have warned that deregulation may, by prompting increased competition, cause coal-fired plants to use more coal in order to produce more electricity, which would send more pollutants into the air. The Natural Resources Defense Council cautioned that deregulation could lead to as many as 500,000 tons of increased emissions of nitrogen oxides per year. Responding to those concerns EPA administrator Carol Browner agreed that the open access rule could lead to future increases in carbon dioxide emissions and said that the EPA has to work closely with the Energy Department and the states to monitor the results of open access.

FOSSIL FUEL USE IN THE DEVELOPING WORLD

Environmental pollution is worldwide and environmental problems of the future are expected to become increasingly regional and global. Evidence mounts that human activities—especially from the production of gases from the combustion of coal, oil, and natural gas—may be causing atmospheric warming worldwide. Developing countries stand on the brink of economic growth that they hope will equal that of the developed world. That explosion in growth will undoubtedly be fueled by fossil fuels, as was the case in America and Europe decades earlier. The filthiest smoke and water generally arise in the early stages of industrialization.

China especially faces a dilemma—coal harms the environment but it surely fuels economic growth. China's heavy reliance on coal, along with its inefficient and wasteful patterns of energy use, will make it the largest single producer of carbon dioxide by 2020, surpassing even the United States. Between 1970 and 1990 energy consumption in China rose 208 percent compared with an average rise of 28 percent in developed countries during the same period. More than five million Chinese participate in coal extraction, feeding China's enormous and growing appetite for energy.

Five of China's largest cities are among the world's ten most polluted cities. Polluted air reportedly kills 178,000 Chinese people prematurely each year, primarily from emphysema and bronchitis. Children with sooty faces dodge traffic; rain brings rivers of black flowing down city streets.

China's situation is repeated, on a lesser scale, in India, Brazil, and the rest of the developing world where meeting environmental goals is considered a rich country's luxury. Chinese officials believe, as do officials in many other developing nations, that developed countries cause 80 percent of the world's pollution. Chinese leadership believes the developed nations should be held responsible for the problems and, as a result, should help pay for cleaner coal-burning technologies in the third world as well as financing hydroelectric plants, nuclear power stations, and alternative energy sources.

In India the country's residents have adjusted to living in a haze of dust and fumes. The World Health Organization reports the level of microscopic particles in the air is two to five times the amount the organization deems healthy. On the streets pedestrians use handkerchiefs or saris to cover their faces in protection from airborne pollutants. In 1994 the Indian Supreme Court ruled that cars sold in the largest cities must run on unleaded fuel and have catalytic converters. But unleaded fuel is expensive and hard to find—less than 1 percent of the fuel sold in New Delhi, India's capital, is unleaded. Inefficient two-wheeled scooters spew out much pollution and, with 400 new scooters added every day in New Delhi alone, the air quality continues to deteriorate.

Forest Fires—A Clear Example of Transborder Pollution

The burning of forests worldwide, either intentionally or accidentally, has consequences for nations many miles away. In 1997 fires in Indonesian forests purposefully started by humans practicing slash-and-burn agriculture (a method of cutting down and burning vegetation to clear the land) darkened the skies and blotted out the sun in seven Southeast Asian nations for many months. The haze caused the closure of airports, loss of work, the shutdown of mines and factories, loss of investments and tourism, and illness among hundreds of thousands of people. Reduced sunlight slowed the growth of fruits and vegetables and caused reductions in corn, cocoa, and rice crops. Hundreds of people died from respiratory ailments, starvation, dysentery, and influenza.

In Mexico in 1998 scores of people died and at least 50 million were left choking in smoke from nearly 10,000 fires (most set for agricultural purposes). A cloud of haze and cinders hung over most of Mexico, Central America, and much of the United States for many weeks. Cities declared the air so unhealthful they even closed airports.

CHAPTER 6

WATER ISSUES

Water is the most essential resource for sustaining life. A living cell is mostly water. An adult human's body is about 65 percent water; blood is 90 percent water. It is the most common substance on Earth, covering three-fourths of Earth's surface.

Civilizations originated and declined based upon the availability of water. While the water *supply* in the United States is a serious problem, water *quality* is also threatened. The days of an unlimited bounty of water are over.

THE VALUE OF WATER

The U.S. economy depends upon water and studies have repeatedly shown a positive relationship between strong environmental standards and economic growth. Good water quality is important to local and national economic development.

Water is a powerful attraction for people. Companies that want to attract workers often locate in areas noted for parks and open spaces where air and water quality are good and recreational opportunities are abundant. The EPA's *Liquid Assets 2000: America's Water Resources at a Turning Point* (May 2000) estimates that the travel, tourism, and recreation industries support jobs for more than 6.8 million people and generate annual sales in excess of $450 billion. One-third of Americans visit the coast each year, spending about $44 billion.

All species of animals and plants depend on adequate supplies of clean water. According to the U.S. Fish and Wildlife's 1996 *National Survey of Fishing, Hunting, and Wildlife-Associated Recreation*, fishing, hunting, and wildlife-associated recreation produced revenues of $101 billion in 1996 and generated about 206,000 jobs, although not all of that amount can be attributed directly to healthy water bodies.

More than 95 percent of U.S. foreign trade, which is expected to triple by 2020, passes through U.S. harbors and ports. Water is an absolute necessity for ensuring agricultural production. It is necessary for irrigating crops, raising livestock, and enables American farmers to produce and sell $197 billion worth of food and fiber annually. Each year the Great Lakes, the Gulf of Mexico, and many other coastal areas produce more than 10 billion pounds of fish and shellfish. The National Marine Fisheries Service estimated the value of U.S. commercial fishing at about $3.1 billion in 1998. The U.S. commercial fleet includes nearly 75,000 vessels; almost 5,000 processing and wholesale plants employ more than 83,000 people.

According to the U.S. Geological Survey (USGS), manufacturers use about nine trillion gallons of fresh water every year. The Center for Marine Conservation reports that federal, state, and local governments maintain more than 25,000 recreational facilities along U.S. coasts, while private organizations operate another 20,000.

THE HYDROLOGIC CYCLE

The amount of water on Earth remains constant, simply passing from one stage to another in a circular pattern known as the hydrologic cycle. Water in the atmosphere condenses and falls to Earth as precipitation such as rain, sleet, or snow. Precipitation seeps into the ground saturating the soil and refilling underground aquifers; it is drawn from the soil by vegetation for growth and returned into the air through plant leaves by the process of transpiration; and some precipitation flows into surface waters such as rivers, streams, lakes, wetlands, and oceans. From surface water moisture evaporates back into the atmosphere to repeat the cycle. (See Figure 6.1.)

WATER USE

Water must be considered as a finite resource that has limits and boundaries to its availability and suitability for use.
—United States Geological Survey, 1998

FIGURE 6.1

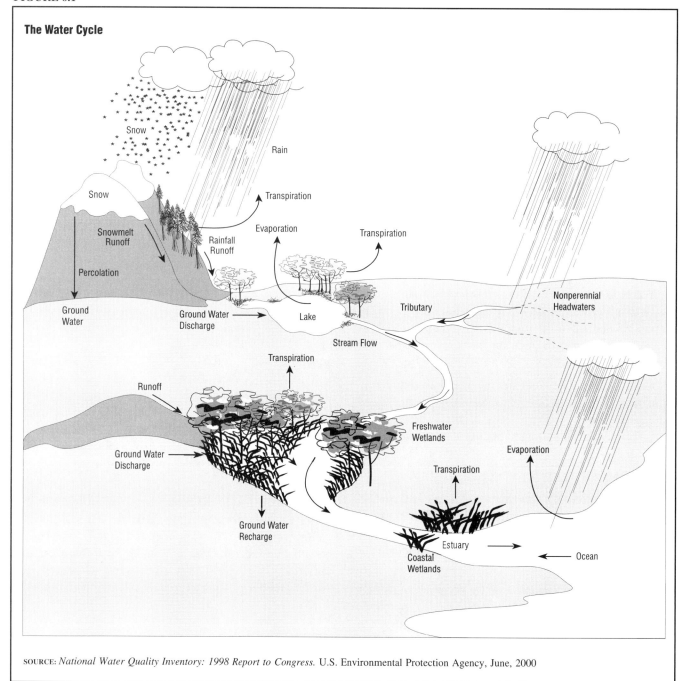

The Water Cycle

SOURCE: *National Water Quality Inventory: 1998 Report to Congress.* U.S. Environmental Protection Agency, June, 2000

Water is a fundamental need in every society. Families use water for drinking, cooking, and cleaning. Industry needs it to make chemicals, prepare paper, and clean factories and equipment. Cities use water to fight fires, clean streets, and fill public swimming pools. Farmers water their livestock, clean barns, and irrigate crops. Hydroelectric power stations use water to drive generators, while thermonuclear power stations need it for cooling. Very few human activities do not require the use of water. Water use in the United States is monitored and reported by the USGS in its *Estimated Use of Water in the United States,* published at five-year intervals since 1950.

For reporting purposes, water use in the United States is classified as *offstream* or *instream.* In off-stream use water is diverted from a surface source or withdrawn from a ground-water source and conveyed to the place where it is actually used. Off-stream use is divided into eight categories:

- Public supply
- Domestic
- Commercial
- Irrigation
- Livestock
- Industrial

TABLE 6.1

Total Offstream Water Use, by State, 1995

[Figures may not add to totals because of independent rounding. Mgal/d = million gallons per day; gal/d = gallons per day]

STATE	POPULA-TION, in thou-sands	PER CAPITA USE, fresh-water, in gal/d	WITHDRAWALS, in Mgal/d (includes irrigation conveyance losses)									RECLAIMED WASTE-WATER, in Mgal/d	CONVEY-ANCE LOSSES, in Mgal/d	CONSUMP-TIVE USE, fresh-water, in Mgal/d
			By source and type						Total					
			Ground water			Surface water								
			Fresh	Saline	Total	Fresh	Saline	Total	Fresh	Saline	Total			
Alabama	4,253	1,670	436	9.1	445	6,650	0	6,650	7,090	9.1	7,100	0.1	0	532
Alaska	604	350	58	75	132	154	43	196	211	117	329	0.	1	25
Arizona	4,218	1,620	2,830	12	2,840	3,980	2.3	3,990	6,820	14	6,830	180	1,030	830
Arkansas	2,484	3,530	5,460	0	5,460	3,310	0	3,310	8,770	0	8,770	0	416	760
California	32,063	1,130	14,500	185	14,700	21,800	9,450	31,300	36,300	9,640	45,900	334	1,670	500
Colorado	3,747	3,690	2,260	17	2,270	11,600	0	11,600	13,800	17	13,800	11	3,770	230
Connecticut	3,275	389	166	0	166	1,110	3,180	4,290	1,280	3,180	4,450	0	0	97
Delaware	717	1,050	110	0	110	642	743	1,390	752	743	1,500	0	0	71
D.C.	554	18	.5	0	.5	9.7	0	9.7	10	0	10	0	0	15
Florida	14,166	509	4,340	4.6	4,340	2,880	11,000	13,800	7,210	11,000	18,200	236	32	2,780
Georgia	7,201	799	1,190	0	1,190	4,560	64	4,630	5,750	64	5,820	.6	0	1,170
Hawaii	1,187	853	515	16	531	497	906	1,400	1,010	922	1,930	6.2	98	542
Idaho	1,163	13,000	2,830	0	2,830	12,300	0	12,300	15,100	0	15,100	0	5,480	4,340
Illinois	11,830	1,680	928	25	953	19,000	0	19,000	19,900	25	19,900	2.0	0	857
Indiana	5,803	1,570	709	0	709	8,430	0	8,430	9,140	0	9,140	0	0	505
Iowa	2,842	1,070	528	0	528	2,510	0	2,510	3,030	0	3,030	0	0	290
Kansas	2,565	2,040	3,510	0	3,510	1,720	0	1,720	5,240	0	5,240	6.8	143	3,620
Kentucky	3,860	1,150	226	0	226	4,190	0	4,190	4,420	0	4,420	0	.5	318
Louisiana	4,342	2,270	1,350	0	1,350	8,500	0	8,500	9,850	0	9,850	0	166	1,930
Maine	1,241	178	80	0	80	141	105	246	221	105	326	0	0	48
Maryland	5,042	289	246	0	246	1,210	6,270	7,480	1,460	6,270	7,730	70	0	150
Massachusetts	6,074	189	351	0	351	795	4,370	5,160	1,150	4,370	5,510	0	0	180
Michigan	9,549	1,260	858	4.4	862	11,200	0	11,200	12,100	4.4	12,100	0	0	667
Minnesota	4,610	736	714	0	714	2,680	0	2,680	3,390	0	3,390	0	0	417
Mississippi	2,697	1,140	2,590	0	2,590	502	112	614	3,090	112	3,200	0	17	1,570
Missouri	5,324	1,320	891	0	891	6,140	0	6,140	7,030	0	7,030	11	0	692
Montana	870	10,200	204	13	217	8,640	0	8,640	8,850	13	8,860	0	4,410	1,960
Nebraska	1,637	6,440	6,200	4.7	6,200	4,350	0	4,350	10,500	4.7	10,500	2.0	906	7,020
Nevada	1,530	1,480	855	42	896	1,400	0	1,400	2,260	42	2,300	24	473	1,340
New Hampshire	1,148	388	81	0	81	364	877	1,240	446	877	1,320	0	0	35
New Jersey	7,945	269	580	0	580	1,560	3,980	5,530	2,140	3,980	6,110	1.1	0	210
New Mexico	1,686	2,080	1,700	0	1,700	1,800	0	1,800	3,510	0	3,510	0	628	1,980
New York	18,136	567	1,010	1.5	1,010	9,270	6,500	15,800	10,300	6,500	16,800	0	0	469
North Carolina	7,195	1,070	535	2.1	535	7,200	1,550	8,750	7,730	1,560	9,290	1.0	0	713
North Dakota	641	1,750	122	0	122	1,000	0	1,000	1,120	0	1,120	0	5.1	181
Ohio	11,151	944	905	0	905	9,620	0	9,620	10,500	0	10,500	0	.2	791
Oklahoma	3,278	543	959	259	1,220	822	0	822	1,780	259	2,040	0	4.9	716
Oregon	3,140	2,520	1,050	0	1,050	6,860	0	6,860	7,910	0	7,910	0	1,300	3,210
Pennsylvania	12,072	802	860	0	860	8,820	0	8,820	9,680	0	9,680	1.1	0	565
Rhode Island	990	138	27	0	27	109	275	383	136	275	411	0	0	19
South Carolina	3,673	1,690	322	0	322	5,880	0	5,880	6,200	0	6,200	0	0	321
South Dakota	729	631	187	0	187	273	0	273	460	0	460	0	54	249
Tennessee	5,256	1,920	435	0	435	9,640	0	9,640	10,100	0	10,100	.5	0	233
Texas	18,724	1,300	8,370	411	8,780	16,000	4,860	20,800	24,300	5,280	29,600	109	540	10,500
Utah	1,951	2,200	776	14	790	3,530	143	3,670	4,300	157	4,460	14	612	2,200
Vermont	585	967	50	0	50	515	0	515	565	0	565	0	0	24
Virginia	6,618	826	358	0	358	5,110	2,800	7,900	5,470	2,800	8,260	0	2.9	218
Washington	5,431	1,620	1,760	0	1,760	7,060	38	7,100	8,820	38	8,860	0	1,090	3,080
West Virginia	1,828	2,530	146	.5	146	4,470	0	4,470	4,620	.5	4,620	0	0	35
Wisconsin	5,102	1,420	759	0	759	6,490	0	6,490	7,250	0	7,250	0	0	44
Wyoming	480	14,700	317	18	335	6,720	0	6,720	7,040	18	7,060	9.1	2,470	2,80
Puerto Rico	3,755	154	155	0	155	422	2,260	2,680	576	2,260	2,840	0	15	18
Virgin Islands	103	113	.5	.2	.7	11	190	201	12	190	202	0	0	1.9
Total	267,068	1,280	76,400	1,110	77,500	264,000	59,700	324,000	341,000	60,800	402,000	1,020	25,300	100,000

SOURCE: *Estimated Use of Water in the United States in 1995.* U.S. Geological Survey, 1998

• Mining

• Thermoelectric power

The 1995 USGS report (published in 1998) found that an estimated 402 billion gallons of water per day (freshwater and saline water) were withdrawn from surface or groundwater sources for off-stream use in the United States. (See Table 6.1.) This was down from 408 billion gallons in 1990. Per capita use for all off-stream uses in 1995 was 1,500 gallons per day of fresh and

FIGURE 6.2

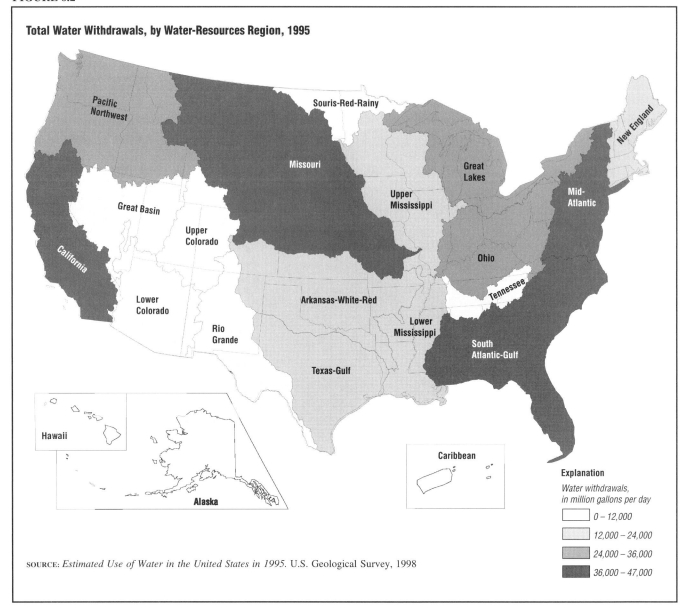

Total Water Withdrawals, by Water-Resources Region, 1995

SOURCE: *Estimated Use of Water in the United States in 1995.* U.S. Geological Survey, 1998

saline water combined (1,280 gallons per day of freshwater). California, South Atlantic-Gulf, and Middle Atlantic regions accounted for one-third of the total water withdrawn in the nation. (See Figure 6.2.) Thermoelectric activities used the most off-stream water, followed by irrigation.

In-stream use includes water that is used at its source, usually a river or stream, for example, for the production of hydroelectric power.

Trends in Water Use—1950 to 1995

After continual increases in total water withdrawals in the United States since the USGS began reporting in 1950, water use peaked in 1980, declined, and has remained generally constant. From 1990 to 1995, a period that experienced a 6 percent increase in population, total off-stream water use declined 2 percent. While use

in most categories declined, water use for public supply (4 percent) and livestock (13 percent) increased.

In-stream use (hydroelectric power) also peaked in 1980 and declined 4 percent between 1990 and 1995. Use of reclaimed water increased 36 percent from 1990 to 1995. (See Table 6.2.)

Experts believe the general increase in water use from 1950 to 1980 and the decrease from 1980 to 1995 can be attributed to several factors:

• The expansion of irrigation systems and increases in energy development from 1950 to 1980 increased the demand for water.

• The development of center-pivot irrigation systems. In some western areas the application of water directly to the roots of plants has replaced sprayer arms that

TABLE 6.2

Estimated Water Use Trends in the U.S., 1950—1995

[The water-use data are in thousands of million gallons per day and are rounded to two significant figures for 1950-80, and to three significant figures for 1985-95; percentage change is calculated from unrounded numbers]

	Year										Percentage change
	[1]1950	[1]1955	[2]1960	[2]1965	[3]1970	[4]1975	[4]1980	[4]1985	[4]1990	[4]1995	1990-95
Population, in millions	150.7	164.0	179.3	193.8	205.9	216.4	229.6	242.4	252.3	267.1	+6
Offstream use:											
Total withdrawals	180	240	270	310	370	420	[5]440	399	408	402	-2
Public supply	14	17	21	24	27	29	34	36.5	38.5	40.2	+4
Rural domestic and livestock	3.6	3.6	3.6	4.0	4.5	4.9	5.6	7.79	7.89	8.89	+13
Irrigation	89	110	110	120	130	140	150	137	137	134	-2
Industrial:											
Thermoelectric power use	40	72	100	130	170	200	210	187	195	190	-3
Other industrial use	37	39	38	46	47	45	45	30.5	29.9	29.1	-3
Source of water:											
Ground:											
Fresh	34	47	50	60	68	82	[5]83	73.2	79.4	76.4	-4
Saline	([6])	.6	.4	.5	1	1	.9	.652	1.22	1.11	-9
Surface:											
Fresh	140	180	190	210	250	260	290	265	259	264	+2
Saline	10	18	31	43	53	69	71	59.6	68.2	59.7	-12
Reclaimed wastewater	([6])	.2	.6	.7	.5	.5	.5	.579	.750	1.02	+36
Consumptive use	([6])	([6])	61	77	[7]87	[7]96	[7]100	[7]92.3	[7]94.0	[7]100	+6
Instream use:											
Hydroelectric power	1,100	1,500	2,000	2,300	2,800	3,300	3,300	3,050	3,290	3,160	-4

[1]48 States and District of Columbia.

[2]50 States and District of Columbia.

[3]50 States and District of Columbia, and Puerto Rico.

[4]50 States and District of Columbia, Puerto Rico, and Virgin Islands.

[5]Revised

[6]Data not available.

[7]Freshwater only.

SOURCE: *Estimated Use of Water in the United States in 1995.* U.S. Geological Survey, 1998

project the water into the air, where much is lost to wind and evaporation.

• Higher energy prices in the 1970s and a decrease in groundwater levels in some areas increased the cost of irrigation water.

• A downturn in the farm economy reduced demands for irrigation water.

• New industrial technologies requiring less water, as well as improved efficiency, increased water recycling, higher energy prices, and changes in the law to reduce pollution, decreased the demand for water.

• Increased awareness by the general public and active conservation programs reduced the demand for water.

Predicted Water Use

Even though water-use data are good indicators of where and how the nation consumes water, they are not necessarily good predictors of future water-use trends. Water use is also dependent on prices, technology, customs, and regulations.

For much of the country the era of free and easily developed water supplies has ended; in some areas, water use is approaching or has exceeded the available supply. However, in most areas, the nation is not running out of water. Hydrologists (scientists who study the properties, distribution, and circulation of water), engineers, and economists must provide a wider knowledge of how to use water more efficiently to fit regional demands. Although water itself can neither be created nor destroyed, its usefulness and availability can. Water users must treat water as a valuable natural resource that cannot be carelessly wasted.

A WATER CRISIS LOOMING IN THE WEST? Most of the fastest growing states are in the West. (See Figure 6.3.) Census projections report that the nation's population growth in the next 30 years will be concentrated in western and southern states, particularly California, Texas, Florida, Georgia, Washington, Arizona, and North Carolina, which are expected to gain more than two million people each by 2025. California's population alone is expected to leap by 18 million over that period—by far the nation's biggest gain—which would boost that state

FIGURE 6.3

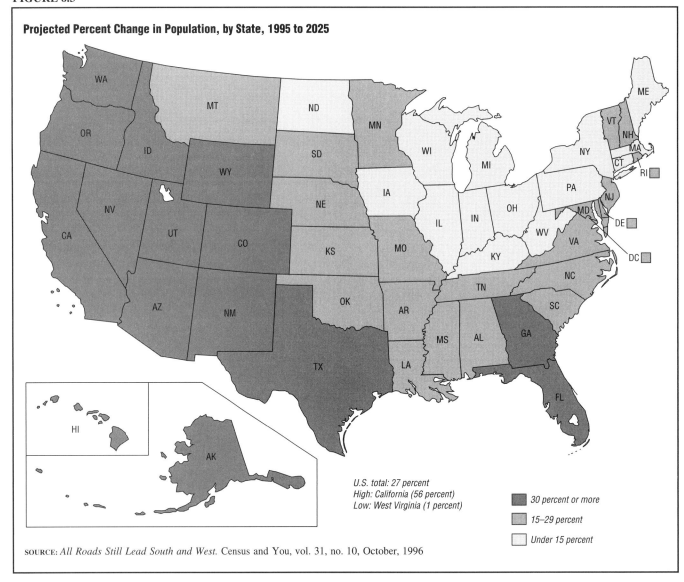

Projected Percent Change in Population, by State, 1995 to 2025

U.S. total: 27 percent
High: California (56 percent)
Low: West Virginia (1 percent)

30 percent or more
15–29 percent
Under 15 percent

SOURCE: *All Roads Still Lead South and West.* Census and You, vol. 31, no. 10, October, 1996

from 12 to 15 percent of the nation's population. With the casino building boom in Las Vegas and surrounding areas, Nevada has become the fastest growing state in the 1990s and Las Vegas the fastest growing metropolitan area. This population growth is expected to put enormous pressure on natural resources, including water, and to force huge changes in water consumption practices and prices.

The USGS predicts that, as population pressures increase, water withdrawals for public supply and domestic uses will most likely rise, although higher water prices and conservation programs may reduce per capita use rates. Agricultural water use is expected to level off or decline, since the amount of land under irrigation has declined slightly. With rising population and urbanization some conflicts between urban and rural areas, especially in the West, could arise. For the foreseeable future industrial water use will continue to decline, although not as sharply as in the past.

When Water Is Scarce—or Humans Intervene

Scarcity of water in an area can produce a number of conditions, including desertification, subsidence, and salinization. Humans are altering the environment by directly or indirectly intervening in the natural water cycle in order to increase or divert water supplies for household use, irrigation, and hydroelectric power. Modern technological developments allow massive quantities of water to be pumped out of the ground. Half of all the drinking water in the United States comes from groundwater, which is also used extensively for irrigation.

When large amounts of water are removed from the ground, underground aquifers can become depleted much more quickly than they can naturally be replenished. On almost every continent, many major aquifers are being drained faster than their natural rate of recharge. Depletion is most severe in India, China, the United States, North Africa, and the Middle East. In some areas this has led to the subsidence, or sinking, of the ground above major aquifers. Farmers in California's San Joaquin Valley began tapping the area's aquifer in the late nineteenth century. Since that time dehydration of the aquifer has caused the soil to subside by as much as 29 feet, cracking foundations, canals, and roadways. Removal of groundwater also disturbs the natural filtering process that occurs as water travels through rocks and sand.

Building dams has also changed the water cycle. The huge dams built in the United States just before and after World War II have substantially changed the natural flow of rivers. By reducing the amount of water available downstream and slowing stream flow, a dam not only affects a river but the river's entire ecological system.

Deforestation and overgrazing have destroyed thousands of acres of vegetation that play a vital role in controlling erosion. Erosion leads to soil runoff into rivers and streams, causing disruption of stream flow. Destruction of vegetation reduces the amount of water released into the atmosphere by transpiration and less water in the atmosphere can mean less rainfall, which can—in turn—lead to desertification (transformation to desert) of once-fertile regions.

Many scientists believe that global warming, or the "greenhouse effect" (thought to be caused largely by burning fossil fuels by factories and automobiles), will increase global temperatures, causing ice caps and glaciers to melt thus raising ocean levels which will destroy vital coastal marine habitats and flood cities located near the shoreline.

The most severe form of land degradation—desertification—is most acute in arid regions. Where land degradation has begun the hydrologic cycle is disrupted, leaving water tables depleted and causing the sinking and drying of the land. Although desertification was long thought to be the result of droughts, there is much more involved in the process of degradation and desertification of grazing land, including:

- Vegetation loss

- Water erosion

- Wind erosion

- Salinization

- Compaction of the land by machinery

- Accumulation of toxic substances such as lead, chromium, pesticides, and industrial waste

A few hundred million years ago oceanic waters were still fresh enough to drink. It is the earth that contains the mineral salts that one tastes in seawater. These salts are leached from soil and rock by runoff water. The runoff concentrations in rivers end up in the oceans or in salt lakes such as Mono Lake in California and the Great

Salt Lake in Utah. These lakes are seven times saltier than the sea. Once in these bodies of water the salts have nowhere to go. Continuous runoff and evaporation of water leave increasingly higher concentrations of salt, gradually causing the oceans or lakes to grow saltier. What is changing, however, is the drastic increase in concentrations of salt in the nation's rivers and on some of its prime agricultural land.

In much of the American West millions of acres of profitable land overlie a shallow and impermeable clay layer, the residual bottom of an ancient sea, that is sometimes only a few feet below the earth's surface. During the irrigation season temperatures in much of the region fluctuate between 90 and 110 degrees. The good water evaporates and polluted and saline water seep downward. Very little of this water seeps through the clay. As the water supplies are replenished with rainfall, the water table—now high in concentrations of salts and pollutants—rises back up through the root zone (the area containing plant roots) soaking the land and killing crops. (In general high salt concentrations obstruct germination and impede the absorption of nutrients by plants.)

Several thousand acres in the West have already gone out of production and salt covers the ground like a dusting of snow; not even weeds can grow there. In the coming decades, as irrigation continues, that acreage is expected to increase dramatically. It is this process rather than drought that is believed to have resulted in the decline of ancient civilizations such as Mesopotamia, Assyria, and Carthage.

Dams—Unexpected Consequences

Some 100,000 dams regulate America's rivers and creeks. Nationwide, reservoirs encompass an area equivalent to New Hampshire and Vermont combined. Of all the major rivers (more than 600 miles in length) in the lower 48 states, only the Yellowstone River still flows freely. America is second only to China in the use of dams. Worldwide, dams collectively store 15 percent of Earth's annual renewable water supply. Globally, water demand has more than tripled since the mid-twentieth century, and the rising demand has been met by building ever more and larger water supply projects.

Being a world leader in dams was a point of pride for the United States during the golden age of dam building, a 50-year flurry of architectural innovation that began with the construction of the massive Hoover Dam on the lower Colorado River in the 1930s and ended in approximately 1980. In the early years the Army Corps of Engineers built most dams for flood control; later projects served narrower interests, such as developers who wanted floodplain land.

Dams epitomized progress, Yankee ingenuity, and humanity's mastery of nature. However, the very success of the dam-building endeavor accounted, in part, for its decline: by 1980 nearly all the nation's good sites—and many dubious ones—had been dammed. There were few appropriate places left in the United States to build a major dam.

Three other factors, however, accounted for most of the decline: public resistance to the enormous costs; a growing belief that politicians were foolishly spending taxpayers' money on "pork barrel" (local) projects including dams, and a developing public awareness of the profound environmental degradation that dams can cause.

WHERE HAVE ALL THE RIVERS GONE? Dams provide a source of energy generation; flood control; irrigation; recreation for pleasure boaters, skiers, and anglers; and locks for the passage of barges and commercial shipping vessels. But dams alter rivers as well as the land abutting them, the water bodies they join, and the aquatic life they contain. All this results in profound changes in water systems and the ecosystems they support.

Many regions have fallen into a zero-sum game in which increasing the water supply to one user means taking it away from another. More water devoted to human activities means serious and potentially irreversible harm to natural systems. Many experts believe that the manipulation of river systems is wreaking havoc on the aquatic environment and its biological diversity. Hundreds of species or subspecies of fish are threatened or endangered because of habitat destruction. When rivers are dammed and water flow is stopped or reduced, wetlands dry up, species die, and nutrient loads carried by rivers into the sea are altered, with many negative consequences. Some rivers, including the large Colorado River, no longer reach the sea at all except in years of very high precipitation.

Concern for damage to the environment led the Congress to pass the Grand Canyon Protection Act of 1992. The act directed the Secretary of State to protect the Grand Canyon basin and its life forms and to monitor the effects of damming the Colorado River. Out of concern for any damage possibly being done to the canyon, in 1996—for a two-week period—the Bureau of Reclamation conducted a controlled flood of the canyon by releasing water from Glen Canyon dam (up-canyon). The flooding created dozens of new beaches in the Grand Canyon, cleared out many harmful non-native species, and invigorated fish habitat. The EPA reported that the release of water was significant in that "it was the first time in [U.S.] history that the economic agenda of a large water project was put aside purely for the good of the ecological resources downstream."

THE CLEAN WATER ACT

On June 22, 1969, the Cuyahoga River in Cleveland burst into flames, the result of oil and debris that had accumulated on the river's surface. This episode thrust the problem of water pollution into the public consciousness. Many people became aware—and wary—of the nation's polluted waters and, in 1972, Congress passed the Federal Water Pollution Control Act (PL 92-500) commonly known as the Clean Water Act.

The objective of the Clean Water Act was to "restore and maintain the chemical, physical, and biological integrity of the nation's waters." It called for ending the discharge of all pollutants into the navigable waters of the United States and to achieve "wherever possible, water quality which provides for the protection and propagation of fish, shellfish, and wildlife and provides for recreation in and on the water." The second provision was that waters be restored to "fishable/swimmable" condition.

Section 305(b) of the Clean Water Act required that each state prepare and submit to the Environmental Protection Agency (EPA) a report documenting (1) the water quality of all navigable waters in the state, (2) the extent to which the waters provide for the protection and propagation of marine animals and allow recreation in and on the water, (3) the extent to which pollution has been eliminated or is under control, and (4) the sources and causes of the pollution. The act stipulated that these reports must be submitted to the EPA every two years.

Both political conservatives and environmentalists credit the Clean Water Act with reversing, in a single generation, what had been a decline in the health of the nation's water since the mid-nineteenth century. In the 1990s, however, some politicians proposed legislation to change the Clean Water Act, giving more authority to the states and more weight to economic considerations. These politicians and their supporters (coalitions of industry, agriculture, and state and local governments) argue that enough has been accomplished and that now it is time to make the law more flexible. They claim that the huge cost of maintaining clean water risks making the United States noncompetitive in the international market. Government regulations, they think, demand more than is necessary to maintain drinkable water.

SURFACE WATER—LAKES, RIVERS, AND STREAMS

Most great civilizations began and flourished on the banks of lakes and rivers. Throughout human history societies have depended on these surface water resources for food, drinking water, transportation, commerce, power, and recreation.

The withdrawal of surface water varies greatly depending on its location. In New England, for example, where rainfall is plentiful, less than 1 percent of the annual renewable water supply is used. In contrast almost the

FIGURE 6.4

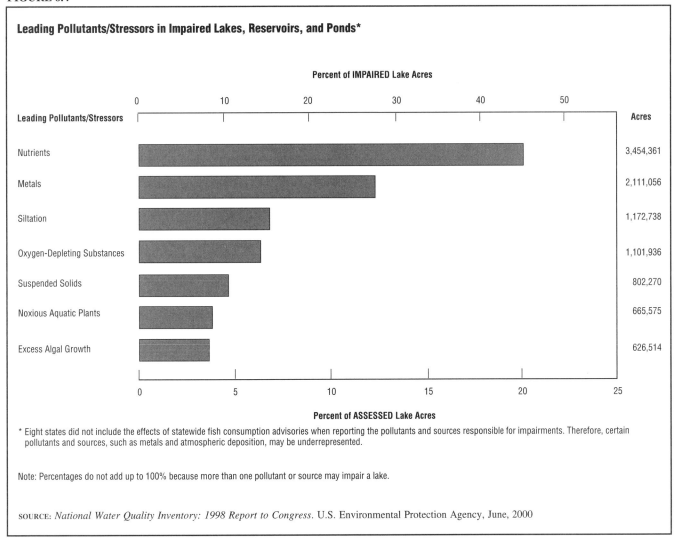

Leading Pollutants/Stressors in Impaired Lakes, Reservoirs, and Ponds*

Percent of IMPAIRED Lake Acres

Leading Pollutants/Stressors		Acres
Nutrients		3,454,361
Metals		2,111,056
Siltation		1,172,738
Oxygen-Depleting Substances		1,101,936
Suspended Solids		802,270
Noxious Aquatic Plants		665,575
Excess Algal Growth		626,514

Percent of ASSESSED Lake Acres

* Eight states did not include the effects of statewide fish consumption advisories when reporting the pollutants and sources responsible for impairments. Therefore, certain pollutants and sources, such as metals and atmospheric deposition, may be underrepresented.

Note: Percentages do not add up to 100% because more than one pollutant or source may impair a lake.

SOURCE: *National Water Quality Inventory: 1998 Report to Congress.* U.S. Environmental Protection Agency, June, 2000

entire annual supply is consumed in the area of the arid Colorado River Basin and the Rio Grande Valley.

Every two years the EPA releases its *National Water Quality Inventory: Report to Congress* report, which is prepared from state assessments. The 1998 report, released in June 2000, found that the states had assessed the quality of water of 42 percent of their lake acres. Among lakes, pond, and reservoirs 55 percent had "good" water quality, which means they fully supported their designated uses (including drinking, swimming, and fishing), while 45 percent did not. The leading pollutants in impaired lakes were nutrients, metals, siltation, and oxygen-depleting substances. (See Figure 6.4.)

The states assessed the quality of 23 percent of their river and stream miles. Of the miles assessed 65 percent were determined to be good quality while 35 percent were deemed impaired. The leading pollutants in impaired rivers were siltation, bacteria, nutrients, and oxygen-depleting substances. (See Figure 6.5.) Agriculture was the leading source of contaminants that polluted

river and stream miles. Hydromodification (damming rivers and altering the flow of water), urban runoff/storm sewers, and municipal point sources also produced contaminants. (See Figure 6.6.)

Siltation alters aquatic habitats by suffocating fish eggs and burying habitat areas of bottom-dwelling species. The loss of those species then impacts fish and other species that feed on them. Silt in the water can also interfere with drinking-water treatment processes and recreational use of surface water bodies. (See Figure 6.7.)

Bacteria reach surface waters through urban runoff, storm sewers, sewage treatment facilities, and septic systems. These organisms sometimes accumulate in shellfish, prompting officials to issue warnings against eating shellfish taken from certain waters. (See Figure 6.8.)

Excessive nutrients (usually attributed to runoff of fertilizers) pollute water bodies by spurring an overgrowth of algae. This results in oxygen depletion and the decomposition of plant matter. Fish suffocate, and unpleasant odor and taste result. (See Figure 6.9.)

FIGURE 6.5

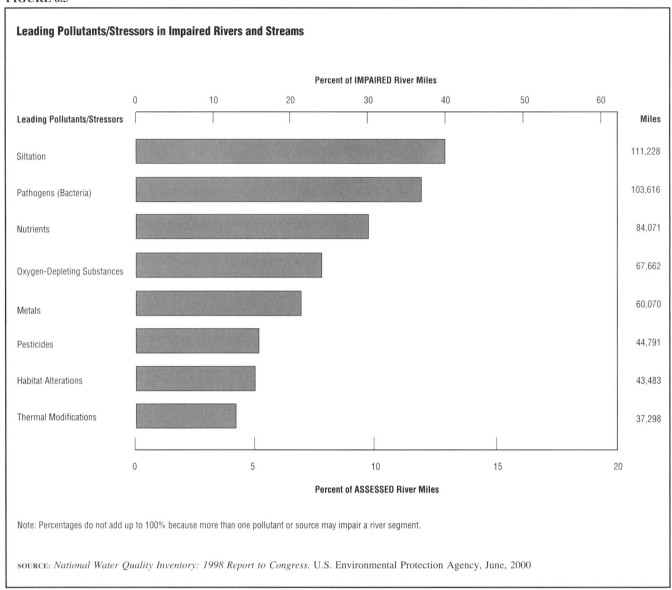

Leading Pollutants/Stressors in Impaired Rivers and Streams

Note: Percentages do not add up to 100% because more than one pollutant or source may impair a river segment.

SOURCE: *National Water Quality Inventory: 1998 Report to Congress.* U.S. Environmental Protection Agency, June, 2000

Point and Nonpoint Sources of Pollution

The main reason that a body of water cannot support its designated uses is that it has become polluted. There are a vast number of pollutants that can make water "impaired," but in order to control a specific pollutant, it is necessary to find out where it is coming from. Although there are many ways in which contaminants can enter waterways, sources of pollution are generally categorized as *point sources* and *nonpoint sources.*

Point sources are those that disperse pollutants from a specific source or area, such as a sewage drain or an industrial discharge pipe. Pollutants commonly discharged from point sources include bacteria (from wastewater treatment plants and sewer overflow), toxic chemicals, and heavy metals from industrial plants.

Nonpoint sources are those that are spread out over a large area and have no specific outlet or discharge

point. These include agricultural and urban runoff, runoff from mining and construction sites, and accidental spills. Some spills, unfortunately, are not accidental but rather the deliberate dumping of toxic wastes, usually at night and in relatively remote areas, by individuals or companies that fail to pay to have their wastes disposed of safely or whose wastes are the result of operations already in violation of environmental regulations. Nonpoint source pollutants can include pesticides, fertilizers, toxic chemicals, and asbestos and salts from road construction. (See Figure 6.10.) The EPA estimates that as much as 65 percent of surface water pollutants come from nonpoint sources.

Do the Nation's Surface Waters Meet the Fishable/Swimmable Goal?

A goal of the Clean Water Act was to return U.S. waters to a fishable/swimmable condition. Meeting the

FIGURE 6.6

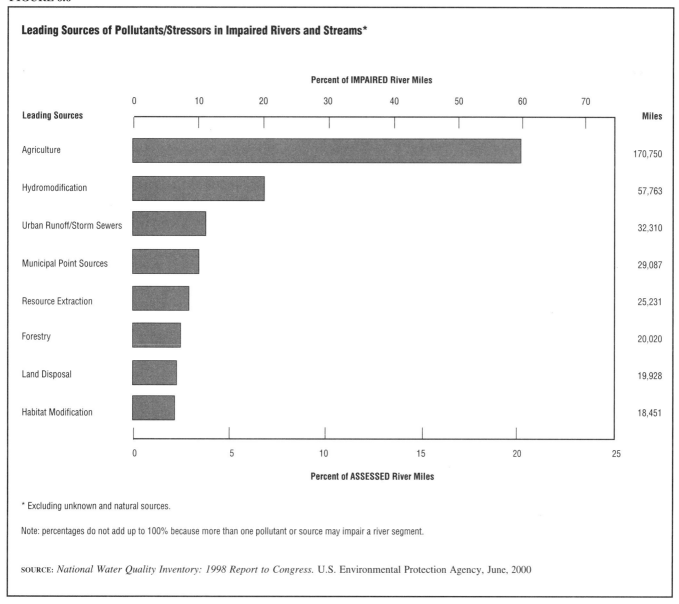

Leading Sources of Pollutants/Stressors in Impaired Rivers and Streams*

* Excluding unknown and natural sources.

Note: percentages do not add up to 100% because more than one pollutant or source may impair a river segment.

SOURCE: *National Water Quality Inventory: 1998 Report to Congress.* U.S. Environmental Protection Agency, June, 2000

fishable goal means providing a level of water quality that protects and promotes the population of fish, shell-fish, and wildlife. As a result of polluted waters fish often become contaminated. When humans eat these fish they can suffer health effects from the toxins. In 1998 the EPA reported that the number of fish advisories rose by 205 to a total of 2,506 from 1997, a 9 percent increase. (See Figure 6.11.) The number of water bodies under advisory represents almost 16 percent of the nation's total lake acres and almost 7 percent of its river miles. In addition 100 percent of the Great Lakes waters and 59 percent of U.S. coastal waters are also under advisory.

In July 2000 the EPA reported in its 1999 BEACH Watch Program that there were about 459 beach closings and swimming advisories issued in 1999, up from 353 in 1998 and 230 in 1997. They occurred at about 24 percent of U.S. beaches, most of which are located in coastal areas.

GROUNDWATER

Groundwater is water that fills pores or cracks in sub-surface rocks. When rain falls or snow melts on the earth's surface, water may run off into lower land areas or lakes and streams. Some is caught and diverted for human use. What is left is absorbed into the soil where it can be used by vegetation; seeps into deeper layers of soil and rock; or evaporates back into the atmosphere.

Below the topsoil is an area called the unsaturated zone where, in times of adequate rainfall, the small spaces between rocks and grains of soil contain at least some water while the larger spaces contain mostly air. After a major rain the zone may become saturated—that is, all the open spaces fill with water. During a drought, the area may become drained and almost completely dry.

With excessive rainfall water will drain through the unsaturated zone (which has now absorbed as much

FIGURE 6.7

Effects of Siltation in Rivers and Streams

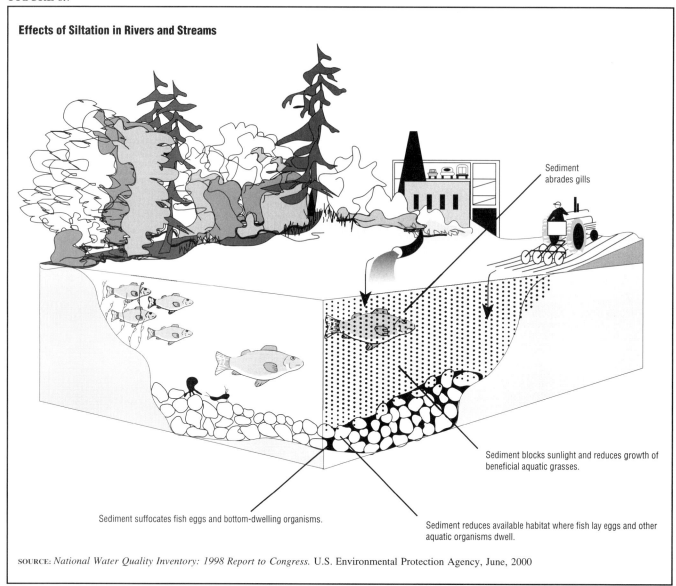

Sediment abrades gills

Sediment blocks sunlight and reduces growth of beneficial aquatic grasses.

Sediment suffocates fish eggs and bottom-dwelling organisms.

Sediment reduces available habitat where fish lay eggs and other aquatic organisms dwell.

SOURCE: *National Water Quality Inventory: 1998 Report to Congress.* U.S. Environmental Protection Agency, June, 2000

water as it can hold) to the saturated zone. The saturated zone is always full of water—all the spaces between soil and rocks, and the rocks themselves, contain water. In the saturated zone water is under higher-than-atmospheric pressure. Thus, when a well is dug into the saturated zone, water flows from the area of higher pressure (in the ground) to the area of lower pressure (in the hollow well), and the well fills with water to the level of the existing water table (the level of groundwater). A well dug just into the unsaturated zone will not fill with water because the water in the unsaturated zone is at atmospheric pressure.

The water table is the level at which the unsaturated zone and the saturated zone meet. The water table is not fixed but may rise or fall, depending on water availability. In areas where the climate is fairly consistent the level of the water table may vary little; in areas subject to extreme flooding and drought it may rise and fall substantially.

An aquifer is an underground formation that contains enough water to yield significant amounts of water when a well is sunk. The formation of an aquifer is actually a path of porous or permeable material through which substantial quantities of water flow relatively easily. The word "aquifer" comes from the Latin *aqua* ("water") and *ferre* ("to bear or carry"). An aquifer can be a layer of gravel or sand, a layer of sandstone or cavernous limestone, a rubble zone between lava flows, or even a large body of massive rock, such as fractured granite.

Aquifers vary from a few feet thick to tens or hundreds of feet thick. They can be located just below the earth's surface or thousands of feet beneath it, and one aquifer may be only a part of a large system of aquifers that feed into one another. They can cover a few acres of land or many thousands of square miles. Because runoff water can easily seep down to the water table, aquifers are susceptible to contamination.

FIGURE 6.8

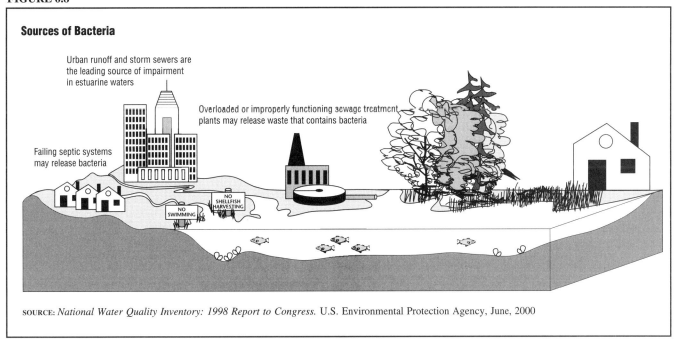

Sources of Bacteria

Urban runoff and storm sewers are the leading source of impairment in estuarine waters

Overloaded or improperly functioning sewage treatment plants may release waste that contains bacteria

Failing septic systems may release bacteria

NO SWIMMING

NO SHELLFISH HARVESTING

SOURCE: *National Water Quality Inventory: 1998 Report to Congress.* U.S. Environmental Protection Agency, June, 2000

FIGURE 6.9

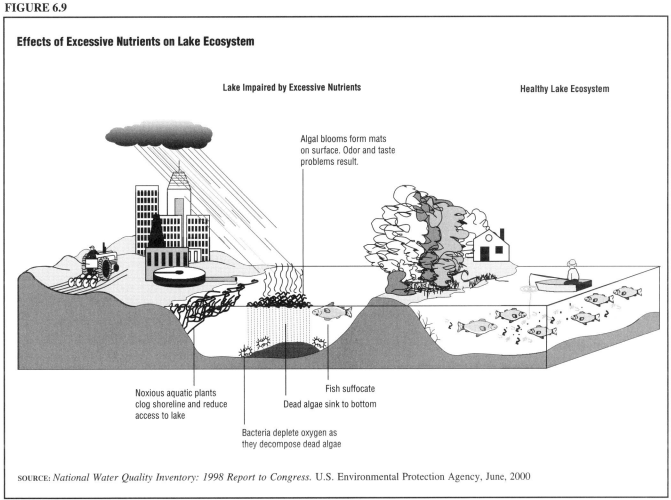

Effects of Excessive Nutrients on Lake Ecosystem

Lake Impaired by Excessive Nutrients

Healthy Lake Ecosystem

Algal blooms form mats on surface. Odor and taste problems result.

Noxious aquatic plants clog shoreline and reduce access to lake

Fish suffocate

Dead algae sink to bottom

Bacteria deplete oxygen as they decompose dead algae

SOURCE: *National Water Quality Inventory: 1998 Report to Congress.* U.S. Environmental Protection Agency, June, 2000

FIGURE 6.10

Examples of Point and Nonpoint Sources of Pollution

Airborne pollution

Tributaries

Animal agriculture

Factory

River

Crop agriculture

Wastewater treatment plant

Boating

Forestry

Power plant

Town

Roads

Examples of point source pollution are indicated on the left side of the river.

Examples of nonpoint source pollution are indicated on the right side of the river.

SOURCE: *Water Quality: Key EPA and State Decisions Limited by Inconsistent and Incomplete Data.* U.S. General Accounting Office, March, 2000

Groundwater is the primary source of public and domestic water supply, irrigation, livestock watering, industrial manufacturing and commercial enterprises, mining, and thermoelectric power. (See Figure 6.12.) As Payal Sampat, researcher for World Watch Institute, reported, "Hydrologists [scientists who study water] now know that healthy aquifers are essential to life above ground—that they play a vital role not just in providing water to drink, but in replenishing rivers and wetlands and, through their ultimate effects on rainfall and climate, in nurturing the life of the land and air as well."

The EPA's *National Water Quality Inventory: 1998 Report to Congress* (June 2000) reported that 26 states assessed water quality in aquifers in their states and identified sources of contamination. The leading sources were underground storage tanks containing toxic chemicals, septic systems, landfills, large industrial facilities, fertilizers, and oil spills. (See Figure 6.13.)

The Federal Role in Protecting Groundwater

Parts of several federal laws help to protect groundwater. (See Table 6.3.) The 1972 Clean Water Act provides guidance and money to the states to help develop groundwater programs. The Safe Drinking Water Act of 1974 (SDWA; PL 93-523) and the Safe Drinking Water Act Amendments of 1996 (PL 104-182) require commu-

FIGURE 6.11

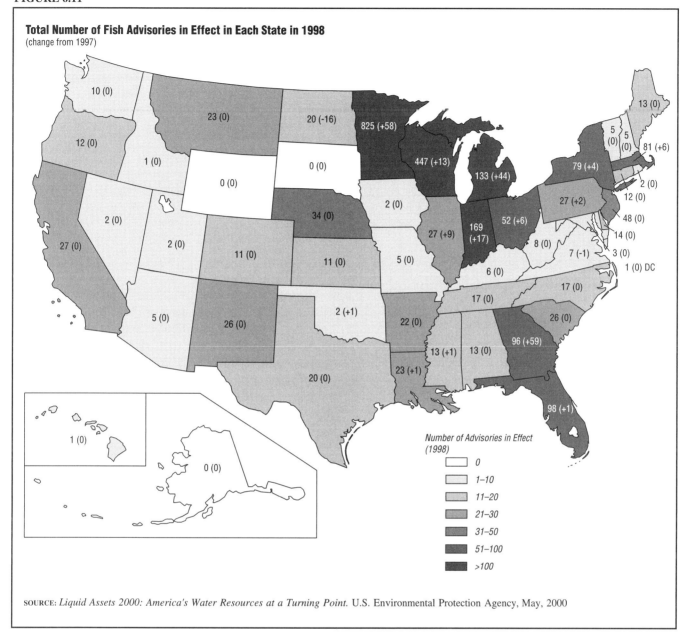

Total Number of Fish Advisories in Effect in Each State in 1998
(change from 1997)

Number of Advisories in Effect
(1998)

☐ 0
☐ 1–10
☐ 11–20
☐ 21–30
☐ 31–50
☐ 51–100
☐ >100

SOURCE: *Liquid Assets 2000: America's Water Resources at a Turning Point.* U.S. Environmental Protection Agency, May, 2000

nities to test their water to make sure it is safe and to help communities finance projects needed to comply with SDWA regulations. The 1976 Resource Conservation and Recovery Act includes many programs designed to clean up hazardous waste, landfills, and underground storage tanks. New storage tanks must be made of strong plastics that will not rust or leak contaminants into the water table.

The Comprehensive Environmental Response, Compensation, and Liability Act of 1980 (CERCLA; PL 96-510) and the Superfund Amendments and Reauthorization Act of 1986 (PL 99-499; PL 99-563; PL 100-202) require the cleanup of hazardous wastes that can seep into the groundwater. The two laws also require that cities and industry build better-managed and better-constructed garbage dumps and landfills for

hazardous materials so that groundwater will not be polluted in the future.

The Federal Insecticide, Fungicide, and Rodenticide Act (FIFRA; 61 Stat 163, amended 1988; PL 100-532) regulates dangerous chemicals used on farms. The act requires the EPA to register the pesticides farmers use against insects, rats, mice, etc. If the EPA thinks pesticides might be dangerous to the groundwater, it can refuse to register them.

OCEANS AND COASTAL WATERS

Throughout history humans have used the oceans virtually as they pleased. Ocean waters have long served as highways and harvest grounds. Now, however, humankind is at a threshold. Marine debris (garbage created by humans) is a problem of global proportions and is

FIGURE 6.12

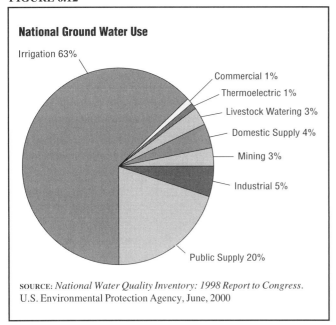

National Ground Water Use

Irrigation 63%

Commercial 1%

Thermoelectric 1%

Livestock Watering 3%

Domestic Supply 4%

Mining 3%

Industrial 5%

Public Supply 20%

SOURCE: *National Water Quality Inventory: 1998 Report to Congress.* U.S. Environmental Protection Agency, June, 2000

extremely evident in countries like the United States where there is extensive recreational and commercial use of coastal waterways. The ocean has become overfished and badly polluted. An estimated half of fish species in the United States are overexploited—more are being caught than can be replenished by natural reproduction. Coastal wetlands are decreasing. Louisiana, for example, loses about 50 square miles of estuaries each year. Sewage, wastewater, and runoff foul the remaining wetlands and ocean waters. Toxic chemicals contaminate fish caught for food. And trash is piling up on many shorelines.

Only 15 states have coastal waters within their jurisdiction. The *National Water Quality Inventory: 1998 Report to Congress* reported that 10 states assessed their coastal waters in 1998. Of the shoreline miles assessed 88 percent were judged to be of good quality while 12 percent were impaired. The three leading pollutants were pathogens (bacteria), turbidity (high particle content in the water, causing muddy or cloudy appearance), and nutrients. (See Figure 6.14.) The sources of those contaminants were urban runoff/storm sewers, land disposal, municipal point sources, and spills (usually of oil).

Every year since 1990 the Natural Resources Defense Council (NRDC), a private environmental protection organization, has studied the condition of U.S. beaches. In *Testing the Waters 2000—A Guide to Water Quality at Vacation Beaches* (August 2000), the NRDC reported 6,160 days of closing and advisories in 1999. Another 23 days were extended closings (6–12 weeks), and 28 days were permanent closings (more than 12 weeks). The total came to more than 7,214 closings and advisories. This was 50 percent higher than in 1997 but a 13 percent decline from 1998. The NRDC found that

about two-thirds of closings were due to excess bacteria levels. Part of the increases were also attributed to an increase in beach monitoring by authorities.

The NRDC reported that only 11 states—California, Connecticut, Delaware, Hawaii, Illinois, Indiana, New Hampshire, New Jersey, North Carolina, Ohio, and Pennsylvania—regularly monitored most or all of their beaches and notified the public when hazards existed. Even those states, however, did not necessarily close the beaches when hazards occurred.

The major causes of beach closings and advisories were elevated levels of bacteria found in monitoring tests (70 percent), known pollution events (14 percent), heavy rains carrying pollution (10 percent), and other (6 percent).

Coastal Erosion

Erosion has become a problem in much of the world in areas that are overfarmed or where topsoil cannot be protected, such as on coasts which are often overdeveloped. The H. John Heinz III Center for Science, Economics, and the Environment, a nonprofit research organization, in *Evaluation of Erosion Hazards*, a study prepared for the Federal Emergency Management Agency (FEMA) in April 2000, found that approximately 25 percent of structures within 500 feet of the U.S. coastline will suffer the effects of coastal erosion within 60 years. (See Figure 6.15.) Especially hard-hit will be areas along the Atlantic and Gulf of Mexico coasts, which are expected to suffer 60 percent of nationwide losses.

The nation's highest average erosion rates—up to six feet or more per year—occur along the Gulf of Mexico. The average erosion rate on the Atlantic coast is two to three feet per year. A major storm can erode 100 feet of coastline in a day. The Heinz Center estimates that roughly 10,000 structures are within the estimated 10-year erosion zone closest to the shore. This does not include structures in the densest areas of large coastal cities, such as New York, Chicago, Los Angeles, and Miami, which are heavily protected against erosion.

The powerful effects of erosion were dramatized by the predicament of the Cape Hatteras lighthouse in North Carolina. When it was constructed in 1870 the lighthouse was 1,500 feet from the shore. By 1987 the lighthouse stood only 160 feet from the sea and was in danger of collapsing. In 1999 the National Park Service, at a cost of $9.8 million, successfully moved the lighthouse back 2,900 feet.

The International Convention for the Prevention of Pollution from Ships

Established in 1973, the International Convention for the Prevention of Pollution from Ships (MARPOL) regulates numerous materials that are dumped at sea. The

FIGURE 6.13

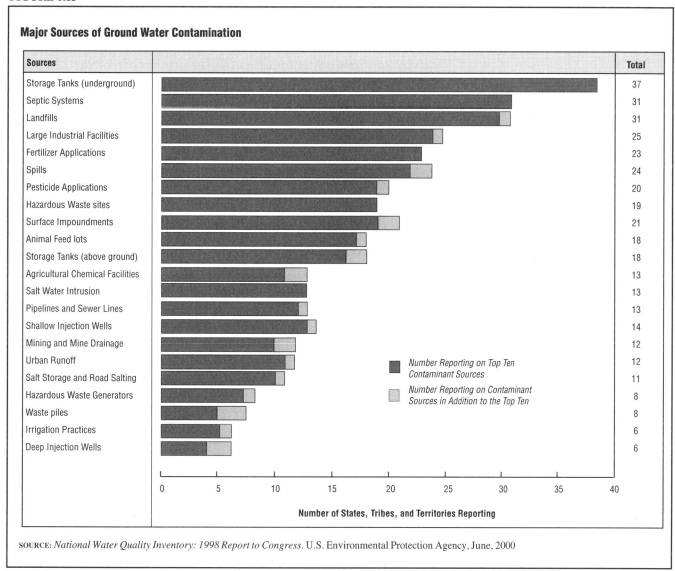

Major Sources of Ground Water Contamination

Sources	Total
Storage Tanks (underground)	37
Septic Systems	31
Landfills	31
Large Industrial Facilities	25
Fertilizer Applications	23
Spills	24
Pesticide Applications	20
Hazardous Waste sites	19
Surface Impoundments	21
Animal Feed lots	18
Storage Tanks (above ground)	18
Agricultural Chemical Facilities	13
Salt Water Intrusion	13
Pipelines and Sewer Lines	13
Shallow Injection Wells	14
Mining and Mine Drainage	12
Urban Runoff	12
Salt Storage and Road Salting	11
Hazardous Waste Generators	8
Waste piles	8
Irrigation Practices	6
Deep Injection Wells	6

■ Number Reporting on Top Ten Contaminant Sources
▢ Number Reporting on Contaminant Sources in Addition to the Top Ten

Number of States, Tribes, and Territories Reporting

SOURCE: *National Water Quality Inventory: 1998 Report to Congress.* U.S. Environmental Protection Agency, June, 2000

international treaty has been in effect in the United States only since its ratification in 1998. Although 83 countries have ratified the treaty they have not necessarily complied, as evidenced by the current level of marine debris.

The Ocean Dumping Act

Congress enacted the Marine Protection, Research, and Sanctuaries Act in 1972 (MPRSA; PL 92-532) to regulate intentional ocean disposal of materials and to authorize research. Title I of the act, known as the Ocean Dumping Act, contains permit and enforcement provisions for ocean dumping. Four federal agencies have authority under the act—the EPA, the U.S. Army Corps of Engineers, the National Oceanic and Atmospheric Administration (NOAA), and the U.S. Coast Guard. Title I prohibits all ocean dumping, except that allowed by permits, in any ocean waters under U.S. jurisdiction by any U.S. vessel or by any vessel sailing from a U.S. port. The act bans dumping of radiological, chemical, and biological

warfare agents, high-level radioactive waste, and medical wastes. In 1997 Congress amended the act to ban dumping of municipal sewage sludge and industrial waste.

The act authorizes the EPA to assess civil penalties up to $50,000 for each violation, as well as criminal penalties (seizure and forfeiture of vessels). For dumping of medical wastes the act authorizes civil penalties up to $125,000, criminal penalties up to $250,000 and five years in prison, or both.

In July 1999 the world's second largest cruise line pleaded guilty in federal court to criminal charges of dumping oil and hazardous chemicals in U.S. waters and lying about it to the Coast Guard. Royal Caribbean agreed to pay a record $18 million fine, the largest ever paid by a cruise line for polluting waters, in addition to the $9 million in criminal fines the company agreed to pay in a previous plea agreement. Six other cruise lines have pleaded guilty to illegal waste dumping since 1993 and have paid

TABLE 6.3

EPA Statutory Authorities With Ground Water Protection Provisions

Resource Conservation and Recovery Act	-Provides the authority to prevent hazardous wastes from leaching into groundwater from hazardous waste facilities and sources such as municipal landfills, impoundments, and underground storage tanks.
Comprehensive Environmental Response, Compensation, and Liability Act, or Superfund	-Provides the authority to clean up abandoned hazardous waste sites that present a major threat to human health or the environment.
Federal Insecticide, Fungicide, and Rodenticide Act	-Provides the authority to control the availability and use of harmful pesticides, including those with a potential to leach into groundwater.
Toxic Substances Control Act	-Provides the authority to control the availability and use of harmful toxic substances, including those with a potential to contaminate groundwater.
Safe Drinking Water Act	-Provides the authority for (1) setting and enforcing drinking water standards for surface and groundwater public drinking water supplies, (2) controlling underground injection practices, and (3) establishing state wellhead protection programs.
Clean Water Act.	-Provides authority for federal grant programs to assist states in developing groundwater protection strategies and nonpoint source pollution programs.

SOURCE: *Water Pollution: More Emphasis Needed on Prevention in EPA's Efforts to Protect Groundwater.* U.S. General Accounting Office, 1991

fines ranging up to $1 million. The cases have focused attention on the difficulties of regulating the fast-growing cruise line industry in which most major ships sailing out of American ports are registered in foreign countries.

The Oil Pollution Act

In 1989 the oil freighter *Exxon Valdez* ran into a reef in Prince William Sound, Alaska, spilling more than 11 million gallons of oil into one of the richest and most ecologically pristine areas in North America. An oil slick the size of Rhode Island killed wildlife and marine species. A $5 billion damage penalty was levied against Exxon, whose ship captain was found to be at fault in the wreck.

In response to the *Valdez* oil spill Congress passed the Oil Pollution Act of 1990 (PL 101-380) that went into effect in 1993. The law requires companies involved in storing and transporting petroleum to have standby plans for cleaning up oil spills on land or in water. Under the Act a company that does not adequately take care of a spill is vulnerable to almost unlimited litigation and expense. The law makes the Coast Guard responsible for approving cleanup plans and procedures for coastal and seaport oil spills, while the EPA oversees cleanups on land and in inland waterways. The law also requires that oil tankers be built with double hulls, to better secure the oil in event of a hull breach.

WHERE WATER IS POWER—INTERNATIONAL WATER WARS?

As it becomes rarer because of the growing population and the pollution of water supplies, usable water is expected to become a commodity, like iron or oil, leading some experts to predict that, at some point, water will be more expensive than oil. Of the 200 largest river systems in the world, 120 flow through two or more countries. All are potential objects of world political power struggles over this critical resource.

Three areas of the world are particularly short of water—Africa, the Middle East, and South Asia. Other dry areas include the American Southwest, parts of South America, and large areas of Australia.

Middle East countries are especially threatened as growing nations compete for a shrinking water supply. Freshwater has never come easily to this area. Rainfall occurs only in winter and drains quickly through the parched land. Most Middle Eastern countries are joined by common aquifers—underground layers of porous rock that contain water. The United Nations has cautioned that future wars in the Middle East could be fought over water.

The oil-rich Middle Eastern nation of Kuwait has little water but has the money to secure it. In order to use seawater Kuwait has constructed six large-scale, oil-powered water desalination plants. Saudi Arabia, farther down the Arabian Peninsula, leads the world in water desalination. Its 22 plants produce 30 percent of all the desalinated water in the world. It is also a leader in the pumping of fossil water—water accumulated in an earlier geologic age lying deep in aquifers beneath Africa and the Middle East.

In 1998 the Ontario Environment Ministry in Canada authorized a company in Sault Ste. Marie, Ontario, to pump up to 158 million gallons of water a year from Lake Superior—which Canada shares with Minnesota, Wisconsin, and Michigan—into cargo ships for export to drought-stricken Asian countries. Americans protested and Canada blocked the sale—at least temporarily. Environmentalists and politicians on both sides of the border worry that such a plan would undermine United States/Canada treaties and encourage other companies to undertake large-scale water exports.

FIGURE 6.14

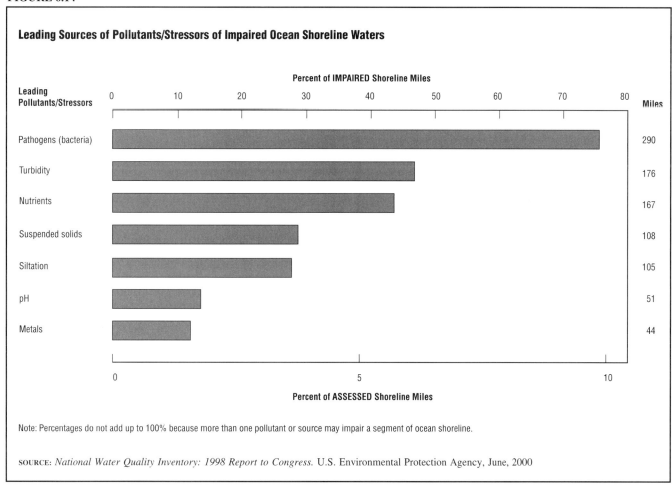

Leading Sources of Pollutants/Stressors of Impaired Ocean Shoreline Waters

Note: Percentages do not add up to 100% because more than one pollutant or source may impair a segment of ocean shoreline.

SOURCE: *National Water Quality Inventory: 1998 Report to Congress.* U.S. Environmental Protection Agency, June, 2000

DRINKING WATER

Sources of Drinking Water—Public and Private Supply

Public water supply systems (that is, utility systems that provide water to the public) serve 90 percent of the American population. The remainder get their water from private wells. Half of America's drinking water comes from surface water such as lakes, rivers, and reservoirs. The other half comes from wells or springs that are filled by groundwater from aquifers. Urban areas draw their supplies largely from surface-water sources, while 90 percent of the drinking water in rural areas is drawn from groundwater. (See Figure 6.16.)

While most Americans rely on drinking water from public water systems, millions receive their water from individually owned and operated sources such as personal wells, cisterns, and springs. Individual wells serving four or fewer housing units provide the water for approximately 15 percent of homes. Most of these are "drilled" wells less than 1.5 feet wide. The remainder are "dug" wells, usually wider than 1.5 feet and generally hand-dug. Only 1 percent of the nation's homes get their water directly from creeks, rivers, and lakes. System owners are solely responsible for the quality of the water provided from private sources.

The EPA and state health or environmental departments regulate public water supplies. Public supplies are required to ensure that the water meets certain government-defined health standards. The Safe Drinking Water Act of 1974 (SDWA; PL 93-523) governs this regulation. The law mandates that all public suppliers test their water on a regular basis to check for the existence of contaminants, and treat their water supplies constantly to take out or reduce certain pollutants to levels that will not harm human health.

On the other hand private water supplies, usually wells, are not regulated under the SDWA, although many states have programs designed to help well owners protect their water supplies. Usually, these state-run programs are not regulatory but provide safety information. This type of information is vital because private wells often are shallower than those used by public suppliers. The shallower the well the greater is the potential for contamination.

Chemicals and Contaminants in Drinking Water

All drinking water contains minerals dissolved from the earth. In small amounts some of these are acceptable because they often enhance the quality of the water (for

FIGURE 6.15

Average Annual Erosion Rates in Selected U.S. Counties

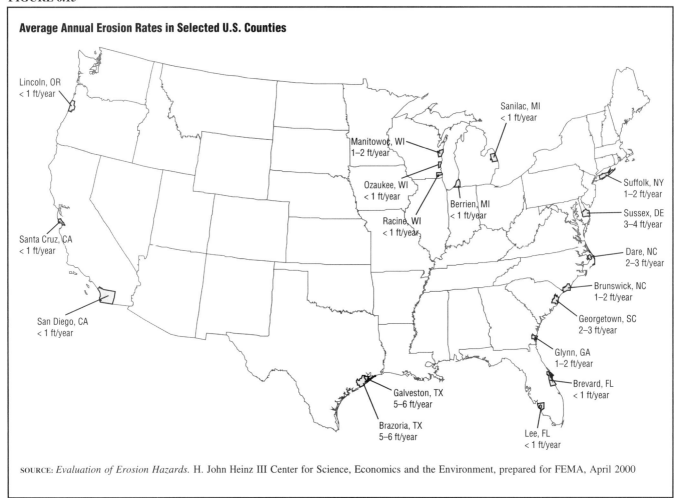

SOURCE: *Evaluation of Erosion Hazards.* H. John Heinz III Center for Science, Economics and the Environment, prepared for FEMA, April 2000

example, giving it a pleasant taste). A few, such as zinc and selenium, in very small amounts, contribute to good health. Other naturally occurring minerals are not desirable because they may cause a bad taste or odor (as excessive amounts of iron, manganese, or sulfur often do) or because they may be harmful to health.

The health effects from drinking contaminated water can occur either over a short or long period of time, depending on the type of pollutant. Short-term, or acute, reactions are those that occur within a few hours or days after drinking tainted water. Long-term, or chronic, effects occur after water with relatively low doses of a pollutant has been consumed for several years or even over a lifetime. Fortunately, the ability to detect contaminants has improved over the past few decades. Scientists can now identify specific pollutants in terms of one part contaminant in one billion parts of water. In some cases contaminants can be measured in the trillionths.

Chemical Pollutants

Water supplies may contain a wide variety of contaminants that can cause serious health risks. While bacteri-

al infections generally make their presence known quickly by causing illness with fairly obvious symptoms, the effects of noxious chemicals may not be apparent for months, or even years, after exposure. Some pollutants are known carcinogens (cancer-causing agents), while others are suspected of causing birth defects, miscarriages, and heart disease. In many cases the effects occur only after prolonged exposure, but no one can say for sure what is a safe level of exposure.

LEAD. When water leaves the treatment plant it is relatively free of lead, but it can pick up the metal from lead pipes (and pipes with lead solder) in distribution systems or homes, resulting in tap water containing significant amounts of this highly toxic substance. Unlike many water contaminants lead has been extensively studied for its prevalence and effects on human health and for ways to eliminate it from the water supply.

Most people understand that drinking water contaminated with lead is very dangerous. When ingested lead causes extremely serious health problems, especially in children, whose developing bodies absorb and retain more lead than adults' bodies do. According to EPA reports even very low level exposures can result in low-

FIGURE 6.16

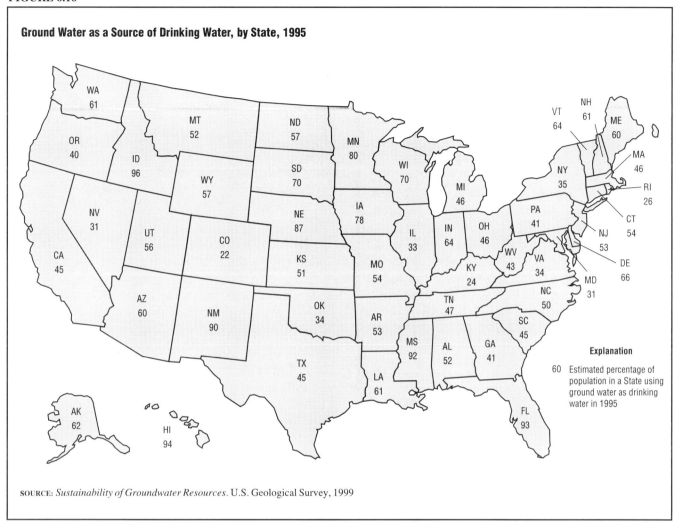

Ground Water as a Source of Drinking Water, by State, 1995

Explanation

60 Estimated percentage of population in a State using ground water as drinking water in 1995

SOURCE: *Sustainability of Groundwater Resources.* U.S. Geological Survey, 1999

ered I.Q., impaired learning and language skills, loss of hearing, reduced attention spans, and poor school performance. High levels damage the brain and central nervous system interfering with both learning and physical development. While drinking water is not the only means of lead contamination (lead exists in some paints, soils, and older food containers), the EPA estimates that it accounts for between 10 and 20 percent of total lead poisoning in young children.

Pregnant women are another high-risk group. Lead is believed to cause miscarriages, premature births, and impaired fetal development. It has also been linked to high blood pressure, fatigue, and hearing loss.

Lead is rarely found in either surface water or groundwater that are the sources of drinking water for most Americans. Lead usually enters the water supply after it leaves the treatment plant. The industries that release the most lead into the environment include lead smelting and refining, copper smelting, steelworks, manufacturers of storage batteries and china plumbing fixtures, iron foundries, and copper mining.

Normally, areas served by lead service lines or residences containing lead interior piping or copper piping with lead solder installed after 1982, are considered high risk. Since 1986 it has been illegal to use lead solder that contains more than 0.2 percent lead. But ripping up and replacing lead piping is extremely expensive. As an alternative some cities are using chemicals to make the water less acidic so that it picks up less lead.

In 1991 the EPA's Lead and Copper Rule set the Maximum Contaminant Level of lead at 0 milligrams per liter (mg/L), with water systems required to take action if lead levels reach 0.015 mg/L. The EPA believes this is the lowest level to which water systems can reasonably be required to control the contaminant. If a water supply is found to exceed that amount in more than 10 percent of the homes served, it must be tested twice a year. If levels remain above the standard the supplier must take steps to reduce those levels. The water supplier must also notify the public via newspapers, radio, TV, and other means, and must inform its consumers of additional measures they may take to reduce the levels in their homes.

NITRATES AND NITRITES. Nitrates and nitrites are nitrogen-oxygen chemicals that combine with organic and inorganic compounds. Once taken into the body, nitrates are converted into nitrites. Nitrates are most frequently used as fertilizer. Primary sources include human sewage and livestock manure, especially from feedlots. Since they are soluble nitrates can easily migrate into groundwater.

Nitrates in drinking water are an immediate threat to small children. In some babies high levels of nitrate react with red blood cells to cause an anemic condition commonly known as "blue baby." The maximum contaminant level (MCL) for nitrates has been set at 10 parts per million (ppm) and for nitrites at 1 ppm. If contaminant levels exceed these standards a water provider must take steps to reduce the levels—either through ion exchange, reverse osmosis, or electrodialysis—and must notify the public of their presence.

MERCURY. Mercury is unique among metals in that it can evaporate when released into water or soil. Large amounts of mercury are released naturally from the earth's crust. Metal smelters, cement manufacture, landfills, sewage, and combustion of fossil fuels are also important sources of mercury release. Mercury is especially dangerous when released into water because it tends to accumulate in the tissues of fish. When tainted fish are eaten by humans mercury poisoning is often the result. The MCL for mercury has been set at 2 ppb.

MICROBIOLOGICAL ORGANISMS AS CONTAMINANTS. Many kinds of biological organisms exist in drinking water. These include certain types of bacteria, viruses, or parasites. These tiny organisms get into the water supplies when the water is contaminated with human or animal wastes. The bacteria, viruses, and parasites that contaminate drinking water can cause flu-like symptoms, including headaches, vomiting, diarrhea, abdominal pain, and dehydration. Although usually not life-threatening they can be debilitating and uncomfortable for their victims.

One type of microscopic parasite that is a common cause of illness is *Giardia lamblia.* Once thought to be harmless it is now believed to be the most frequent cause of waterborne epidemics in the United States. Its symptoms include mild-to-severe gastrointestinal pain, vomiting, and diarrhea. It is estimated that up to 4 percent of Americans carry *Giardia* in the upper intestine. *Giardia* is particularly threatening because it can enter the water supply in any number of ways, including as sewage overflow. It is also easily transmitted from person to person making places with high concentrations of people (day care centers and schools, for example) particularly susceptible to outbreaks.

Cryptosporidium is a one-celled, infectious parasite that frequently contaminates the water supply. It is pre-

sent in 65 to 87 percent of surface water samples tested throughout the United States. Although there are tests for *Cryptosporidium*, current testing methods cannot determine with certainty whether *Cryptosporidium* detected in drinking water is alive or whether it can affect humans. In addition, the technology often requires several days to get results, by which time the tested water has already been used by the public and is no loner in the community's water pipes. Consequently, water utilities do not routinely test to detect its presence. (A water utility may voluntarily test for the microorganism, and it is also possible that a state may require water systems to test for it. Otherwise, it is unlikely that a given water system tests for *Cryptosporidium*.)

Because *Cryptosporidium* is highly resistant to chlorination, disinfection of water is not a reliable method of preventing exposure to it. The Centers for Disease Control and Prevention (CDC) and the EPA report that the organism can be killed by boiling water for one minute. Outbreaks of *Cryptosporidium* have generally occurred when turbidity reached 0.9–2.0 NTU (nephelometric turbidity units).

Coliform bacteria from human or animal wastes can also pose serious health problems. Waterborne diseases such as typhoid, cholera, infectious hepatitis, and dysentery have all been traced to untreated drinking water.

MODERN DRINKING WATER TREATMENT

Community water systems are those that have at least 15 service connections used by year-round residents or that serve at least 25 year-round residents. As of 1999 there were approximately 55,000 community water systems in the United States and approximately 114,000 non-community water systems. Non-community systems serve travelers and intermittent users; for example, campgrounds, rest stops, and other public accommodations. Some 25,000 businesses and institutions also have water supply systems that are regulated by federal water standards.

The water treatment process begins with choosing the highest quality source available. Raw water must be transported from the source to the treatment plant while groundwater is usually pumped directly into the plant. In many cases the only treatment needed before the water is distributed to consumers is disinfection. Groundwater is naturally filtered as it seeps through layers of rock and soil. However, sometimes it must be treated to remove contaminants that may have percolated down from the surface to the aquifer. In addition some groundwater must have certain minerals or gases removed to make the water less "hard" (high in natural minerals). Hard water can clog pipes, stain fixtures, and make soap hard to lather.

Surface water is sent to the water treatment plant through aqueducts or pipes. An initial screen at the

FIGURE 6.17

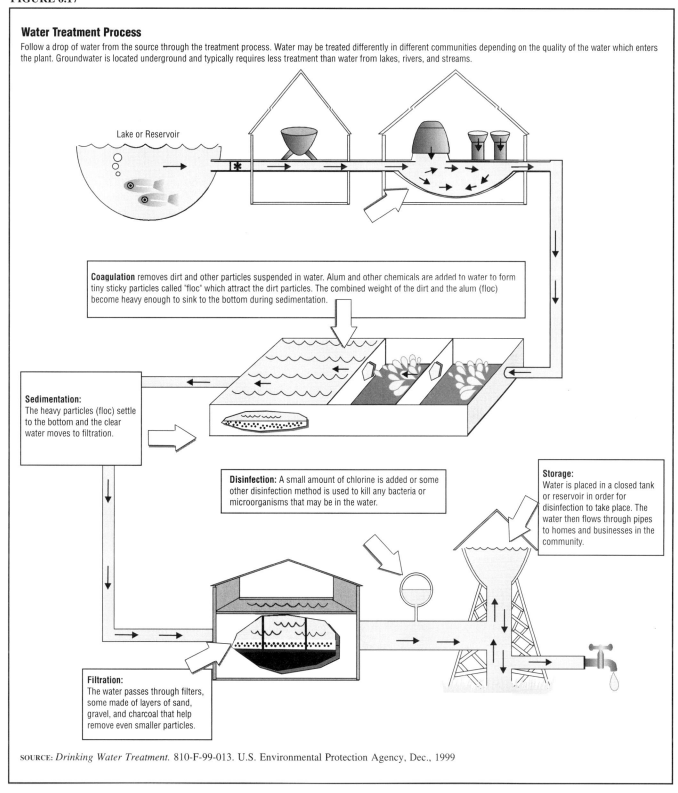

Water Treatment Process

Follow a drop of water from the source through the treatment process. Water may be treated differently in different communities depending on the quality of the water which enters the plant. Groundwater is located underground and typically requires less treatment than water from lakes, rivers, and streams.

Lake or Reservoir

Coagulation removes dirt and other particles suspended in water. Alum and other chemicals are added to water to form tiny sticky particles called "floc" which attract the dirt particles. The combined weight of the dirt and the alum (floc) become heavy enough to sink to the bottom during sedimentation.

Sedimentation:
The heavy particles (floc) settle to the bottom and the clear water moves to filtration.

Disinfection: A small amount of chlorine is added or some other disinfection method is used to kill any bacteria or microorganisms that may be in the water.

Storage:
Water is placed in a closed tank or reservoir in order for disinfection to take place. The water then flows through pipes to homes and businesses in the community.

Filtration:
The water passes through filters, some made of layers of sand, gravel, and charcoal that help remove even smaller particles.

SOURCE: *Drinking Water Treatment.* 810-F-99-013. U.S. Environmental Protection Agency, Dec., 1999

intake pipe removes large objects. The water is then aerated to eliminate gases and add oxygen. Once the water is inside the plant chemicals may be added both to clean the water and also to make it more palatable. If the water is hard lime or soda is added to remove the calcium and magnesium. Chlorine or other such disinfectants may be used as well.

The water is mixed well with the various chemicals then sent to sedimentation basins where the heavy particles (floc) settle to the bottom and are removed. The water is then sent to filtration beds for "polishing"—the removal of any remaining small particles and disease-causing protozoa, bacteria, and viruses. (Although filtration removes some viruses most pass through the filtration process.)

Additional treatment may be required if the raw water contains high levels of toxic chemicals. The pollutants that are the most difficult to detect and remove are often the ones with the greatest potential for severe health effects for large portions of the population.

At various points in the treatment process the water is monitored by computers and other technological procedures. As the water leaves the treatment plant chlorine is added as a disinfectant to keep it free of organisms as it travels to customers.

The water then goes to reservoirs where it is stored until needed. These reservoirs may be elevated towers, where gravity brings the water to the consumer without unnecessary energy expense, or ground-level containers that require pumps to move the water. (See Figure 6.17.) The water that flows from the tap should be *clear*, *tasteless*, and *safe* to drink.

Chemicals Deliberately Added to Drinking Water

Water purification facilities deliberately add certain chemicals to drinking water to destroy contaminants that may cause illness and to improve the taste, smell, and look of the water.

CHLORINE. The most extensively used disinfectant in the United States is chlorine, which is used to kill infectious microorganisms and parasites. Disinfection with chlorine or similar chemicals can prevent outbreaks of salmonellosis, dysentery, and *Giardia*. Chlorination first began in the early 1900s as an attempt to eliminate cholera and typhoid.

Chlorination is not risk free, however. Chlorine reacts with organic chemicals to form trihalomethanes (THMs) that have been shown to cause cancer in laboratory animals. In 1979 the EPA established regulations limiting the amount of THMs to 0.1 mg/L for water supplies serving 10,000 people or more. In the 1980s several epidemiology studies reported an increased risk of colon, bladder, and rectal cancer from chlorinated drinking water. However, studies with animals were negative. The EPA and the medical community continue to study the effects of chlorine. Most concerns about chlorine involve possible damage to the earth's ozone layer. Some water systems now use ozone gas instead of chlorine. Ozone, bubbled through water, can kill more microorganisms than chlorine and may present less risk.

FLUORIDE. Fluoride was first added to drinking water in 1945 to prevent tooth decay. Since that time most community water systems in the United States have introduced water fluoridation. Because the fluoridation of drinking water proved effective in reducing dental cavities, researchers also developed other methods to deliver fluoride to the public (toothpastes, rinses, dietary supplements). The widespread use of these products has assured that virtually all Americans have been exposed to fluoride. The American Dental Association estimated in 1992 that each $1 expenditure for water fluoridation results in a savings of approximately $80 in dental treatment costs.

There have been concerns about the effects of fluoridation since it was first introduced. The Public Health Service has recommended further assessment of potential problems, although it notes that fluoridation is believed to be greatly beneficial for a number of bone-related conditions.

DRINKING WATER LEGISLATION

Almost any legislation concerning water affects drinking water, either directly or indirectly. The following pieces of legislation are aimed specifically at providing safe drinking water for the nation's residents.

The Safe Drinking Water Act of 1974

The Safe Drinking Water Act of 1974 (SDWA; PL 93-523) mandated that the EPA establish and enforce minimum national drinking water standards for all public water systems—community and non-community—in the United States. The law also required the EPA to develop guidelines for water treatment and to set testing, monitoring, and reporting requirements.

To address pollution of surface water supplies to public systems, the EPA established a permit system requiring any facility that discharges contaminants directly into surface waters (lakes and rivers) to apply for a permit to discharge a set amount of materials—and that amount only. It also created groundwater regulations to govern underground injection of wastes.

Congress intended that, after the EPA had set regulatory standards, each state would run its own drinking water program. Since 1974, 54 states and territories have been granted "primacy"; that is, they have been given the primary responsibility for enforcing the requirements of the SDWA. In order to be granted primacy a state must adopt drinking water standards at least as stringent as the national standards (those established by the EPA), and it must be able to conduct monitoring and enforcement programs that meet federal standards.

The EPA established the Primary Drinking Water Standards by setting maximum containment levels (MCLs) for contaminants known to be detrimental to human health. All public water systems in the United States are required to meet primary standards. Only two contaminants regulated at this time by primary standards—bacteria and nitrates—pose an immediate health risk when the MCL standards are exceeded. Secondary standards cover non-health-threatening aspects of drinking water such as odor, taste, staining properties, and color. Secondary standards are recommended but not required.

SOLE SOURCE AQUIFERS. Under the SDWA, the EPA has the authority to designate certain groundwater supplies as the sole source of drinking water for a community (referred to as "sole source aquifers") and to determine if federal financially assisted projects may contaminate these aquifers. If the EPA determines that contamination could occur, no commitment of federal financial assistance—grants, contracts, loan guarantees, etc.—can be made for that project.

As of October 1999 the EPA had designated 70 sole source aquifers nationwide. To be designated as a sole source aquifer for an area, at least 50 percent of the population in a given area must depend on the aquifer for drinking water; a significant public health hazard would result if the aquifer were contaminated, and no reasonable alternative drinking water supplies exist.

1986 AMENDMENTS TO THE SAFE DRINKING WATER ACT. The 1986 amendments to the Safe Drinking Water Act (PL 99-359) required that the EPA set MCLs for an additional 53 contaminants by June 1989, 25 more by 1991, and 25 every three years thereafter. The amendments also required the EPA to issue a maximum contaminant level goal (MCLG) along with each MCL. An MCLG is a health goal equal to the maximum level of a pollutant not expected to cause any health problems over a lifetime of exposure. The EPA is mandated by law to set MCLs as close to MCLGs as technology and economics will permit.

The 1986 amendments banned the use of lead pipe and lead solder in new public drinking water systems and in the repair of existing systems. In addition the EPA had to specify criteria for filtration of surface water supplies and set standards for disinfection of all surface and groundwater supplies. The EPA was required to take enforcement action, including filing civil suits against violators of drinking water standards, even in states granted primacy, if those states did not adequately enforce regulations. Violators became subject to fines up to $25,000 daily until violations were corrected.

The Water Quality Control Act of 1987

Section 304 (1) of the revised Clean Water Act of 1987 (PL 100-4) determines the state of the nation's water quality and reviews the effectiveness of the EPA's regulatory programs designed to protect and improve that water quality. Section 308—known as the Water Quality Control Act—requires that the administrator of the EPA report annually to Congress on the effectiveness of the water quality improvement program.

The main purpose of the Water Quality Control Act is to identify water sources that need to be brought up to minimum standards and to establish more stringent controls where needed. States are now required to develop lists of contaminated waters as well as lists of the sources and amounts of pollutants causing toxic problems. In addition each state is required to develop "individual control strategies" dealing with these pollutants.

Lead Contamination Control Act of 1988

The Lead Contamination Control Act of 1988 (PL 100-572) strengthened the controls on lead contamination set out in the 1986 amendments to the SDWA. It requires the EPA to provide guidance to states and localities in testing for and remedying lead contamination in drinking water in schools and day care centers. The act also contains requirements for the testing, recall, repair, and/or replacement of water coolers with lead-lined storage tanks or parts containing lead. It attaches civil and criminal penalties to the manufacture and sale of water coolers containing lead.

The ban on lead states that plumbing must be lead-free. In addition each public water system must identify and notify anyone whose drinking water may be contaminated with lead, and the states must enforce the lead ban through plumbing codes and the public-notice requirement. The federal government gave the EPA the power to enforce the lead ban law by authorizing the agency to withhold up to 5 percent of federal grant funds to any state that does not comply with the new rulings.

Reinventing Drinking Water Law—The 1996 Amendments to the Safe Drinking Water Act

In 1996 Congress passed a number of significant amendments to the Safe Drinking Water Act (PL 104-182). The law changed the relationship between the federal government and the states in administering drinking water programs, giving states greater flexibility and more responsibility.

The centerpiece of the law is the State Revolving Fund (SRF), a mechanism for providing low-cost financial aid to local water systems to build the treatment plants necessary to meet state and federal drinking water standards. Congress appropriated $1.275 billion in fiscal year 1997 to fund the SRFs and authorized $1 billion for each succeeding fiscal year through 2003. States must come up with a 20 percent match. The revenue can be used to make loans, purchase or finance government debt, or buy local bond insurance to help cover the costs of constructing or repairing water treatment facilities. To qualify for grants, states must have a drinking water SRF in place.

The law also requires states to train and certify operators of drinking water systems. If they do not, states risk losing up to 20 percent of their federal grants. The law requires states to approve the operation of any new water supply system, making sure it complies with the technical, managerial, and financial requirements. The 1996 SDWA gives the EPA discretion in regulating only those contaminants that may be harmful to health, and requires the EPA to select at least five contaminants every five years for consideration for new standards. A further change is that the EPA, when proposing a regulation, now must determine—and publish—whether or not the benefits of a new standard justify the costs.

Furthermore, the law affirms Americans' "right to know" the quality of their drinking water, and mandates notification. Water suppliers must promptly (within 24 hours) alert consumers if water becomes contaminated by something that can cause illness and must advise as to what precautions can be taken. In 1998 states began to compile information about individual systems, which the EPA now summarizes in an annual compliance report. As of October 1999 water systems have been required to make that data available to the public. Large suppliers will have to mail their annual safety reports to customers, while smaller systems can post the reports in a central location or publish it in a local newspaper. (Information on individual water systems is available on the EPA web site at http://www.epa.gov.)

HOW CLEAN IS OUR DRINKING WATER?

Safe drinking water is a cornerstone of public health. Fortunately, the nation's drinking water is generally safe. The vast majority of U.S. residents receive water from systems that have no reported violations of MCLs and no flaws in treatment techniques, monitoring, or reporting. Nevertheless, numerous studies have found some deficiencies in those systems, and recent measurements suggest areas of potential danger.

In 1994, in a 3,500-page report (Brian Cohen and Eric Olson, *Victorian Water Treatment Enters the 21st Century*) the Natural Resources Defense Council (NRDC), a private environmental protection group, documented some 250,000 violations of the SDWA in 1991 and 1992 alone. The council found that 43 percent of the water systems, serving about 120 million people, had committed violations. The NDRC contended that:

- Most water systems did little to prevent contamination of source water. Two-thirds of large water systems did nothing to prevent encroachment of contamination into well or raw water.

- Fewer than 10 percent of large community systems were using modern techniques such as carbon or ozone—both widely used in European drinking water systems.

- More than 50 large groundwater systems provided no water treatment at all, claiming that groundwater protected from surface contamination does not require treatment because passage of the water through the aquifer is enough to remove many pollutants.

- Aged, crumbling distribution systems, some as old as 100 years, were neglected, could permit cross-contamination, and might be made of lead pipe.

This indictment of drinking water suppliers caught the attention of the general public, the water industry, and the scientific community nationwide.

The 1995 NRDC Study of Water Contamination

In its 1995 study, *You Are What You Drink*, the NRDC reported 180,276 total drinking water violations in the United States—15,761 violations of health standards, 8,267 violations of treatment techniques, and 150,487 monitoring and reporting violations. Of that number, 2,575 state enforcement actions and 849 federal enforcement actions resulted. A total of 92 million people served by 25,102 water systems nationwide were affected by those violations. The NRDC study also reported that:

- There was widespread *Cryptosporidium* in the nation's water systems. Tens of millions of Americans drank water that contained the organism.

- About one in five Americans drank water supplied by systems that did not comply with the EPA's basic health or treatment standards. According to best estimates hundreds of thousand, or possibly millions, of Americans became sick as a result. The EPA set a "point estimate" of 7.1 million illnesses and perhaps as many as 1,200 deaths annually.

- Persons with immune system deficiencies were at risk from tap water and should boil their water for one minute, use certified bottled water, or use a water filter certified for cyst (a stage in the life cycle of *Cryptosporidium*) removal.

- Consumers needed to be wary of poorly regulated or unregulated common chemical contaminants such as arsenic and aluminum.

The NRDC also contended that many water utilities were hiding facts about contamination. Only about half responded to questions about their water quality and many of the rest were either evasive or made excuses about why they could not provide the requested data.

The 1998 National Public Water System Compliance Report

In accordance with the 1996 SDWA amendments public water systems are mandated to submit compliance reports on the quality of their drinking water, and the EPA prepares an annual national compliance report from that data. According to the EPA's *1998 National Summary of Public Water Systems Compliance Report* (April 2000), 94 percent of public water systems in the United States reported no violations in 1998. The overwhelming majority—89 percent—of Americans received water from systems without reported violations. (See Figure 6.18.) Most violations were reported in small systems that served fewer than 3,300 people.

1997 GAO Study of Private Wells

The U.S. General Accounting Office (GAO), in its 1997 report *Information on the Quality of Water Found at Community Water Systems and Private Wells,* surveyed the quality of water from 5,500 private wells in nine states. The survey measured coliform bacteria, *E. coli,* nitrate, and the herbicide atrazine. The wells studied ranged in age from one to 200 years and represented many different construction types. More than 41 percent of wells tested positive for coliform bacteria, 11.2 percent contained *E. coli,* and 13.4 percent contained nitrate levels above EPA standards. In 4,847 wells in eight of the nine states atrazine above the MCL was detected in 0.2 percent of cases.

Several factors influenced the quality of well drinking water. The report found that source water quality was key. Community systems were much more likely to treat their water than were private well owners. Construction standards were much more stringent for community systems than for private wells. In only two of the states surveyed (California and Illinois) were permits required prior to construction. All the states required registration of new wells, although officials estimate that only 60 percent of wells are reported. Only two of the states tested water quality at new wells and two states inspected new wells.

Milwaukee—"The Nation's Worst Drinking Water Disaster"

In April 1993, 403,000 residents of Milwaukee became victims of what is considered the nation's worst drinking water disaster. *Cryptosporidium* flourished in the city water supply which had been turbid for several days. For a week more than 800,000 residents were without potable (drinkable) tap water. By the end of the disaster more than 40 people lost their lives because of the outbreak. In addition to the human suffering the disease cost an estimated $37 million in lost wages and productivity.

Among the possible causes for the outbreak were the advanced age and flawed design of the Milwaukee water plant which returned dirty water back to the reservoir. Other explanations included plant personnel failing to react quickly when turbidity levels rose; critical monitoring equipment was broken at the time that turbidity levels peaked; the water intake point was vulnerable to contamination; and a slaughterhouse, feedlot, and sewage treatment plant were located upriver from the plant. Water treatment experts blamed the complacency of officials and false assumptions based on a history of quality water dispersal.

As a result of the disaster Milwaukee launched one of the most aggressive drinking water programs in the coun-

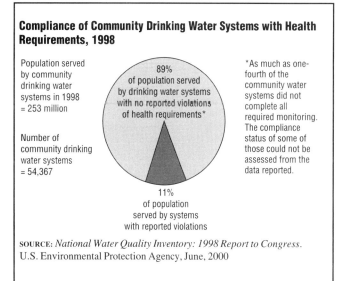

FIGURE 6.18

Compliance of Community Drinking Water Systems with Health Requirements, 1998

Population served by community drinking water systems in 1998 = 253 million

Number of community drinking water systems = 54,367

89% of population served by drinking water systems with no reported violations of health requirements*

11% of population served by systems with reported violations

*As much as one-fourth of the community water systems did not complete all required monitoring. The compliance status of some of those could not be assessed from the data reported.

SOURCE: *National Water Quality Inventory: 1998 Report to Congress.* U.S. Environmental Protection Agency, June, 2000

try. Each week the city monitors for *Cryptosporidium* and has set a zero standard for the parasite. It has also adopted a turbidity standard five times tougher than federal regulations. Turbidity, although harmless in itself, is often a precursor to the presence of organisms such as *Cryptosporidium.*

Incidence of Disease Caused by Tainted Water— CDC/EPA Surveillance Report

It is difficult to know how many illnesses are caused by contaminated water. People may not know the source of many illnesses and may attribute them to food (which may also have been in contact with polluted water), chronic illness, or other infectious agents. The EPA has noted that some researchers think that the actual number of drinking-water-related diseases may be 25 times the reported number. They believe most are not reported because victims believe them to be "stomach upsets" and simply treat themselves.

Since 1971 the CDC and the EPA have collected and reported data that relate to waterborne-disease outbreaks. The latest report, *Surveillance for Waterborne-Disease Outbreaks—United States, 1997–1998* (May 2000), includes data about outbreaks associated with both drinking water and recreational water.

DRINKING WATER. In 1997–98, 13 states reported a total of 17 outbreaks of disease associated with drinking water. Those outbreaks caused an estimated 2,038 persons to become ill. No deaths were reported. Florida reported the most outbreaks (four), followed by New Mexico (two). Eleven other states reported one outbreak each. The number of waterborne-illness outbreaks increased in the summer and early fall months, especially May, June, and July. (See Figure 6.19.)

FIGURE 6.19

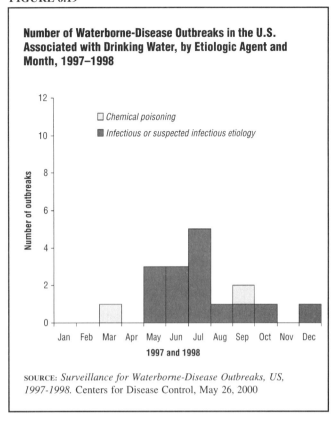

Number of Waterborne-Disease Outbreaks in the U.S. Associated with Drinking Water, by Etiologic Agent and Month, 1997–1998

SOURCE: *Surveillance for Waterborne-Disease Outbreaks, US, 1997-1998.* Centers for Disease Control, May 26, 2000

FIGURE 6.20

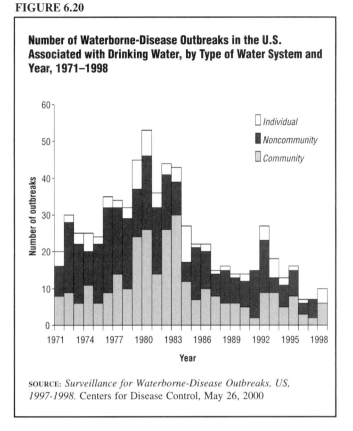

Number of Waterborne-Disease Outbreaks in the U.S. Associated with Drinking Water, by Type of Water System and Year, 1971–1998

SOURCE: *Surveillance for Waterborne-Disease Outbreaks, US, 1997-1998.* Centers for Disease Control, May 26, 2000

TABLE 6.4

Waterborne-Disease Outbreaks in the U.S. Associated with Drinking Water, by Etiologic Agent and Type of Water System, 1997–1998

| | Type of water system* | | | | | | | |
| | Community | | Noncommunity | | Individual | | Total | |
Etiologic agent	Outbreaks	Cases	Outbreaks	Cases	Outbreaks	Cases	Outbreaks	Cases
AGI[†]	1	10	3	148	1	5	5	163
Copper	2	37	0	0	0	0	2	37
Cryptosporidium parvum	1	1,400	0	0	1	32	2	1,432
Escherichia coli O157:H7	1	157	1	4	1	3	3	164
Giardia intestinalis	2	57	1	100	1	2	4	159
Shigella sonnei	1	83	0	0	0	0	1	83
Total (%)	8	1,744	5	252	4	42	17	2,038
	(47.1%)	(85.6%)	(29.4%)	(12.4%)	(23.5%)	(2.1%)	(100.0%)	(100.0%)

*Community and noncommunity water systems are public water systems that serve ≥15 service connections or an average of ≥25 residents for ≥60 days/year. A community water system serves year-round residents of a community, subdivision, or mobile home park with ≥15 service connections or an average of ≥25 residents. A noncommunity water system can be nontransient or transient. Nontransient systems serve ≥25 of the same persons for >6 months of the year (e.g., factories or schools), whereas transient systems do not (e.g., restaurants, highway rest stations, or parks). Individual water systems are small systems not owned or operated by a water utility that serve <15 connections or <25 persons.

[†]Acute gastrointestinal illness of unknown etiology.

SOURCE: *Surveillance for Waterborne-Disease Outbreaks, US, 1997-1998.* Centers for Disease Control, May 26, 2000

The number of waterborne-disease outbreaks reported for 1996–98 was lower than that reported for any period since 1971. Outbreaks peaked during 1979–83. (See Figure 6.20.) The largest number of people (1,744) made ill by a waterborne disease contracted *Cryptosporidium* from a community water system. Another 148 people experienced acute gastrointestinal illness (AGI) by ingesting water from a noncommunity water system,

while 100 others contracted *Giardia* from a noncommunity system. (See Table 6.4).

RECREATIONAL WATER. People also become ill from exposure to contaminated water while involved in various kinds of recreational activities. During 1997–98 a total of 18 states reported 32 outbreaks associated with recreational water. These outbreaks affected a total of 2,128 people.

TABLE 6.5

Opinion Regarding Safety of Drinking Water

Do you think the drinking water which comes from the tap in your home is safe to drink or not safe to drink? Based on half sample; n = 494; margin of error = +/- 5 percentage points.

98 Jun 5-7	Total	Urban	Suburban	Rural
Yes, safe	68%	50%	72%	79%
No, not safe	30%	46%	26%	20%
No opinion	2%	5%	2%	*
	100%	100%	100%	100%

* Less than 0.5%

SOURCE: The Gallup Poll Monthly, June 1998. Gallup Organization, 2000

By year this was the highest number of outbreaks associated with recreational water since 1991. (See Figure 6.21.)

Of the cases where the cause was known, 56.6 percent occurred in treated water while 44.4 percent occurred in fresh water. In treated water most cases (80 percent) were caused by *Cryptosporidium* and the remainder by *E. coli* and *Shigella* (10 percent each). In fresh water AGI accounted for 37.5 percent of the outbreaks, *E. coli* 25 percent, Norwalk-like virus 25 percent, and *Cryptosporidium* 12.5 percent. (See Figure 6.22.)

Americans Are Concerned about Their Drinking Water

Drinking water safety is a growing public concern. A 1998 Gallup Poll asked if drinking water from the tap was safe. While 68 percent of respondents believed it was, 30 percent did not. Only 50 percent of people living in urban areas thought their tap water was safe, compared with 79 percent of people living in rural areas and 72 percent of people in suburban areas. (See Table 6.5.) Seventy-nine percent of respondents reported they had never received a warning about safety problems with their drinking water supply. (See Table 6.6.)

Tampa's Plan

In 1997 many of the citizens of Tampa, Florida, were shocked to find out about the city's plan to recycle wastewater—treating wastewater from toilets, garbage disposals, and showers, and returning it as drinking water. (Before this, treated wastewater drained into Tampa Bay.) This was not, however, a novel concept. Prince William County, Virginia, has been recycling its own treated

FIGURE 6.21

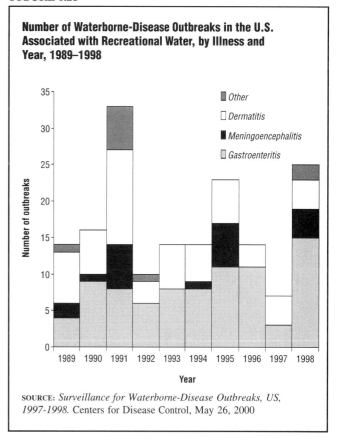

Number of Waterborne-Disease Outbreaks in the U.S. Associated with Recreational Water, by Illness and Year, 1989–1998

SOURCE: *Surveillance for Waterborne-Disease Outbreaks, US, 1997-1998.* Centers for Disease Control, May 26, 2000

TABLE 6.6

Prevalence of Safety Problem Alerts Concerning Drinking Water Supplies

Have you ever received a notice or heard a community alert concerning safety problems with your drinking water supply?

Yes	21%
No	79
No opinion	0
	100%

SOURCE: The Gallup Poll Monthly, June 1998. Gallup Organization, 2000

wastewater since the 1970s. In 1999 San Diego began construction of a $154 million treatment plant that would convert 30 million gallons of wastewater a day into drinking water. The treatment process forces wastewater through membranes so fine that no known viruses, bacteria, or other pathogens can get through. As a backup the water would also be disinfected.

FIGURE 6.22

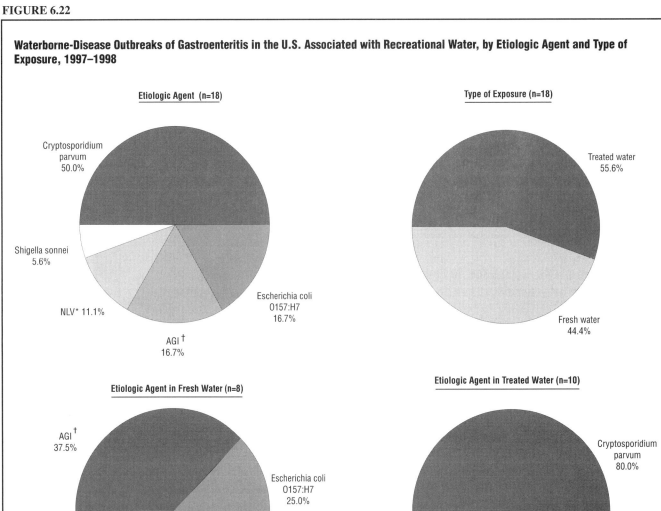

Waterborne-Disease Outbreaks of Gastroenteritis in the U.S. Associated with Recreational Water, by Etiologic Agent and Type of Exposure, 1997–1998

Etiologic Agent (n=18)

Cryptosporidium parvum 50.0%

Shigella sonnei 5.6%

NLV* 11.1%

AGI † 16.7%

Escherichia coli O157:H7 16.7%

Type of Exposure (n=18)

Treated water 55.6%

Fresh water 44.4%

Etiologic Agent in Fresh Water (n=8)

AGI † 37.5%

Cryptosporidium parvum 12.5%

NLV* 25.0%

Escherichia coli O157:H7 25.0%

Etiologic Agent in Treated Water (n=10)

Cryptosporidium parvum 80.0%

Escherichia coli O157:H7 10.0%

Shigella sonnei 10.0%

*Norwalk-like virus

†Acute gastrointestinal illness of unknown etiology

SOURCE: Surveillance for Waterborne-Disease Outbreaks, US, 1997-1998, CDC, May 26, 2000

CHAPTER 7
ACID RAIN

WHAT IS ACID RAIN?

Acid rain is any form of precipitation that contains an abnormally high amount of acid. Industrial plants and automobiles burning fossil fuels emit chemicals such as sulfur dioxide and nitrogen oxides. When these chemicals combine with moisture in the atmosphere, they form sulfuric acid and nitric acid. These toxic compounds eventually fall to the earth in rain, snow, fog, or dust—all of which are called acid rain. Acid rain can create dangerously high levels of acidic impurities in our water, soil, and plants. In nature, the combination of rain and oxides is part of a natural balance that nourishes plants and aquatic life. However, when pollution-causing human activity interferes, the results on humans and the environment can be harmful and destructive. (See Table 7.1.)

The concept of "acid rain" dates from 1872 when the term was coined by English chemist Robert Angus Smith to refer to acidic precipitation in and around the city of Manchester, England. Scientific research on acid rain was sporadic and largely focused on local problems until the late 1960s when Scandinavian scientists began more systematic studies. Acid precipitation in North America was not identified until 1972 when scientists found that precipitation was acidic in eastern North America, especially in the Northeast, and eastern Canada. The 1975 First International Symposium on Acid

TABLE 7.1

Effect of Acid Rain on Human Health and Selected Ecosystems and Anticipated Recovery Benefits

Human health and ecosystem	Effects	Recovery benefits
Human health	In the atmosphere, sulfur dioxide and nitrogen oxides become sulfate and nitrate aerosols, which increase morbidity and mortality from lung disorders, such as asthma and bronchitis, and impacts to the cardiovascular system.	Decrease emergency room visits, hospital admissions, and deaths.
Surface waters	Acidic surface waters decrease the survivability of animal life in lakes and streams and in the more severe instances eliminate some or all types of fish and other organisms.	Reduce the acidic levels of surface waters and restore animal life to the more severely damaged lakes and streams.
Forests	Acid deposition contributes to forest degradation by impairing trees' growth and increasing their susceptibility to winter injury, insect infestation, and drought. It also causes leaching and depletion of natural nutrients in forest soil.	Reduce stress on trees, thereby reducing the effects of winter injury, insect infestation, and drought, and reduce the leaching of soil nutrients, thereby improving overall forest health.
Materials	Acid deposition contributes to the corrosion and deterioration of buildings, cultural objects, and cars, which decreases their value and increases costs of correcting and repairing damage.	Reduce the damage to buildings, cultural objects, and cars, and reduce the costs of correcting and repairing future damage.
Visibility	In the atmosphere, sulfur dioxide and nitrogen oxides form sulfate and nitrate particles, which impair visibility and affect the enjoyment of national parks and other scenic views.	Extend the distance and increase the clarity at which scenery can be viewed, thus reducing limited and hazy scenes and increasing the enjoyment of national parks and other vistas.

SOURCE: *Acid Rain Emissions and Effects.* RCED-00-47. U.S. General Accounting Office

FIGURE 7.1

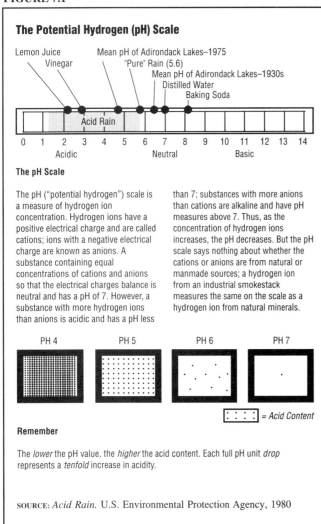

The Potential Hydrogen (pH) Scale

The pH Scale

The pH ("potential hydrogen") scale is a measure of hydrogen ion concentration. Hydrogen ions have a positive electrical charge and are called cations; ions with a negative electrical charge are known as anions. A substance containing equal concentrations of cations and anions so that the electrical charges balance is neutral and has a pH of 7. However, a substance with more hydrogen ions than anions is acidic and has a pH less than 7; substances with more anions than cations are alkaline and have pH measures above 7. Thus, as the concentration of hydrogen ions increases, the pH decreases. But the pH scale says nothing about whether the cations or anions are from natural or manmade sources; a hydrogen ion from an industrial smokestack measures the same on the scale as a hydrogen ion from natural minerals.

Remember

The *lower* the pH value, the *higher* the acid content. Each full pH unit *drop* represents a *tenfold* increase in acidity.

SOURCE: *Acid Rain.* U.S. Environmental Protection Agency, 1980

FIGURE 7.2

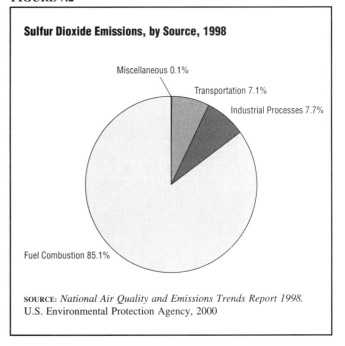

Sulfur Dioxide Emissions, by Source, 1998

SOURCE: *National Air Quality and Emissions Trends Report 1998.* U.S. Environmental Protection Agency, 2000

Precipitation and the Forest Ecosystem (Columbus, Ohio) helped scientists define the acid rain problem and initiated further research.

Acid deposits, both dry and in precipitation, are the most common human-produced causes of water acidification. In cold parts of the country, for example, pollutants become concentrated in upper layers of the snowpack. During spring snowmelt, runoff containing large amounts of acidic particles accumulated over the winter can flow into a lake or river, causing acid shock to its aquatic inhabitants. High levels of acidity can also cause the release of aluminum and manganese particles stored in a lake or river bottom.

The states bordering, and east of, the Mississippi River contain approximately 17,000 lakes and 112,000 miles of streams. An estimated 25 percent of the land contains soil and bedrock that allow acidity to travel through underground water to these lakes and streams. Approximately half of these bodies of water have such a limited ability to neutralize acid that acid-laden pollutants will eventually cause acidification.

Measuring Acid Rain

The acidity of any solution is measured on a pH (potential hydrogen) scale numbered from 0 to 14, with a pH value of 7 considered neutral. Values higher than 7 are considered more alkaline or basic (the pH of baking soda is 8); values lower than 7 are considered acidic (the pH of lemon juice is 2). Pure, distilled water has a pH level of 7. Rainfall, which normally has a pH value of 5.65, is not pure because it accumulates naturally occurring sulfur oxides (SO_x) and nitrogen oxides (NO_x) as it passes through the atmosphere. The pH scale is a logarithmic measure. This means that every pH drop of 1 is a *tenfold* increase in acid content. Therefore, a decrease from pH 6 to pH 5 is a tenfold increase in acidity; a drop from pH 6 to pH 4 is a hundredfold increase in acidity; and a drop from pH 6 to pH 3 is a thousandfold increase. (See Figure 7.1.)

Although pH levels vary considerably from one body of water to another, a "normal" pH value for the lakes and rivers in the United States is approximately 6. Acid rain has a pH of about 4 although it may be higher—as it is in Scandinavia where it is about 4.4.

SOURCES OF SULFATE AND NITRATE IN THE ATMOSPHERE

Natural Causes

Natural causes of sulfate (sulfur oxides) in the atmosphere include ocean spray, volcanic emissions, and readily oxidized hydrogen sulfide released from the decomposition of organic matter found in the earth. Natural sources of nitrogen or nitrates include nitrogen oxides produced by microorganisms in soils, by lightning during thunderstorms, and by forest fires. Scientists generally specu-

FIGURE 7.3

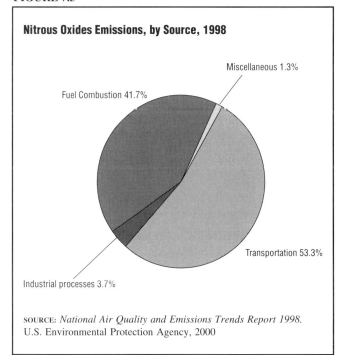

Nitrous Oxides Emissions, by Source, 1998

Miscellaneous 1.3%

Fuel Combustion 41.7%

Transportation 53.3%

Industrial processes 3.7%

SOURCE: *National Air Quality and Emissions Trends Report 1998.* U.S. Environmental Protection Agency, 2000

FIGURE 7.4

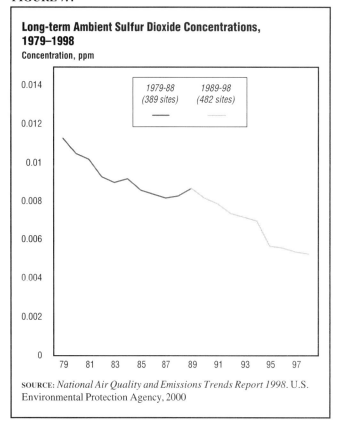

Long-term Ambient Sulfur Dioxide Concentrations, 1979–1998

Concentration, ppm

1979-88 (389 sites)   1989-98 (482 sites)

SOURCE: *National Air Quality and Emissions Trends Report 1998.* U.S. Environmental Protection Agency, 2000

late—a rough estimate since there is no way to measure this—that one-third of the sulfur and nitrogen emissions in the United States comes from these natural sources.

Sources Caused by Human Activity

Most human-produced emissions of sulfur oxides (SO_x) and nitrogen oxides (NO_x)—the two main components of acid rain—are the result of burning fossil fuels (oil and gas) for energy. This includes fossil-fueled electric utilities and industrial plants, motor vehicles using gasoline or diesel fuel, and commercial or residential heating. Non-energy sources of emission include metal smelters that emit sulfur compounds, and nitrogen compounds from agricultural fertilizers carried by the wind.

Nitrogen pollution of the waters had been blamed primarily on surface runoff from fertilizer, animal wastes, sewage, and industrial wastes. It is now believed that airborne nitrates produced by human activities account for one-fourth of all nitrogen, the second most prevalent cause after fertilizers. Not only has the Chesapeake Bay area been affected, but preliminary findings from the Long Island Sound and the lower Neuse River in North Carolina indicate airborne nitrates have caused similar problems in those eastern estuaries.

Levels of pollutants are measured in two ways—emissions and concentrations. Emissions are pollutants expelled into the air by a particular source. Fuel combustion accounted for 85.1 percent of total SO_2 emissions in 1998 while transportation and industrial processes accounted for just over 7 percent each. (See Figure 7.2.) Transportation caused 53.3 percent of NO_x emissions,

fuel combustion 41.7 percent, and industrial processes only 3.7 percent. (See Figure 7.3.)

Concentrations are the total saturation of a contaminant over time. There has been great progress in reducing SO_2 concentration. Between 1979 and 1998 the average annual mean concentration of SO_2 was lowered by 53 percent. (See Figure 7.4.) Annual mean NO_2 concentrations declined in the early 1980s, were relatively stable during the mid- to late 1980s, and resumed their decline in the 1990s. Concentrations of NO_2 have declined 25 percent since 1979. (See Figure 7.5.) Annual emissions of both peaked in the early 1970s.

FACTORS CONTRIBUTING TO ACID RAIN

Several factors contribute to the impact of acid rain on an area. Sulfur and nitrogen oxides carried in the air or precipitation can seriously affect the water quality of a region. (See Figure 7.6.)

Climate

In drier climates such as those experienced in the western United States, windblown alkaline dust moves more freely through the air and tends to neutralize atmospheric acidity. (The effects of acid rain can be greatly reduced by the presence of basic (also called alkali) substances. Sodium, potassium, and calcium are examples of basic chemicals. When a basic and an acid chemical come into contact

FIGURE 7.5

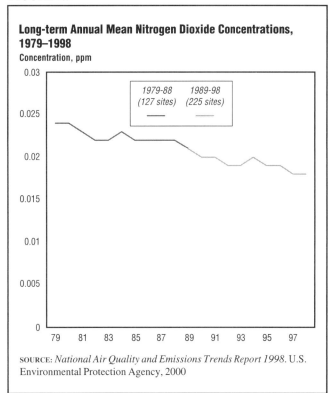

Long-term Annual Mean Nitrogen Dioxide Concentrations, 1979–1998

Concentration, ppm

| 1979-88 | 1989-98 |
| (127 sites) | (225 sites) |

SOURCE: *National Air Quality and Emissions Trends Report 1998*. U.S. Environmental Protection Agency, 2000

they react chemically and neutralize each other.) On the other hand, in more humid climates where there is less dust—such as along the eastern seaboard—precipitation is more acidic. The season of the year also determines the extent of acid rain damage. For instance acids will have a more harmful effect if rainfall occurs during periods when fish are spawning or seed is germinating.

Topography/Geology

Areas most sensitive to acid rain contain hard, crystalline bedrock and very thin surface soils. When no alkaline-buffering particles are in the soil, runoff from rainfall directly affects surface waters such as mountain streams. In contrast a thick soil covering or soil with a high buffering capacity, such as flat land, neutralizes acid rain better. Lakes tend to be most susceptible to acid rain because of low alkaline content in lakebeds. Lake depth, its watershed (the area draining into the lake), and the amount of time the water has been in the lake, are also factors.

TRANSPORT—THE LINK BETWEEN EMISSIONS AND DEPOSITION

Transport systems—primarily the movement of air— distribute acid emissions in definite patterns around the planet. The movement of air masses transports emitted pollutants many miles, during which time these pollutants are transformed into sulfuric and nitric acid by mixing with clouds of water. This process is known as "transport and transformation." (See Figure 7.7.)

In the United States a typical transport pattern occurs from the Ohio River Valley to the northeastern United States and southeastern Canada, since prevailing winds tend to move from west to east and from south to north. About one-third of the total sulfur compounds deposited over the eastern United States originates from sources in the Midwest more than 300 miles away.

The lakes and forests in and near the Adirondack Mountains in upstate New York are an example of what occurs in areas that do not have carbonate rock to quickly neutralize the acid. Approximately half the lakes above the altitude of 2,000 feet have a pH of less than 5.0. Ninety percent of these lakes contain no aquatic life.

In Europe a typical transport pattern carries pollutants from the smokestacks of the United Kingdom over Sweden. In southwestern Germany many trees of the famed Black Forest are dying from the effects of acid rain transported to the region by wind. Germans have coined a word for the phenomenon, "waldsterben," which means "forest death."

Acid Rain in Asia

Acid rain is a growing problem in Asia. According to International Institute for Applied Systems Analysis (ILASA), SO_2 emissions in Asia are surpassing those in Europe and North America. In China acid rain is implicated in large die-offs in southwestern forests. Crops are at risk as well. A study by China's National Environment Protection Agency found that farmland also is affected by acid rain. The World Bank, in *Clear Water, Blue Skies: China's Environment in the New Century*, (Washington, D.C., 1997), estimated annual forest and crop losses at $5 billion. Researchers believe it will be difficult to control pollution from China, the world's biggest consumer of coal, as that nation goes through an accelerated economic expansion that involves growing coal consumption.

Atmospheric acidity levels are highest in the winter and early spring. During this time huge air masses from continental Asia move to Japan propelled by the prevailing monsoons. As in other countries forests in Japan have experienced abnormally high death rates, particularly in stands of red pine and Japanese cedar.

Scientists estimate that more than one-third of Japan's sulfur depositions comes from China. Japanese laws governing the emission of gases that acidify rain are among the strictest in the world. Nonetheless, Japan's rain is increasingly acidic. In Kawasaki, where NO_x levels are posted every day outside City Hall, the rain is sometimes as acidic as grapefruit juice.

EFFECTS OF ACID RAIN ON AQUATIC SYSTEMS

Low pH levels kill fish eggs, frog eggs, and fish food organisms. The degree of damage depends upon the

FIGURE 7.6

Direct and Indirect Pollutant Deposition

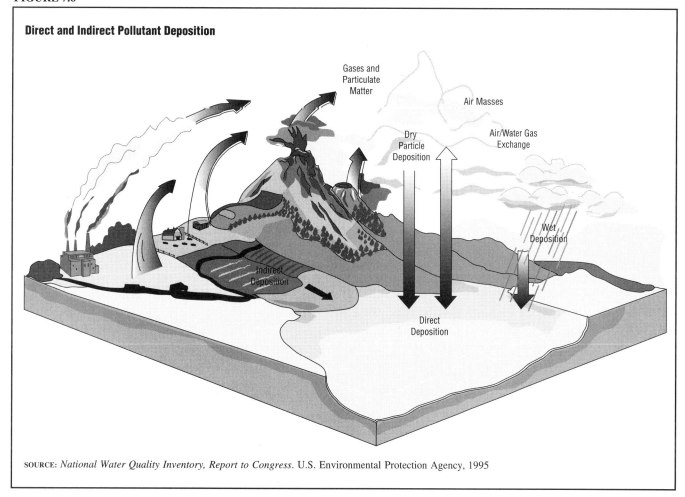

SOURCE: *National Water Quality Inventory, Report to Congress.* U.S. Environmental Protection Agency, 1995

buffering capacity of the watershed soil; the higher the alkalinity, the more slowly the lakes and streams acidify. The death of fish in acidified freshwater lakes and streams has been intensely studied. Since the 1970s aquatic life has increasingly suffered because of:

• Sudden, short-term shifts in pH levels resulting in acid shock for the fish population. For example spring snowmelts release acidic materials accumulated during the winter. At pH levels below 4.9 damage occurs to fish eggs. At acid levels below 4.5 some species of fish die. Below 3.5 most fish die within hours. (See Table 7.2.)

• Gradual decreases of pH levels over time affecting fish reproduction and spawning. Moderate levels of acidity in water can confuse a salmon's sense of smell, which it uses to find the stream from which it came. Atlantic salmon are unable to find their home streams and rivers because of acid rain. In addition excessive acid levels in female fish cause low amounts of calcium preventing the production of eggs. Even if eggs are produced their development is often abnormal. Over time the fish population decreases while the remaining fish population becomes older and larger.

In 1988 the Environmental Defense Fund (EDF), an environmental watch-group, sounded one of the first alarms that the coastal waters of the eastern United States were receiving large inputs of nitrogen. The nitrogen led to an excessive growth of algae on the surface of the water. This resulted in the loss of oxygen and light to the water and the long-term decline of marine life. The EDF concluded that the major sources of the nitrogen were human activities—the runoff of fertilizer, animal waste from farms, and discharge from sewage treatment plants and industrial facilities. Researchers also noted that the significant decline of the Chesapeake Bay and other estuaries could also be attributed to the increase in nitrogen oxides from automobiles and electric power plants, along with toxic chemicals, pesticides, and wetland destruction.

When spring thaws begin or rainstorms sweep an area, residual acid is washed into streams and lakes. Mountainous streams in New York, North Carolina, Pennsylvania, Tennessee, and Arkansas have shown increased acidity during rainstorms and snowmelts of between three to 20 times that experienced during the rest of the year. Because many species of fish hatch in the spring, even mild increases in acidity can harm or kill the new life. If levels are high enough even adult fish can die. Temporary

FIGURE 7.7

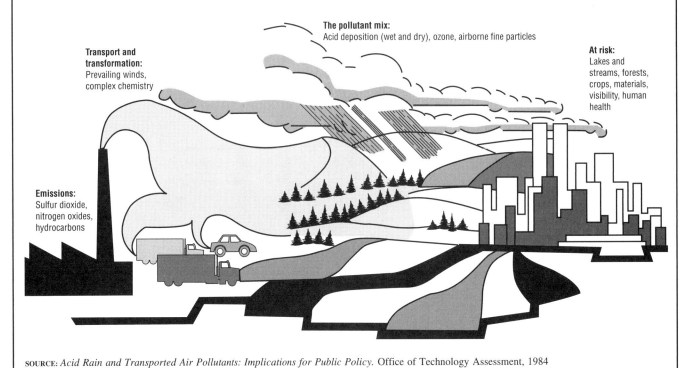

Transported Air Pollutants: Emissions to Effects

The transported pollutants considered in this study result from emissions of three pollutants: sulfur dioxide, nitrogen oxides, and hydrocarbons. As these pollutants are carried away from their sources, they form a complex "pollutant mix" leading to acid deposition, ozone, and airborne fine particles. These transported air pollutants pose risks to surface waters, forests, crops, materials, visibility, and human health.

The pollutant mix:
Acid deposition (wet and dry), ozone, airborne fine particles

Transport and transformation:
Prevailing winds, complex chemistry

At risk:
Lakes and streams, forests, crops, materials, visibility, human health

Emissions:
Sulfur dioxide, nitrogen oxides, hydrocarbons

SOURCE: *Acid Rain and Transported Air Pollutants: Implications for Public Policy.* Office of Technology Assessment, 1984

TABLE 7.2

Generalized Short-Term Effects of Acidity on Fish

pH range	Effect
6.5-9	No effect
6.0-6.4	Unlikely to be harmful except when carbon dioxide levels are very high (>1000 mg l^{-1})
5.0-5.9	Not especially harmful except when carbon dioxide levels are high (>20 mg l^{-1}) or ferric ions are present
4.5-4.9	Harmful to the eggs of salmon and trout species (salmonids) and to adult fish when levels of Ca^{2+}, Na^+ and Cl^- are low
4.0-4.4	Harmful to adult fish of many types which have not been progressively acclimated to low pH
3.5-3.9	Lethal to salmonids, although acclimated roach can survive for longer
3.0-3.4	Most fish are killed within hours at these levels

SOURCE: *National Water Quality Inventory, Report to Congress.* U.S. Environmental Protection Agency, 1998

---

increases in acidity also affect insects and other invertebrates such as snails and crayfish that the fish feed on.

During the 1990s acid and polluted waters caused the disappearance of many aquatic species, leaving gaping holes in the food chain and diminishing forever the bio-

logical balance and diversity that keeps Earth genetically healthy. According to the American Fisheries Society and the Environmental Protection Agency (EPA), many species of freshwater fish have become extinct since the late 1970s, and additional species have become endangered, threatened, or listed as "of special concern" for their ultimate survival.

OTHER DAMAGE CAUSED BY ACID RAIN

Soil and Vegetation

Acid precipitation can cause leafy plants such as lettuce to hold increased amounts of potentially toxic substances such as the mineral cadmium. Research has also found a decrease in carbohydrate production in the photosynthesis process of some plants exposed to acid conditions. Some researchers believe that acid rain disrupts soil regeneration, which is the recycling of chemical and mineral nutrients through plants and animals back to the earth. They also believe acids suppress decay of organic matter. Valuable nutrients like calcium and magnesium are normally bound to soil particles and are, therefore, protected from being rapidly washed into ground water. Acid rain, however, may accelerate the process of breaking these bonds to rob the soil of these nutrients.

Research is underway to determine whether acid rain could ultimately lead to a permanent reduction in tree growth, food crop production, and soil quality. Effects on soils, forests, and crops are difficult to measure because of the numerous species of plants and animals, the slow rate at which ecological changes occur, and the complex interrelationships between plants and their environment.

A national survey of soils conducted for the government of the United Kingdom by the Institute of Terrestrial Ecology in Britain in 1994 pinpointed the "critical loads" of acid rain for the main types of soil in England. Maps of ecological damage to British soils showed the worst-hit areas to be in southern England, including the New Forest in Hampshire and Ashdown Forest in Sussex. Britain is committed by a European directive to reduce its emissions of sulfur dioxide from power stations to 60 percent below the 1980 level by the year 2003.

TREES. The effect of acid rain on trees is influenced by many factors. Some trees adapt to environmental stress better than others; the type of tree, its height, and its leaf structure (deciduous or evergreen) influence how well it will adapt to acid rain. Acid rain may affect trees in at least two ways: in areas with high evaporation rates, acids will concentrate on leaf surfaces; in regions where a dense leaf canopy does not exist, more acid may seep into the earth to affect the soil around the tree's roots.

The study of the effects of acid fog on trees is a relatively new scientific endeavor. In fog conditions the concentration of acid and sulfate in fog droplets is much greater than in rainfall. In areas of frequent fog, such as London, significant damage has occurred to trees and other vegetation because the fog condenses directly on the leaves.

A 1994 joint report of the European Commission and the United Nations Economic Commission for Europe (UNECE) surveyed 102,300 trees at 26,000 sampling plots in 35 European countries and found that almost one-quarter of the trees in Europe were defoliated by more than 25 percent. The report showed that forest damage is a problem in virtually all European countries. The most severely affected country was the Czech Republic, where 53 percent of all trees suffered moderate or severe defoliation or have died. The least affected was Portugal where 7.3 percent of trees were damaged.

Thomas Cahill and Robert A. Eldred, scientists at the University of California at Davis, studied air quality in 12 U.S. national parks from 1982 to 1992. They found, from more than 12,000 measurements of air samples, that levels of sulfate in the air over Eastern parks increased by nearly 40 percent over ten years. This, despite the fact that sulfur emissions remained relatively constant over the preceding decade, suggested that more stringent controls were needed. The Great Smoky Mountains National Park (North Carolina and Tennessee) experienced the largest increase

at 39 percent. Other parks showing increases were Shenandoah National Park (Virginia), Glacier National Park (Montana), Bryce Canyon National Park (Utah), Big Bend National Park (Texas), Grand Canyon National Park (Arizona), Yosemite National Park (California), and Guadalupe Mountains National Park (Texas).

In March 1999 the U.S. Geological Survey (USGS), in *Soil Calcium Depletion Linked to Acid Rain and Forest Growth in the Eastern United States*, reported that calcium levels in forest soils had declined at locations in ten states in the eastern United States. Calcium is necessary to neutralize acid rain and is an essential nutrient for tree growth. Sugar maple and red spruce trees, in particular, showed reduced resistance to stresses such as insect defoliation and low winter temperatures. Although the specific relationships among calcium availability, acid rain, and forest growth are uncertain, Gregory Lawrence, scientist and co-author of the report, speculated, "Acid rain releases aluminum from the underlying mineral soil layer, which is followed by the upward transport of the aluminum into the forest floor by root uptake and water movement. The result is that aluminum replaces calcium, and the trees have a harder time trying to get the needed calcium from the soil layer."

Birds

Increased freshwater acidity harms some species of migratory birds. Experts believe the dramatic decline of the North American black duck population since the 1950s is due to decreased food supplies in the acidified wetlands. The U.S. Fish and Wildlife Service reports that ducklings in wetlands created by humans in Maryland are three times more likely to die before adulthood if raised in acidic waters.

Acid rain leaches calcium out of the soil and robs snails of the calcium they need to form shells. Because titmice and other species of songbirds get most of their calcium from the shells of snails, the birds are also perishing. The eggs they lay are defective—thin and fragile. The chicks either do not hatch or have bone malformations and die.

EFFECTS OF ACID RAIN ON OBJECTS BUILT BY HUMANS

Investigations into the effects of acid rain on objects such as stone buildings, marble statues, metals, and paints, only began in the 1990s. A joint study conducted by the EPA, the Brookhaven National Laboratory, and the Army Corps of Engineers in 1993 found that acid rain caused $5 billion worth of damage annually in a 17-state region. Two-thirds of the damage was created by pollution whose source was less than 30 miles away.

Studies by the U.S. National Park Service were well underway by the year 2000 to determine damage to monuments and statues. (See Figure 7.8.) Acid rain is suspect-

FIGURE 7.8

Effects of Acid Rain on Statues and Monuments

1. Acid rain, or dry deposition falls

2. Crust forms

1. Crust washes off

4. Layer of stone is removed

SOURCE: *Acid Rain and Transported Air Pollutants: Implications for Public Policy.* Congress of the United States, Office of Technology Assessment: Washington, D.C. 1984

ed, in part, of damaging the Statue of Liberty and the Egyptian pyramids. Examination of the seven-centuries-old, 37-foot-tall, bronze Great Buddha of Kamakura, an important symbol of Japanese culture, shows pock marks and rust stains, the result of acid rain.

New kinds of protective chemicals that adhere to limestone and marble—consolidants—are helping to save some of the world's decomposing monuments from acid rain and other pollutants. Consolidants were developed in the 1960s in response to widespread water damage to stone buildings in Venice. Among the monuments getting close attention are the Taj Mahal in India; the Parthenon in Rome, Italy; the Lincoln Memorial in Washington,

D.C.; and the Alamo in Texas. Experts report, however, that these chemicals have many limitations. They are toxic and difficult to apply, and their effects are only temporary yet they permanently alter the nature of the stone. Most important is that their long-term effects are uncertain. For those reasons their use was banned on the Acropolis in Athens, Greece.

AUTOMOTIVE COATINGS. Over the past decade reports of damage to automotive coverings have increased. The general consensus within the automobile industry is that the damage is caused by some form of "environmental fallout," the term used in the auto industry. Carmakers suspect acid rain damage to automobile paint, especially to those cars having clear protectant overcoats used on many newer models. Chemical analyses of the damaged areas of some car finishes have showed elevated levels of sulfate, implicating acid rain.

The auto industry began using clear coat finishes in the mid-1980s. Although the new high gloss paints look better, complaints are mounting over marred surfaces, especially on dark-colored or metallic cars in the northeastern and southeastern United States. Automakers believe that, when acid rain falls on autos, the moisture evaporates leaving a permanent blemish caused by sulfuric acid and nitric acid—the composition of acid rain. Some car dealers now offer optional protective sealants at added expense to buyers. Higher-priced cars often include protective sealants in the purchase price.

Human Health

Extremely small particles of sulfates in air pollution can threaten human health. These small particles can travel long distances and, when inhaled, penetrate deep into the lungs. Studies of death rates across the United States have found some correlation between elevated mortality levels and high sulfate levels. Acid rain and the pollutants that cause it cause bronchitis and asthma in children. It is also believed responsible for increasing health risks to those over age 65; those with asthma, chronic bronchitis, and emphysema; pregnant women; and those with histories of heart disease.

THE POLITICS OF ACID RAIN

The early acid rain debate centered almost exclusively in the eastern portion of the United States and Canada. The controversy was often defined as a problem of property rights. The highly valued production of electricity in coal-fired utilities in the Ohio River Valley caused acid rain to fall on land in the Northeast and Canada. An important part of the acid rain controversy in the 1980s was the adversarial relationship between U.S. and Canadian federal governments over emission controls of SO_2 and NO_2. More of these pollutants crossed the border into Canada than the reverse. Canadian officials very quickly came to a consensus over the need for more stringent controls, while this consensus was lacking in the United States.

Throughout the 1980s the major lawsuits involving acid rain all came from eastern states, and the states that passed their own acid rain legislation were from the eastern part of the United States. There has been a clear difference in the intensity of interest between the eastern and western states regarding acid rain.

Legislative attempts to restrict emissions of pollutants were often defeated after strong lobbying by the coal industry and utility companies. Those industries advocated further research for pollution-control technology rather than placing restrictions on utility company emissions.

A National Debate

In 1975 the First International Symposium on Acid Precipitation and the Forest Ecosystem convened in Columbus, Ohio, to define the acid rain problem. Scientists used the meeting to propose a precipitation-monitoring network in the United States to cooperate with the European and Scandinavian networks and to set up protocols for collecting and testing precipitation.

In 1977 the Council on Environmental Quality was asked to develop a national acid rain research program. Several scientists drafted a report that eventually became the basis for the National Acid Precipitation Assessment Program (NAPAP). Carter's initiative eventually translated into legislative action with the Energy Security Act (PL 96-264) in June 1980. Title VII of the Act (the Acid Precipitation Act of 1980) produced a formal proposal that created NAPAP and authorized federally financed support.

The first international treaty aimed at limiting air pollution was the United Nations Economic Commission for Europe (UNECE) Convention on Long-Range Transboundary Air Pollution, which went into effect in 1983. It was ratified by 38 of the 54 UNECE members which included not only European countries but also Canada and the United States. The treaty targeted sulfur emissions requiring that parties reduce emissions 30 percent from 1980 levels—the so-called "30 percent club."

HOW BAD IS ACID RAIN?

National Acid Precipitation Assessment Program (NAPAP)—1988

Following the creation of the NAPAP in 1980 to study the acid rain phenomenon and assess its damage, two thousand scientists worked with an elaborate multimillion-dollar computer model in an eight-year, $570 million undertaking. In 1988 NAPAP produced an overwhelming 6,000-page report on its findings, including:

- Acid rain had adversely affected aquatic life in about 10 percent of eastern lakes and streams.

- Acid rain had contributed to the decline of red spruce at high elevations by reducing that species' cold tolerance.

- Acid rain had contributed to erosion and corrosion of buildings and materials.

- Acid rain and related pollutants had reduced visibility throughout the Northeast and in parts of the West.

The report concluded, however, that the incidence of serious acidification was more limited than originally

FIGURE 7.9

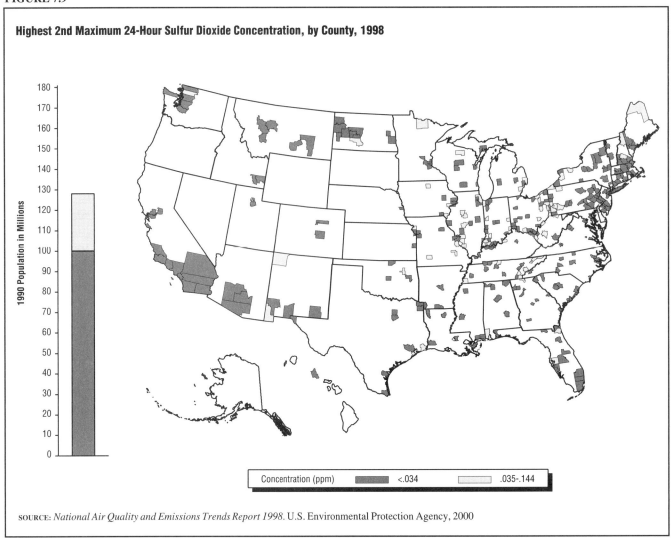

Highest 2nd Maximum 24-Hour Sulfur Dioxide Concentration, by County, 1998

Concentration (ppm)   <.034   .035-.144

SOURCE: *National Air Quality and Emissions Trends Report 1998.* U.S. Environmental Protection Agency, 2000

feared. The Adirondacks area of New York was the only region showing widespread, significant damage from acid at that time.

Update—The U.S. Geological Service Report

The U.S. Geological Service (USGS), in *Trends in Precipitation Chemistry in the United States, 1983–1994: An Analysis of the Effects in 1995 of Phase I of the Clean Air Act Amendments of 1990, Title IV* , found that rainwater tested at 109 test sites across the United States was less acidic in 1995 than in 1983, particularly along the Ohio River Valley and in the Mid Atlantic region. Sulfates had declined at 92 percent of the sites. The USGS attributed the improvement to the standards put in place by the Clean Air Act Amendments (CAAA) Title IV program. For nitrates approximately as many sites showed decreased levels as did the number of sites reporting increased levels. Overall, nitrate levels rose slightly, with the largest increases occurring in the western states.

In its 1998 *National Air Quality and Emissions Trends Report* (2000), the EPA found that approximately

TABLE 7.3

Number of Areas Exhibiting National Air Quality Standards Nonattainment Status, By Pollutant, 1999

Pollutant	Original # areas	1999 # areas	Population 1999 (in 1000s)
CO	43	20	33,230
Pb	12	8	1,116
NO_2	1	0	0
O_3	101	32	92,505
PM_{10}	85	77	29,880
SO_2	51	31	4,371

SOURCE: *National Air Quality and Emissions Trends Report 1998.* U.S. Environmental Protection Agency, 2000

4.3 million persons lived in areas with SO_2 concentrations above national standards in 1998. Because NO_2 had declined to acceptable levels, no persons lived in areas where nitrogen oxides exceeded desired standards. (See Table 7.3.) Figures 7.9 and 7.10 show those areas of the United States where concentrations of sulfur dioxide and nitrogen dioxide were highest.

FIGURE 7.10

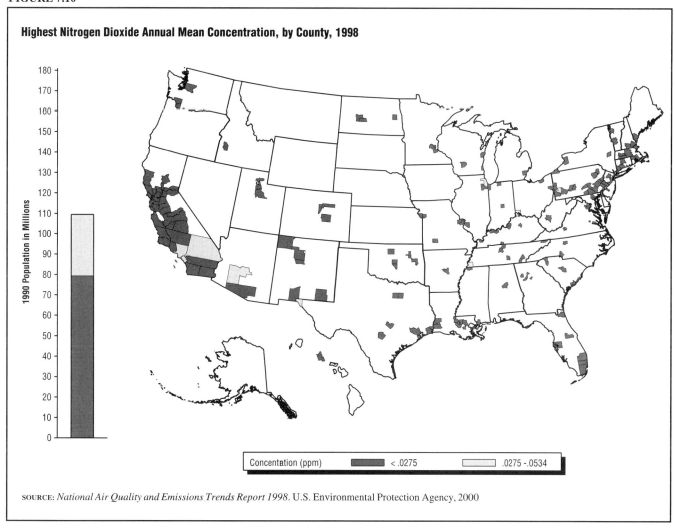

Highest Nitrogen Dioxide Annual Mean Concentration, by County, 1998

Concentration (ppm)   < .0275   .0275 - .0534

SOURCE: *National Air Quality and Emissions Trends Report 1998*. U.S. Environmental Protection Agency, 2000

National Acid Precipitation Assessment Program (NAPAP) 1999—Acid Rain Still a "Serious Problem"

In April 1999 NAPAP released findings from its latest study, *National Acid Precipitation Assessment Program Biennial Report to Congress: An Integrated Assessment*. The study warned that, despite important strides in reducing air pollution, acid rain remained a serious problem in sensitive areas. The report provided more evidence that acid rain is more "complex and intractable than was believed 10 years ago." Among the findings were:

• New York's Adirondack Mountain waterways suffer from serious levels of acid. Even though sulfur levels are declining, nitrogen levels there are climbing. The agency predicted that, by 2040, about half the region's 2,800 lakes and ponds will be too acidic to sustain life.

• The Chesapeake Bay is suffering from excess nitrogen, which is causing algae blooms that suffocate other life forms.

• High elevation forests in Colorado, West Virginia, Tennessee, and Southern California are nearly saturated with nitrogen, a key ingredient in acid rain.

• High elevation lakes and streams in the Sierra Nevadas, the Cascades, and the Rocky Mountains may be on the verge of "chronically high acidity."

The report concluded that further reductions in sulfur and nitrogen would be needed. The report also found, however, that the 1990 CAAA have reduced sulfur emissions and acid rain in much of the United States.

In 1998 the EPA ordered 22 states in the East and Midwest to reduce nitrogen oxide pollution levels which, when accomplished, should lower acid levels even more.

THE ACID RAIN PROGRAM—CLEAN AIR ACT AMENDMENTS, TITLE IV

Title IV of the CAAA (PL 101-549) strives to achieve a 10-million-ton annual reduction in emissions from 1980 levels by the year 2010. Traditionally, environmental regulation has been achieved by the "command and control" approach in which the regulator specifies how to reduce pollution, by what amounts, and what technology to use. Title IV, however, gave utilities flexibility in choosing how to achieve these reductions. For

FIGURE 7.11

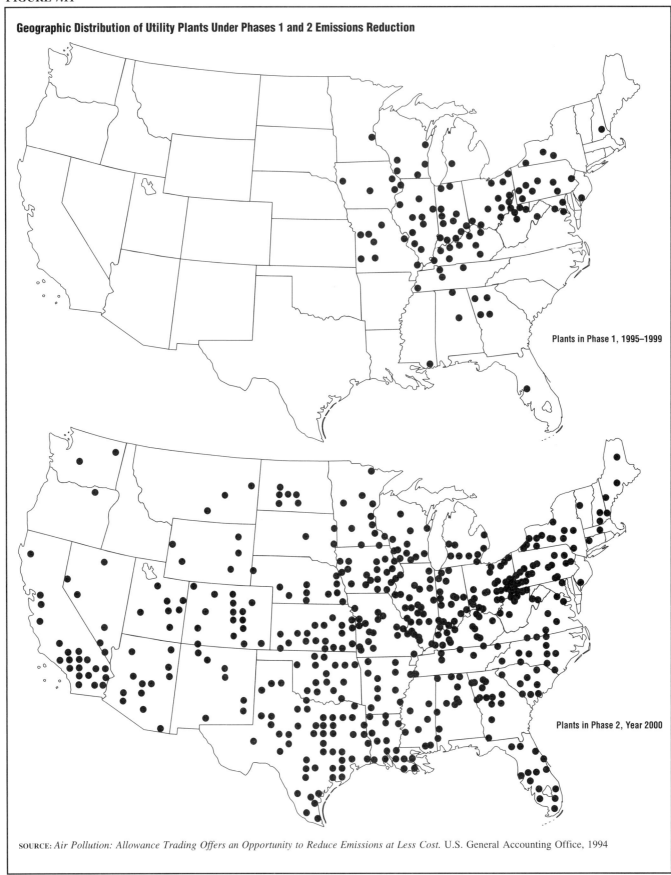

Geographic Distribution of Utility Plants Under Phases 1 and 2 Emissions Reduction

Plants in Phase 1, 1995–1999

Plants in Phase 2, Year 2000

SOURCE: *Air Pollution: Allowance Trading Offers an Opportunity to Reduce Emissions at Less Cost.* U.S. General Accounting Office, 1994

example, utilities may reduce emissions by switching to low-sulfur coal, installing pollution control devices called scrubbers, or shutting down plants.

Targeting the Electric Utilities

Title IV introduced a new regulatory approach to reduce acid rain—allowing electric utilities to trade allowances to emit sulfur dioxide (SO_2). Utilities that reduce their emissions below the required levels can sell their extra allowances to other utilities to help them meet their requirements. Because electric utilities are the source of 70 percent of SO_2 emissions and 30 percent of nitrogen oxide emissions, the act targeted emissions from electric utilities. Of the desired 10-million-ton reduction in SO_2, 8.5 million tons is to come from electric utilities.

The reduction was implemented in two phases. In Phase 1 the 263 units at 110 utility plants in 21 states with the highest levels of emissions were mandated to reduce their annual emissions by 3.5 million tons beginning January 1995. (See Figure 7.11.) An additional 182 units joined Phase 1 voluntarily, bringing the total of Phase 1 units to 445.

Phase 2, which began January 1, 2000, affected an additional 2,000 cleaner and smaller units in all 48 contiguous states and the District of Columbia. (See Figure 7.11.)

Allowance Trading

Title IV allows companies to buy, sell, trade, and bank pollution rights. Utility units are allocated allowances based on their historic fuel consumption and a specific emissions rate. Each allowance permits a unit to emit 1 ton of SO_2 during or after a specific year. For each ton of SO_2 discharged in a given year, one allowance is retired and can no longer be used. Companies that pollute less than the set standards will have allowances left over. They can then sell the difference to companies that pollute more than they are allowed, bringing them into compliance with overall standards. Companies that can clean up their pollution less expensively by changing fuel or persuading their customers to conserve energy would recover some of their costs by selling their pollution rights to other companies. The EPA holds an allowance auction each year. The sale offers allowances at a fixed price. This use of market-based incentives under Title IV is regarded by many as a major new method for controlling pollution.

Utilities also took advantage of their flexibility under Title IV to choose less costly ways to reduce emissions, such as switching from high- to low-sulfur coal, and are achieving sizable reductions in their SO_2 emissions. Fifty-five percent of Phase 1 plants opted to switch to low-sulfur coal. (See Figure 7.12.) Sixteen percent chose to install scrubbers. Only 3 percent initially planned to purchase allowances. Not surprisingly the market for

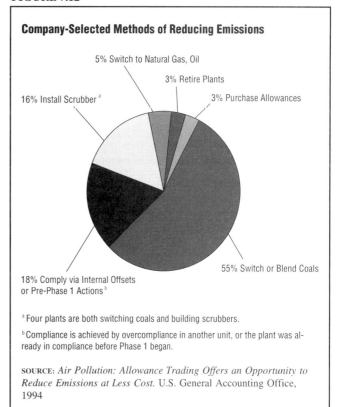

FIGURE 7.12

Company-Selected Methods of Reducing Emissions

5% Switch to Natural Gas, Oil

3% Retire Plants

16% Install Scrubber [a]

3% Purchase Allowances

18% Comply via Internal Offsets or Pre-Phase 1 Actions [b]

55% Switch or Blend Coals

[a] Four plants are both switching coals and building scrubbers.

[b] Compliance is achieved by overcompliance in another unit, or the plant was already in compliance before Phase 1 began.

SOURCE: *Air Pollution: Allowance Trading Offers an Opportunity to Reduce Emissions at Less Cost.* U.S. General Accounting Office, 1994

low-sulfur coal is growing as a result of Title IV, and the market for high-sulfur coal is decreasing.

From 1995 to 1998 there was considerable buying and selling of allowances among utilities. Because the utilities that participated in Phase 1 reduced their sulfur emissions more than the minimum required, they did not use as many allowances as they were allocated for the first four years of the program. Those unused allowances could be used to offset sulfur dioxide emissions in future years. From 1995 to 1998 a total of 30.2 million allowances were allocated to utilities nationwide; almost 8.7 million, or 29 percent, of the allowances were not used.

IS THE ACID RAIN PROGRAM WORKING?

According to the EPA total emissions of sulfur dioxide declined 17 percent from 1990 through 1998, but total emissions of nitrogen oxides changed little during that time. Total deposition (the amount reaching the ground or water) of sulfur decreased 26 percent from 1989 through 1998, while total deposition of nitrogen decreased 2 percent. The EPA and the U.S. General Accounting Office (GAO) concluded, in *Acid Rain: Emissions Trends and Effects in the Eastern United States* (March 2000):

While there has been a reduction in nitrogen oxide emission rates from power generation since 1990, the steady increase in electricity production has resulted in only a small reduction in nitrogen oxide emissions. Now as the over 2-million-ton reduction under Phase 2

TABLE 7.4

Time Periods Needed for Recovery of Selected Ecosystems

Ecosystem	Time period for recovery
Acute human health effects	Hours to weeks
Episodic effects on aquatic resources	Days to months
Chronic effects on aquatic resources	Years to decades
Soil nutrient reserves	Decades to centuries

SOURCE: *Acid Rain Emissions and Effects.* RCED-00-47. U.S. General Accounting Office

of the Acid Rain Program is being implemented, the Agency expects to see the impact of those reductions in the short-term.

The report found that some surface waters in New England were showing signs of recovery, although ecosystems most severely affected—such as the Adirondacks—have not yet shown improvement. The GAO concluded that acidified lakes in the Adirondack Mountains are taking longer to recover than lakes elsewhere and are likely to "recover less or not recover, without further reductions of acid deposition." The recovery of surface waters such as these most likely depends upon improvement in the surrounding soils.

Similarly, in 1998, the NAPAP predicted that acute health effects in humans could require hours to weeks to recover. Episodic changes (from specific ecological events) in aquatic areas may take days to months while chronic damage could require years or decades. Contaminated soil may take decades or even centuries to recover. (See Table 7.4.)

CHAPTER 8

TOXINS IN EVERYDAY LIFE

Many of the substances released into the environment by modern, industrialized society are harmful to humans and other living creatures. These substances may be found in the home or workplace, in the food and water people eat and drink, and even in medications. Most of the chemicals attacked in Rachel Carson's 1962 book, *Silent Spring*, are still used in the United States, and some that have been banned have been replaced by substances even more toxic. The risks posed by environmental contamination may not be blatantly obvious. For example people or animals with impaired immune systems and who are exposed to these contaminants may take longer (or even be unable) to recover from infectious diseases. Tracking this problem to environmental pollutants, however, can be difficult.

TYPES OF TOXINS

A toxin is a substance—bacterial, viral, chemical, metal—that poisons or harms a living organism. A toxin may cause immediate, acute symptoms such as gastroenteritis, or cause harm after long-term exposure such as living in a lead- or radon-contaminated home for many years. Some toxins can have both immediate and long-term effects: Living in an environment with cigarette smoke may trigger an acute asthmatic attack or, after many years' exposure, it may contribute to lung cancer. Although the effects of a toxin may not show up for years, these effects may, nevertheless, be serious. Toxins are often grouped according to their effect on living creatures. They may be called carcinogens, mutagens, and teratogens.

- A carcinogen is any substance that causes cancerous growth.

- A mutagenis is an agent capable of producing genetic change.

- A teratogen is a substance that produces malformations or defective development.

Toxic substances that contribute to these diseases and disorders include heavy metals (lead, cadmium, aluminum, mercury, and manganese), pesticides, petroleum products, industrial wastes, and radionucleids (radium, uranium, and radon). Heavy metals and radionucleids are of particular concern because they reside in the environment for hundreds, and even thousands, of years.

CHEMICALS AND PESTICIDES

Hundreds of chemicals released into the air, water, and the food chain are harmful to the environment—to humans as well as other living things. Under the Federal Insecticide, Fungicide, and Rodenticide Act (61 Stat 163; amended 1988, PL 100-532), the Environmental Protection Agency (EPA) is charged with reviewing chemical studies and taking appropriate action, including banning dangerous substances. The EPA may regulate the manufacture, importation, processing, distribution, use, and disposal of any chemical that poses a risk to human health or the environment. Regulatory tools used by the EPA range from requiring a substance to bear a warning label to placing a total ban on production and importation. The EPA reviews more than 2,000 new chemicals each year, and research and testing continually find new substances to add to that list.

Of the thousands of chemicals currently produced in the United States, many have yet to be tested to see whether they might cause cancer, birth dcfccts, infertility, or abnormal growth in children. Some observers charge that the EPA follow-up on warnings of adverse effects has been slow and, sometimes, has enabled dangerous pesticides to remain on the market.

Pesticides are chemicals and other products used to kill or control insects. These are a unique group of chemicals because they are specifically formulated to be toxic (to some living things) and are deliberately introduced into the environment. Because of these two facts they are closely regulated, and the EPA reviews every pesticide for every particular use. More than half of all pesticide use is on agricultural crops.

While pesticides have important uses, studies show that some cause serious health problems at certain levels of exposure. Among the most common contaminants identified as harmful are DDT, kelthane, lindane, some synthetic pyrethroids, dioxins, polychlorinated biphenyls (PCBs), furans, some heavy metals such as lead and cadmium, and some plastics. Solvents are common in industrial applications, but are also found in such products as furniture polish, bathroom tile cleaners, disinfectants, and shoe polish.

Environmental contaminants—particularly PCBs (used in coolants and lubricants) and pesticide by-products—have been linked to breast cancer in humans. Researchers have found that breast tissue from some women with malignant breast tumors contained more than twice as many PCBs and DDE (a component of the pesticide DDT) than are found in the tissue of women who do not have cancer. Scientists indicate that the carcinogen is stored in body fat, making obesity a risk factor for breast cancer.

In general the level of persistent toxins, including pesticides, has declined in humans and wildlife over recent decades. Newer pesticide compounds are often more toxic than the older types of pesticides but they are generally designed to be less persistent in the environment and tend to cause fewer chronic problems such as birth defects. However, even pesticides originally believed safe are sometimes found to be harmful. In June 2000 researchers announced that recent tests of the pesticide dursban, a very commonly used chemical in residences, found the substance to be harmful, and many applications were withdrawn from the market. Ironically, dursban was often used as a substitute for chlordane, a chemical also withdrawn from use after being discovered to be harmful.

Fear of chemical damage has contributed to a rise in the interest in "organic" foods. The federal government's National Organic Program defines organic agriculture as that which excludes the use of synthetic fertilizers and pesticides. More importantly it strives for low environmental impact and enlists natural biological systems—cover crops, crop rotation, and natural predators—to increase fertility and decrease the likelihood of pest infestation.

Agent Orange

Beginning in 1962 approximately 19 million gallons of herbicides were sprayed over South Vietnam, primarily by aircraft, to defoliate vegetation used as cover for enemy troops in the Vietnam War. In 1969 studies linked chemicals in one of these herbicides, Agent Orange (named after the orange band used to mark the drums it was stored in) to birth defects in laboratory animals. Use of the defoliant was subsequently stopped in 1971. How-

ever, tens of thousands of Americans who served in the war are believed to have been exposed to the chemical, which is a form of dioxin. (In 1994 the EPA affirmed the health danger posed by dioxin.)

Title 38 of the United States Code prohibits veterans from suing the government for injuries suffered while in the military. However, many Vietnam veterans have sued the manufacturers of Agent Orange for damages because of health problems experienced since their return from Vietnam. In 1979 a number of claimants filed a class action, *In re Agent Orange Product Liability Litigation*, which was settled out of court in 1987 for $180 million. The final funds in the case were distributed in 1992. Additional suits against the manufacturers have been attempted but have been prohibited by the courts. The most strongly fought of these legal battles, *Ivy v. Diamond Shamrock*, was supported by the attorneys general of all 50 states. The Supreme Court, however, refused to hear the arguments and the case ended in 1992. The court decreed that the issue was *res judicata*, or "the matter is settled."

The Department of Veterans Affairs sponsored a study of Agent Orange and announced its findings in 1996. The veterans' illnesses were grouped into four categories: (1) those that have a positive association with the herbicide; (2) those for which there is suggestive, but not conclusive, evidence of a link; (3) those for which there is insufficient evidence to make a determination; and (4) those for which there is little or no evidence of an association. The report also emphasized the need for further study. The illnesses that the study positively linked to herbicide use were soft-tissue sarcomas, non-Hodgkin's lymphoma, Hodgkin's disease, and the skin disease chloracne. The second category includes respiratory cancers, prostate cancer, multiple myeloma, acute peripheral neuropathy (nerve numbness or weakness), and spina bifida (a congenital deformity of the spine) in children of veterans. The federal government has generally agreed to pay medical claims on those illnesses.

FERTILIZERS

The U.S. Department of the Interior defines a fertilizer as any substance applied to soil to enhance its ability to produce plentiful, healthy crops. Fertilizers are natural and manufactured chemicals that contain nutrients known to improve the fertility of soils. Nitrogen, phosphorus, and potassium are the three most important of these nutrients; some scientists also consider sulfur a major nutrient for plant health. The EPA estimates that 90 percent of all fertilizers used contain nitrogen, phosphorus, and potassium.

The United States is the largest producer of fertilizers for domestic use and export, and American farmers are the most productive in the world, producing crops for domestic demand as well as export to the rest of the world. Agriculture exists in every state but is concentrat-

FIGURE 8.1

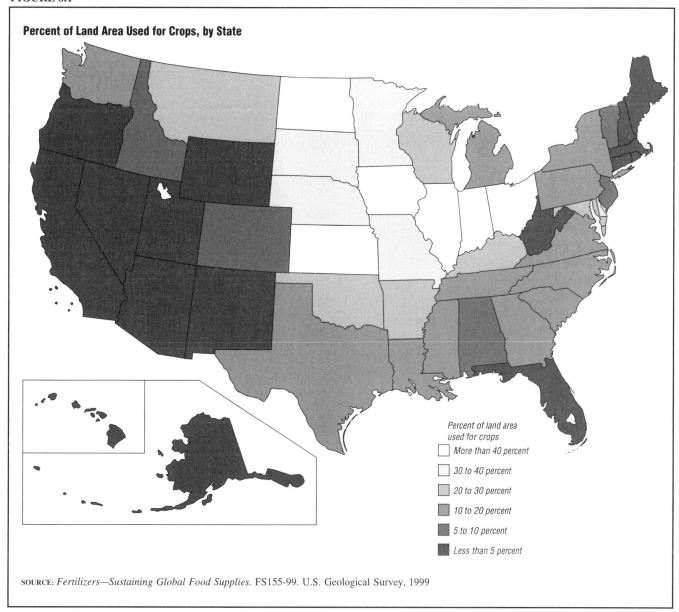

Percent of Land Area Used for Crops, by State

Percent of land area used for crops

- More than 40 percent
- 30 to 40 percent
- 20 to 30 percent
- 10 to 20 percent
- 5 to 10 percent
- Less than 5 percent

SOURCE: *Fertilizers—Sustaining Global Food Supplies.* FS155-99. U.S. Geological Survey, 1999

ed in the Midwest. (See Figure 8.1.) Although many different crops are farmed in the United States, corn, soybeans, and wheat dominate, occupying more than 80 percent of planted cropland. (See Figure 8.2.)

Over-fertilization of crops can cause excess chemicals to leach into surface and groundwater, causing nitrogen accumulation in watersheds. While nitrogen and other nutrients, such as phosphorous, are essential for healthy marine environments, too much can trigger an overabundance of algal growth ("eutrophication"), which then depletes oxygen in the water and kills aquatic species. This condition of oxygen depletion, called hypoxia, has resulted in the mass deaths of marine life ("red tides") in many coastal areas.

The process involved in producing fertilizer is also an environmental concern. For each ton of phosphoric acid produced, five tons of phosphogypsum are created.

This phosphogypsum (which contains small amounts of radium) has accumulated dramatically in some areas, including Florida.

ASBESTOS

First used as a coating for candlewicks by the ancient Greeks, asbestos was developed and manufactured in the twentieth century as an excellent thermal and electrical insulator. Probably the largest single source of asbestos in America is indoor insulation. The physical properties that give asbestos its resistance to heat and decay have long been linked to adverse health effects in humans. Asbestos tends to break into microscopic fibers. These tiny fibers can remain suspended in the air for long periods of time and can easily penetrate body tissues when inhaled. Because of their durability these fibers can lodge and remain in the body for many years. No "safe" exposure threshold for asbestos has been established, but the risk of disease generally increases with the length and amount of exposure.

Asbestos was one of the first substances regulated under Section 112 of the Clean Air Act (CAA) (PL 91-604; 1970) as a hazardous air pollutant. Exposure to asbestos can cause life-threatening diseases including asbestosis, lung cancer, and mesothelioma—a rare cancer of the thin membrane that lines the lungs, chest, abdomen, and heart.

The discovery that asbestos is a strong carcinogen has resulted in the need for its removal or encapsulation (sealing off so that residue cannot escape) from known locations, including schools and public buildings. Many hundreds of millions of dollars have been spent in such cleanups.

Under the CAA asbestos-containing materials must be removed from demolition and renovation sites without releasing asbestos fibers into the environment. Among other safeguards workers must wet asbestos insulation before stripping the material from pipes and must seal the asbestos debris in leak-proof containers while still wet to prevent the release of asbestos dust. The laws of most states have specific requirements for asbestos workers.

A number of legal convictions have resulted from improper and illegal asbestos removal. In many cases the convicted companies had hired homeless people or teenagers to clean up asbestos without advising them that they were dealing with asbestos and without training them in proper handling methods. In response to those cases the Department of Justice and the EPA joined the National Coalition for the Homeless to issue an advisory to be posted in homeless shelters around the United States. The advisory warns about the dangers of asbestos and cautions workers to be on guard for employers who offer work tearing out old asbestos without providing adequate notice, equipment, and training.

Some observers believe that asbestos poses less risk to humans than previously thought, and suggest that asbestos is less harmful than smoking, drug and alcohol abuse, improper diet, or lack of exercise. They contend that Americans can live safely with asbestos—given careful management procedures—and don't need to spend huge sums of money attempting to remove it completely.

The Centers for Disease Control and Prevention (CDC), in *Work-Related Lung Disease Surveillance Report 1999* (2000), found that from 1987 to 1996, 9,614 people died from asbestosis in the United States. In 1987, 710 deaths were attributed to asbestosis; by 1996, 1,176 died from the disease. Most of the deaths were white males. The median age in 1987 was 72 years, increasing

FIGURE 8.2

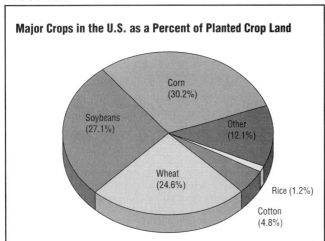

Major Crops in the U.S. as a Percent of Planted Crop Land

Corn (30.2%)
Soybeans (27.1%)
Other (12.1%)
Wheat (24.6%)
Rice (1.2%)
Cotton (4.8%)

NOTE: A tremendous variety of crops are grown in the United States. A few types, however, dominate the total area planted—corn, soybeans, and wheat occupy more than 80 percent of planted crop land.

SOURCE: *Fertilizers—Sustaining Global Food Supplies*. FS155-99. U.S. Geological Survey, 1999

TABLE 8.1

Asbestosis Deaths, by Sex, Race, and Age, U.S. Residents 15 and Over, 1987–1996

Year	Male	Female	White	Black	Other	15-24	25-34	35-44	45-54	55-64	65-74	75-84	85+	Total	Median Age (yrs)
1987	683	27	676	31	3	-	-	3	23	134	283	219	48	710	72.0
1988	742	27	717	48	4	-	-	1	20	141	310	235	62	769	72.0
1989	849	29	807	62	9	-	-	4	18	116	364	307	69	878	73.0
1990	901	47	864	78	6	-	-	4	18	142	359	336	89	948	74.0
1991	908	38	877	63	6	1	-	1	25	114	370	358	77	946	74.0
1992	923	36	898	57	4	-	-	3	13	124	371	355	93	959	74.0
1993	969	30	934	58	7	-	-	1	20	110	365	396	107	999	75.0
1994	1,026	34	993	62	5	-	-	2	21	94	410	422	111	1,060	75.0
1995	1,138	31	1,095	69	5	-	-	3	24	118	411	477	136	1,169	75.0
1996	1,123	53	1,088	84	4	-	-	3	16	104	428	480	145	1,176	75.0
Total	9,262	352	8,949	612	53	1	-	25	198	1,197	3,671	3,585	937	9,614	74.0

- indicates no deaths listed

SOURCE: *Work-Related Lung Disease Surveillance Report.* Centers for Disease Control, 1999

TABLE 8.2

Asbestosis Deaths, by State, U.S. Residents 15 and Over, 1987–1996

State	1987	1988	1989	1990	1991	1992	1993	1994	1995	1996	Total
Alabama	6	10	22	21	23	18	23	29	36	43	231
Alaska	-	-	2	1	1	1	2	2	-	2	11
Arizona	8	7	3	6	6	8	12	9	13	13	85
Arkansas	10	4	1	4	6	5	9	5	9	6	59
California	80	81	87	102	94	95	93	101	113	100	946
Colorado	3	6	3	4	5	4	7	12	6	13	63
Connecticut	9	10	14	11	14	17	8	7	13	18	121
Delaware	3	8	13	6	14	8	8	12	10	10	92
District of Columbia	1	-	-	-	1	-	1	2	-	1	6
Florida	41	59	56	43	54	52	39	60	67	84	555
Georgia	12	8	6	13	10	18	9	12	11	9	108
Hawaii	3	3	6	3	4	4	1	2	5	1	32
Idaho	6	4	5	4	6	3	2	3	4	4	41
Illinois	10	11	26	18	20	21	24	22	21	17	190
Indiana	3	5	5	8	4	4	4	6	6	7	52
Iowa	6	2	2	4	3	7	3	8	8	4	47
Kansas	5	3	5	2	3	7	9	5	10	6	55
Kentucky	3	4	4	1	5	5	11	5	3	9	50
Louisiana	7	12	20	20	20	14	20	15	18	20	166
Maine	9	12	8	17	8	8	13	12	8	6	101
Maryland	18	16	24	36	27	33	35	44	53	50	336
Massachusetts	31	30	43	36	27	48	25	45	40	39	364
Michigan	11	7	11	17	15	16	16	17	27	21	158
Minnesota	4	11	5	8	6	17	19	17	18	7	112
Mississippi	12	15	17	16	25	25	20	25	34	33	222
Missouri	7	11	7	9	11	14	18	13	11	11	112
Montana	1	2	8	6	2	4	4	4	-	4	35
Nebraska	5	2	2	3	3	2	6	4	4	2	33
Nevada	3	1	-	2	3	1	3	6	5	5	29
New Hampshire	6	8	2	2	1	4	8	6	6	4	47
New Jersey	80	83	102	115	93	80	80	81	93	109	916
New Mexico	4	2	5	2	3	1	3	6	8	2	36
New York	30	29	32	44	37	30	26	34	43	42	347
North Carolina	17	12	17	25	21	25	12	32	29	33	223
North Dakota	-	-	1	-	3	-	2	3	-	3	12
Ohio	21	12	23	27	24	32	29	31	35	43	277
Oklahoma	5	5	8	6	6	5	1	5	5	5	51
Oregon	18	21	17	12	22	22	29	26	18	30	215
Pennsylvania	65	79	101	67	83	100	114	90	114	106	919
Rhode Island	-	11	5	7	1	5	4	6	5	2	46
South Carolina	8	14	18	11	8	13	21	13	17	18	141
South Dakota	-	-	-	-	-	-	-	-	-	3	3
Tennessee	4	12	5	6	8	4	8	13	12	14	86
Texas	48	46	49	91	95	54	72	80	93	87	715
Utah	2	1	2	3	-	4	5	2	4	5	28
Vermont	2	1	-	1	-	3	3	-	2	2	14
Virginia	22	35	35	47	36	43	47	38	44	37	384
Washington	40	34	37	34	50	40	59	60	44	48	446
West Virginia	15	15	6	17	20	28	18	20	32	32	203
Wisconsin	4	4	7	9	11	6	10	9	10	3	73
Wyoming	2	1	1	1	4	1	4	1	2	3	20
TOTAL	**710**	**769**	**878**	**948**	**946**	**959**	**999**	**1,060**	**1,169**	**1,176**	**9,614**

- indicates no deaths listed.

SOURCE: *Work-Related Lung Disease Surveillance Report.* Centers for Disease Control, 1999

to 75 years by 1996. (See Table 8.1.) As of 1996 California (946), Pennsylvania (919), New Jersey (916), Texas (715), and Florida (555) had the largest number of asbestosis deaths. (See Table 8.2.)

Of those who died from asbestosis the largest percentage was among plumbers, pipe fitters, and steamfitters (9.2 percent), followed by insulation workers (4.8 percent). (See Table 8.3.) Construction accounted for, by far, the greatest proportion (24.8 percent) of asbestosis deaths; second was ship/boat building and repairing (7.6 percent).

(See Table 8.4.) In 1996 about 13,000 people were treated in short-stay hospitals for asbestosis, up from 300 in 1970.

CHLORINE

Chorine is a gaseous element first isolated in 1774 by chemist Wilhelm Scheele. The gas has an irritating odor and, in large concentrations, is dangerous. It was the first substance used as a poisonous gas in World War I. It can be liquefied under pressure and is usually transported as a liquid in steel bottles or tank cars.

TABLE 8.3

Asbestosis Deaths, Most Frequently Recorded Occupations on Death Certificate, U.S. Residents 15 and Over, Selected States and Years, 1987–1996

COC	Occupation	Number	Percent
585	Plumbers, pipefitters, and steamfitters	237	9.2
593	Insulation workers	124	4.8
019	Managers and administrators, n.e.c.	113	4.4
575	Electricians	106	4.1
567	Carpenters	97	3.8
889	Laborers, except construction	88	3.4
783	Welders and cutters	73	2.8
633	Supervisors, production occupations	72	2.8
453	Janitors and cleaners	67	2.6
637	Machinists	67	2.6
	All other occupations	1,441	56.0
	Occupation not reported	90	3.5
	TOTAL	**2,575**	**100.0**

COC - Census Occupation Code
n.e.c. - not elsewhere classified
NOTE: Percentages may not total to 100% due to rounding.

SOURCE: *Work-Related Lung Disease Surveillance Report.* Centers for Disease Control, 1999

TABLE 8.4

Asbestosis Deaths, Most Frequently Recorded Industries on Death Certificate, U.S. Residents 15 and Over, Selected States and Years, 1987–1996

CIC	Industry	Number	Percent
060	Construction	638	24.8
360	Ship and boat building and repairing	195	7.6
192	Industrial and miscellaneous chemicals	105	4.1
400	Railroads	81	3.1
262	Miscellaneous nonmetallic mineral and stone products	67	2.6
392	Not specified manufacturing industries	66	2.6
901	General government, n.e.c.	51	2.0
460	Electric light and power	45	1.7
270	Blast furnaces, steelworks, rolling and finishing mills	44	1.7
961	Non-paid worker or non-worker	43	1.7
	All other industries	1,145	44.5
	Industry not reported	95	3.7
	TOTAL	**2,575**	**100.0**

CIC - Census Industry Code
n.e.c. - not elsewhere classified
NOTE: Percentages may not total to 100% due to rounding.

SOURCE: *Work-Related Lung Disease Surveillance Report.* Centers for Disease Control, 1999

The use of chlorine to disinfect water supplies is one of the greatest public health success stories of the twentieth century. First used to purify water in the early 1900s, chlorine is, by far, the world's primary water disinfectant and is indisputably valuable in preventing the spread of disease. It is inexpensive, effective, and available nearly everywhere. It is credited with banishing typhoid fever, cholera, and dysentery from the United States and elsewhere. About three-fourths of all U.S. drinking water is chlorinated to kill parasites, viruses, and bacteria, while most of the rest is treated with a combination of chlorine and ammonia.

Researchers, however, are trying to determine if there is any connection between chlorine in drinking water and bladder and rectal cancer in humans. When chlorine is added to water containing organic matter it produces by-products (such as chloroform) that are suspected of causing harm to humans and other species. Environmentalists contend that chlorine is responsible for a thinning ozone layer (see Chapter 3) and for causing reproductive malformations in aquatic species and cancer in humans.

Many sources claim that reducing or eliminating chlorine in water would cost many lives. Although the body of evidence seems to show a slight increase in cancer risk from chlorine by-products, most scientists consider the risk not "statistically significant" and believe that the risks to public health would be far greater if chlorine is reduced or eliminated.

Researchers are investigating alternatives to chlorination. However, by the turn of the twenty-first century, all alternative methods—with the possible exception of ultraviolet radiation—also form by-products. In addition alternative disinfectants do not provide the residual protection of chlorine-based treatment; they must be used in combination with chlorine or chlorine derivatives to provide a complete disinfection system.

ELECTROMAGNETIC FIELDS

Earth produces electromagnetic fields (EMFs) naturally, such as during thunderstorms and deep within Earth's molten core. Electricity also occurs naturally in the human body where it can be measured by electroencephalograms (EEGs) of brain wave activity, or electrocardiograms (ECGs) of heart rhythms.

Electric power is a fact of life in America and the developed world. Many consumer and industrial products use some form of electromagnetic energy. While the danger of electric shock is well known, another concern has arisen about electric power—does it cause certain types of cancer, particularly leukemia and cancer of the central nervous system?

One type of electromagnetic energy that is of increasing importance worldwide is radio frequency (RF) energy, which includes radio waves and microwaves. RF energy is used for a wide array of applications including television broadcasting, cellular telephones, pagers, cordless phones, radio communications for police and fire departments, amateur radio, and satellite transmissions. In the United States the Federal Communications Commission (FCC) authorizes or licenses most RF telecommunications activities. Non-communications uses of RF energy include microwave ovens, radar, and diathermy (the delivery of heat to body tissues for medical or surgical purposes).

FIGURE 8.3

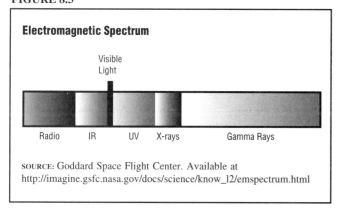

Electromagnetic Spectrum

Visible
Light

Radio   IR   UV   X-rays   Gamma Rays

SOURCE: Goddard Space Flight Center. Available at
http://imagine.gsfc.nasa.gov/docs/science/know_12/emspectrum.html

TABLE 8.6

Office Sources of Magnetic Fields

Distance from Source	6"	1'	2'	4'
AIR CLEANERS				
Lowest	110	20	3	-
Median	180	35	5	1
Highest	250	50	8	2
COPY MACHINES				
Lowest	4	2	1	-
Median	90	20	7	1
Highest	200	40	13	4
FAX MACHINES				
Lowest	4	-	-	-
Median	6	-	-	-
Highest	9	2	-	-
FLUORESCENT LIGHTS				
Lowest	20	-	-	-
Median	40	6	2	-
Highest	100	30	8	4
ELECTRIC PENCIL SHARPENERS				
Lowest	20	8	5	-
Median	200	70	20	2
Highest	300	90	30	30
VIDEO DISPLAY TERMINALS (PCS WITH COLOR MONITORS)				
Lowest	7	2	1	-
Median	14	5	2	-
Highest	20	6	3	-

Magnetic field measurements in units of milligauss (mG).

SOURCE: *Questions and Answers About Electric and Magnetic Fields Associated with the Use of Electric Power.* U.S. Dept. of Energy, 1995

TABLE 8.5

Kitchen Sources of Magnetic Fields

Distance from Source	6"	1'	2'	4'
BLENDERS				
Lowest	30	5	-	-
Median	70	10	2	
Highest	100	20	3	-
CAN OPENERS				
Lowest	500	40	3	-
Median	600	150	20	2
Highest	1500	300	30	4
COFFEE MAKERS				
Lowest	4	-	-	-
Median	7	-	-	-
Highest	10	1	-	-
CROCK POTS				
Lowest	3	-	-	-
Median	6	1	-	-
Highest	9	1	-	-
DISHWASHERS				
Lowest	10	6	2	-
Median	20	10	4	-
Highest	100	30	7	1
FOOD PROCESSORS				
Lowest	20	5	-	-
Median	30	6	2	-
Highest	130	20	3	-

Magnetic field measurements in units of milligauss (mG).

SOURCE: *Questions and Answers About Electric and Magnetic Fields Associated with the Use of Electric Power.* U.S. Dept. of Energy, 1995

The spectrum of electromagnetic waves ranges from extremely low-frequency energy to X-rays and gamma rays which have very high frequencies. (Frequency is the number of waves passing a given point in one second.) (See Figure 8.3.) All humans are exposed to EMFs at some time during the course of a day, either from natural sources or sources produced by humans. Human-produced sources include electrical items found in the typical home, such as kitchen appliances. (See Table 8.5.) Office equipment is another example of EMF produced by humans. (See Table 8.6.) Most people are exposed to some level of EMF daily and that level can differ at different times of the day. (See Figure 8.4.)

Scientists have known for years that exposure to high levels of RF radiation can be harmful because of the ability of RF energy to heat biological tissue rapidly (this is the principle by which microwave ovens work). Two areas of the body, the eyes and the testes, are known to be particularly vulnerable to RF heating. RF radiation, even in low levels, has caused cataracts and sterility in laboratory rabbits.

In 1996 the U.S. Public Health Service reviewed the scientific data on EMFs and determined the following:

In total, the epidemiological data on both residential and occupational exposures show a moderate risk of cancer, generally between 1.1 and 3.0 times, for adults and children exposed to magnetic fields. This is not extremely large relative to other known risks (for example, the smoking-related risk of approximately 10 or the asbestos-related risk of 5). However, what is unusual about magnetic field exposures is that they are universal. Virtually all of us are exposed....This means that (a) the observed risk may be underestimated because we cannot identify a truly unexposed comparison group; (b) because of the widespread exposure, even a small risk may result in a large number of individual cancers.

The Public Health Service concluded that "the cost of mitigation [cleanup] already instituted far exceeds the health protection offered, and mitigation of other environmental risks is more important. From a cost-benefit view only limited, low-cost mitigation should be considered."

FIGURE 8.4

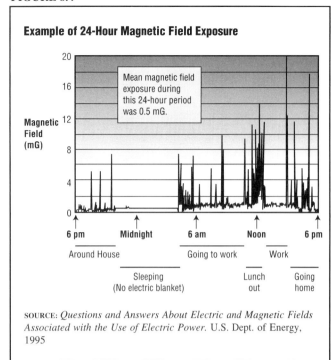

Example of 24-Hour Magnetic Field Exposure

Mean magnetic field exposure during this 24-hour period was 0.5 mG.

SOURCE: *Questions and Answers About Electric and Magnetic Fields Associated with the Use of Electric Power.* U.S. Dept. of Energy, 1995

The FCC, in its 1999 *Questions and Answers about Biological Effects and Potential Hazards of Radiofrequency Electromagnetic Fields*, reported that

> Environmental levels of RF energy routinely encountered by the general public are *far below* levels necessary to produce significant heating and increased body temperature. However, there may be situations, particularly workplace environments near high-powered RF sources, where recommended limits for safe exposure of human beings to RF energy could be exceeded. In such cases, restrictive measures or actions may be necessary to ensure the safe use of RF energy.

Nonetheless, some scientists have continued to suggest that exposure to electric and magnetic fields generated by electric power is responsible for certain cancers (particularly among children), reproductive dysfunction, birth defects, neurological disorders, and Alzheimer's disease. Some activist groups allege the hazards to be so great that they have called for the closure of schools and other public facilities near power lines, and restructuring of the entire power delivery system. EMFs are even cited as causing decreases in property values. Some utilities, with equally strong beliefs, claim there is no proof of risk.

In response to concern about the risks of EMFs, in 1992, under the Energy Policy Act (PL 102-486; 1992), the U.S. Congress authorized the Electric and Magnetic Fields Research and Public Information Dissemination Program (EMF-RAPID Program). The program mandated that the National Institute of Environmental Health Sciences (NIEHS), the National Institutes of Health (NIH), and the Department of Energy (DOE) research

any possible link between EMFs and health. In 1998 an international panel of 30 scientists met to consider the evidence. In 1999 the panel issued its findings in *Health Effects from Exposure to Power-Line Frequency Electric and Magnetic Fields*.

The report found that the evidence was lacking to prove EMFs were a "known human carcinogen" or "probable human carcinogen," and thus EMFs did not warrant aggressive regulatory action. However, a majority of the members concluded that exposure to power-line frequencies is a "possible human carcinogen." The NIEHS concluded that EMF exposure cannot be recognized as entirely safe because of weak scientific evidence that exposure may pose a leukemia hazard. However, because virtually everyone in the United States uses electricity, and is thus exposed, the panel recommended research into ways to reduce exposure. The researchers found no indication of increased cancer incidence in experimental animal studies. The strongest evidence of any connection between EMF exposure and cancer came from observations of human populations, although the scientists said some other factor could explain the findings, especially since the correlation was, at best, weak.

Other Studies

In 1998 the FCC studied EMF emission levels of cellular phones, vehicle-mounted antennas, and fixed transmitting antennas used for such devices. The FCC reported, in its *Information on Human Exposure to Radiofrequency Fields from Cellular and PCS Radio Transmitters*, that a human would have to remain within a few feet of a main transmitting beam for extremely long periods of time to be exposed to levels of RF energy in excess of recognized safety levels. In addition the study noted that, regarding vehicle antennas, vehicle occupants are effectively shielded by the metal auto body, especially when the antennas are mounted in the center of the roof or trunk. The study yielded similar results for handheld cellular devices— exposures vary greatly depending on use. Researchers, nevertheless, continue a number of programs to study electromagnetic fields.

In 2000 the American Medical Association, after conducting a study of the possible link between EMFs and childhood leukemia, found no evidence that exposure to EMFs increased the risk for childhood cancer ("Exposure to Power-Frequency Magnetic Fields and the Risk of Childhood Cancer," *Journal of the American Medical Association*, Feb. 2000).

LEAD

In Paint

Mined along the Eastern Seaboard since 1621, lead created an important industry, providing bullets, piping, and a base for paint. Because of its malleability it was valued as a conduit for water. The use of wallpaper steadily declined with the almost universal use of paint, not only for protecting surfaces but also for interior decorating. "White lead" paint was sold as the best thing to use on interior and exterior surfaces. In cities teeming with millions of new immigrants, the glossy, durable finish of white lead-based paint meant walls could be easily washed. In 1922 a General Motors researcher discovered that the addition of lead to automobile fuel reduced the "knocking" that limited power and efficiency in car engines.

As early as the late 1890s medical reports concerning problems with lead began to appear. In 1914 the first U.S. case of lead poisoning was reported, although the cause was undetermined. Scientists eventually began to link lead poisoning to lead paints and, as World War II ended, began to address the problem. In the mid-1960s, medical reports documented the connection of lead poisoning to both auto emissions and paint.

In 1971 Congress passed the Lead-Based Poisoning Prevention Act (PL 91-695), restricting residential use of lead paint in structures constructed or funded by the federal government. The phasedown of leaded fuel in automobiles began in the 1970s—not to safeguard health but to protect cars' catalytic converters, which were rendered inoperable by lead.

In an effort to protect families from exposure to the hazards of lead-based paint, Congress amended the Toxic Substances Control Act (TSCA; PL 94-469, 1976) in 1992 to add Title IV, entitled "Lead Exposure Reduction." Title IV directs the EPA to address the general public's exposure to lead-based paint through regulations, education, and other activities. A particular concern of Congress and the EPA is the potential lead exposure risk associated with housing renovation. The law directs the EPA to publish lead hazard information and make it available to the general public, especially to those undertaking renovations.

Also in 1992 Congress passed the Residential Lead-Based Paint Hazard Reduction Act (PL 102-550) to stop the use of lead-based paint in federal structures and set up framework to evaluate and remove paint from buildings nationwide. In 1996 Congress once again amended the TSCA, adding Section 402a to establish and fund training programs for lead abatement and to set up requirements and training of technicians and lead-abatement professionals.

Lead Builds Up in the Body

Lead is a cumulative poison. People are exposed to it every day and, over time, it begins to accumulate in the body. At very high levels of exposure (now rare in the U.S.), lead can cause mental retardation, convulsions, and even death. More commonly, exposure occurs at very low levels over an extended period of time. Exposure to lead causes nervous system disorders, fatigue, high blood pressure, and reduced intelligence. Children under the age of six are particularly vulnerable and they can suffer a host of effects including reduced I.Q., impaired growth, hyperactivity, reading and learning disabilities, insomnia, and hearing loss. Many scientists believe that the federal standards for exposure should be lowered, and, in fact, some researchers believe there is no safe level for lead. The CDC believes that effects on the central nervous system of children begin at 10 micrograms per deciliter, and the greater the blood lead level (BLL), the higher the risk. The CDC recommends that if many children in a community have BLLs above 10 micrograms per deciliter, community-wide lead poisoning prevention activities should begin. The Occupational Safety and Health Administration (OSHA) requires that a worker be removed from a workplace if his or her BLL reaches 50 micrograms per deciliter, although two Harvard School of Public Health studies released in 1996 found kidney damage and hypertension correlated with lead levels in bone below 10 micrograms per deciliter.

Lead Can Be Found in Many Places

Many structures are still covered by old lead paint. Nearly three-quarters of all U.S. homes constructed before 1980 contain some lead paint. The EPA reports that lead paint poses little danger if stable. But when renovations are made that involve sanding or stripping paint, the old paint may become hazardous. Certain ceramics and crystalware, especially those made in foreign countries, still contain unacceptable levels of lead. In 1996 researchers announced that ingredients used to manufacture some miniblinds could be toxic to humans because they contain lead. Lead can be found in many other places as well, including weights used for draperies, wheel balances, or fishing lures; seams in stained-glass windows; linoleum; batteries; solder; gun shot; and plumbing. Test kits and laboratories that test for lead can now check questionable items and locations for the presence of the heavy metal.

In Water

The EPA estimates that nearly 40 million Americans drink water that has a lead content of more than 50 parts per billion (ppb), the maximum recommended level, while another 75 million Americans drink water that exceeds 20 ppb.

Most of the lead in water comes from lead pipes and lead solder in plumbing systems. Lead toxicity is the single most significant disease of environmental origin in American children. Exposure to lead causes nervous system disorders, reduced intelligence, fatigue, high blood pressure, inhibited infant growth, and hearing loss, especially in children. A 1992 EPA report revealed that 20 percent of the nation's large cities exceeded government

FIGURE 8.5

Adults With Elevated Blood Lead Levels in Relation to All Adults Tested

Total number of adults* tested[†] and whose blood lead levels (BLLs) were ≥ 25 μg/dL, by quarter — 27 states participating in Adult Blood Lead Epidemiology and Surveillance,[§] 1997–1998

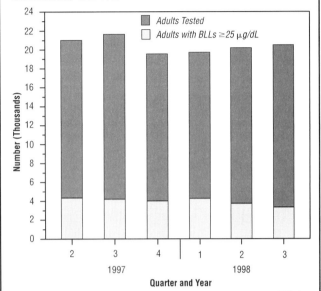

* Persons aged 16–64 years, categorized according to the highest reported BLL for that person during the given quarter. Data for the second and third quarters of 1998 were not available for New Mexico; the corresponding 1997 quarters were used as estimates.

[†] The reporting threshold varies among the participating states; the value includes persons with BLLs <25 μg/dL. However, the following states do not report persons with BLLs <25 μg/dL: Maryland, Massachusetts, New Jersey, North Carolina, and Oregon.

[§] Alabama, Arizona, California, Connecticut, Iowa, Maine, Maryland, Massachusetts, Michigan, Minnesota, New Hampshire, New Jersey, New Mexico, New York, North Carolina, Ohio, Oklahoma, Oregon, Pennsylvania, Rhode Island, South Carolina, Texas, Utah, Vermont, Washington, Wisconsin, and Wyoming.

SOURCE: *Adult Blood Epidemiology and Surveillance.* Morbidity and Mortality Weekly Report, vol. 48, no. 10, March 19, 1999

limits for lead in drinking water. By 1993 all large public water-supply systems were required to add substances such as lime or calcium carbonate to their water lines to reduce the corrosion of older pipes, which releases lead.

Lead Levels Are High in a Significant Number of Americans

The federal government estimates that about 15 percent of all U.S. preschoolers now have unacceptable BLLs. In 1995 the New York State Department of Health assessed the effect of lead exposure resulting from home renovation and remodeling on children under the age of six. Of the 4,608 children studied, 320 (7 percent) had elevated BLLs (over 20 micrograms per deciliter). In 2000 the CDC reported that the level of lead in Americans' blood now averages 2.8 micrograms per deciliter. It also estimated that 1.7 million American children under the age of six have BLLs above 10 micrograms per deciliter, and that approximately 200,000 have BLLs greater than

20 micrograms per deciliter. These levels can cause subtle, but significant, impairment of learning skills.

The CDC conducts quarterly BLL tests among adults in just over half the U.S. states. In the third quarter of 1998 (the latest reported), approximately 16 percent of adults tested had BLLs above 25 micrograms per deciliter, a 21 percent decrease from the third quarter of 1997. (See Figure 8.5.) The study also found that, from 1994 to 1997, states on the East Coast, the Northeast, and the Great Lakes region generally reported higher BLLs among adults. (See Figure 8.6.)

Contribution to Delinquency?

A 1996 University of Pittsburgh Medical Center study of 800 male public school students found that, even after taking into account other predictors of delinquency, such as socioeconomic status, boys with higher lead levels were more likely to engage in antisocial acts. A direct relationship was found between the amount of lead in boys' leg bones and reports from parents, teachers, and the children themselves of criminal or aggressive behavior.

RADON

In 1984 a worker in a nuclear plant triggered a radiation contamination alarm as he entered the plant to work. Since the alarms were intended to check for contamination as workers left the plant, plant officials were amazed. Investigations discovered the source of the worker's contamination was radon present in extraordinarily high amounts in his home.

Radon is an invisible, odorless, radioactive gas formed by the decay of uranium in rocks and soil. This gas seeps from underground rock into the basements and foundations of structures via cracks in foundations, pipes, and sometimes through the water supply. Because it is naturally occurring, it cannot be entirely eliminated from homes.

Although there is no federal regulation addressing radon in indoor air, the EPA recommends that a resident take action if levels reach or exceed 4 picocuries per liter (pCi/L). The average indoor radon level is 1.25 pCi/L; the average outdoor level is 0.4 pCi/L. The EPA estimates that there are at least six million homes—one in 15 nationwide—with levels over 4 pCi/L, and 100,000 with levels over 20 pCi/L. Most homes today can reduce radon content to 2 pCi/L or below by using devices such as specially designed fans that prevent radon from seeping into a house.

In 1999 the EPA estimated that radon causes an estimated 20,000 deaths from lung cancer each year—the second leading cause of lung cancer after smoking. A synergistic effect has been noted: when radon levels are high in a home where a smoker resides, the likelihood of that person contracting lung cancer is greatly increased.

FIGURE 8.6

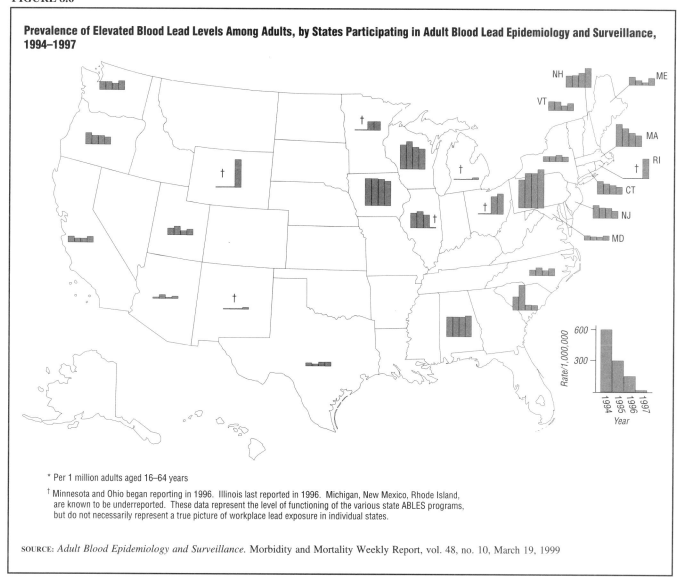

Prevalence of Elevated Blood Lead Levels Among Adults, by States Participating in Adult Blood Lead Epidemiology and Surveillance, 1994–1997

* Per 1 million adults aged 16–64 years

† Minnesota and Ohio began reporting in 1996. Illinois last reported in 1996. Michigan, New Mexico, Rhode Island, are known to be underreported. These data represent the level of functioning of the various state ABLES programs, but do not necessarily represent a true picture of workplace lead exposure in individual states.

SOURCE: *Adult Blood Epidemiology and Surveillance*. Morbidity and Mortality Weekly Report, vol. 48, no. 10, March 19, 1999

The EPA has proposed voluntary guidelines calling for builders to install protective measures in hundreds of thousands of new houses across the country to prevent radon from seeping in. The EPA recommends the testing of homes, which is generally quite inexpensive; home-test kits are even available to owners. Radon contamination is addressed under the Superfund Amendments and Reauthorization Act of 1986 (PL 99-499).

Radon in Drinking Water

In addition to exposure through the air, humans can also contact radon in drinking water. Radon gas can dissolve and accumulate in groundwater such as that found in wells. The EPA estimates that only about 1–2 percent of radon comes from drinking water. However, ingesting radon-contaminated drinking water can lead to the development of internal-organ cancers, primarily stomach cancer—although the risk is smaller than that of developing lung cancer from radon released to the air from tap water. A National Academy of Sciences report, *Risk Assessment of Radon in Drinking Water* (1999), a study mandated by the Safe Drinking Water Act Amendments (PL 104-182, 1996), estimates that radon in drinking water causes about 169 cancer deaths a year; 89 percent of those are caused by breathing radon released into the air from tap water while 11 percent are caused by consuming water that contains radon.

Not all water contains radon. Surface waters—such as rivers, lakes, or reservoirs—usually do not carry radon because the radon tends to evaporate before it has a chance to reach residences. Underground sources such as groundwater, however, may contain radon. Those who get water from a public water system that serves 25 or more year-round residents may obtain an annual water-quality report that tells where the water comes from, what is in that water, and whether or not radon was found.

As of 2000 there is no federally enforced drinking-water standard for radon. The EPA has proposed requir-

FIGURE 8.7

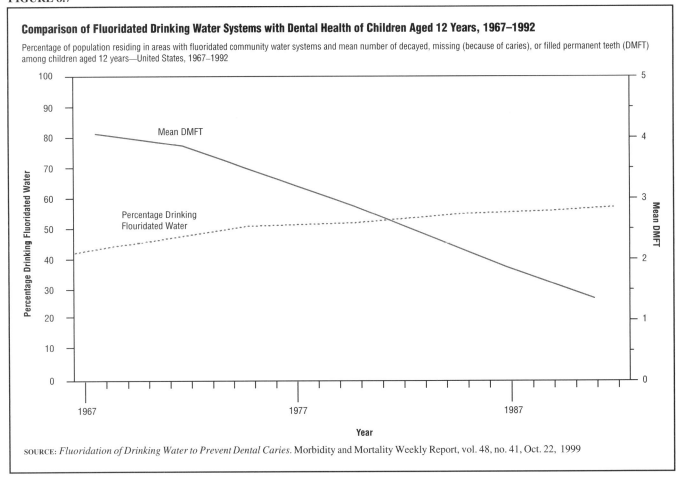

Comparison of Fluoridated Drinking Water Systems with Dental Health of Children Aged 12 Years, 1967–1992

Percentage of population residing in areas with fluoridated community water systems and mean number of decayed, missing (because of caries), or filled permanent teeth (DMFT) among children aged 12 years—United States, 1967–1992

SOURCE: *Fluoridation of Drinking Water to Prevent Dental Caries.* Morbidity and Mortality Weekly Report, vol. 48, no. 41, Oct. 22, 1999

ing water suppliers to provide water with radon levels no higher than 4,000 pCi/L, a level that contributes about 0.4 pCi/L of radon to the air in a home. Such a mandate is expected to be finalized by late 2000. Many states have also developed EPA-approved programs to reduce indoor radon. For states that choose not to develop their own programs, water systems in those states will be required to reduce radon levels in drinking water to 300 pCi/L. For home-owners who get their water from private wells—which are not regulated—radon in the water can be reduced by using a carbon filter, or aeration devices that bubble air through the water and carry radon gas out through an exhaust fan.

FLUORIDATION

Dental caries is an infectious, communicable disease in which bacteria dissolve the enamel surface of a tooth. Left unchecked, the bacteria may then penetrate the underlying dentin and soft tissue, resulting in tooth loss, discomfort, and even acute infection throughout the body. At the beginning of the twentieth century dental caries was common in the United States and most developed countries. Failure to meet the minimum standard of having six opposing teeth was a leading cause of rejection from military service in both world wars.

Fluoride was first added to drinking water in 1945 to control dental caries and prevent tooth decay. As a result dental caries declined greatly during the second half of the twentieth century. (See Figure 8.7.) Since that time most community water systems in the United States have introduced water fluoridation. The CDC estimates that 144 million Americans now have access to drinking water with fluoride levels of at least 0.7 parts per million (ppm).

Fluoridation safely and inexpensively benefits children and adults by preventing tooth decay—regardless of socioeconomic status or access to care. Because fluoridation of drinking water has proved effective in reducing dental cavities, researchers have developed other methods to deliver fluoride to the public (toothpastes, rinses, dietary supplements). The widespread use of these products has assured that virtually everyone in the United States has been exposed to fluoride. The CDC estimates that fluoride has reduced tooth decay in children by as much as 40–70 percent and in adults by 40–60 percent. Water fluoridation is believed to be especially beneficial in low-income areas where residents often have less access to dental-care services and other sources of fluoride. Consequently, the CDC rates the fluoridation of drinking water as one of the top ten greatest public health achievements of the twentieth century. (See Figure 8.8.)

Despite these benefits, however, there have been long-standing concerns about the negative effects of fluoridation. Because of these concerns studies continue into the possible adverse effects of fluoride, although nothing had been uncovered as of 2000.

TOBACCO

Cigarette smoking is the single most preventable cause of premature death in the United States.
—The Centers for Disease Control and Prevention (CDC)

Tobacco causes more death and suffering than any other environmental toxin. In 2000 the CDC reported that smoking kills more people than AIDS, alcohol, drug abuse, car crashes, murders, suicides, and fires combined, and costs more than $50 billion each year in direct medical costs. According to the American Cancer Society nearly one in five deaths in the United States results from the use of tobacco. Every year smoking claims more than 400,000 lives nationwide. Nearly half of all smokers between the ages of 35 and 69 will die prematurely; smoking takes, on average, 20–25 years off the life of a smoker—a total of five million years of life lost each year. Passive smokers (those who inhale the smoke from others' cigarettes) have a 30 percent greater risk of dying of lung cancer than those who do not inhale other people's smoke. Mothers who smoke risk diminishing the physical and mental abilities of both their born and unborn children.

People in industrialized countries smoke twice as much as those in developing nations. Education level and smoking are inversely related; people with only grade-school educations are much more likely to smoke than are those with advanced degrees.

In addition to nicotine, which is very addictive, cigarette smoke contains hundreds of mutagens, carcinogens, and some 4,000 other chemical compounds including carbon monoxide and radioactive polonium. These chemicals not only enter the lungs but also the bloodstream where they circulate into internal organs. Smoking can be responsible for, or contribute to, asthma; heart disease; cancer of the lungs, esophagus, mouth, bladder, pancreas, and pharynx; bronchitis; emphysema; and low-birth-weight babies. One ironic result of the attempt to reduce smoking has been the marked increase in the use of smokeless tobacco. While smokeless tobacco does not produce secondhand smoke, it can cause cancer—especially oral cancer—in those who use it.

The presence of tobacco smoke in the air increases the risk of lung cancer in nonsmoking spouses and of respiratory disease in infants. Until the late 1990s nonsmokers generally had no choice about breathing tobacco smoke in many public buildings, including hospitals. That is no longer the case and, in fact, smokers may find themselves ostracized or, at the very least, required to smoke in designated areas. Many American companies are banning smoking from the workplace; some even refuse to hire smokers. Also, some restaurants and social clubs—places where smoking has historically been common—have begun to ban, or severely curtail, smoking on their premises.

Besides the pain and suffering caused by smoking—both active and passive—smoking is costly to society. Surveys show that inefficiency and ill health as a result of smoking waste about 7 percent of smokers' working time. Smokers also add to insurance and cleanup costs. More significant is the cost of medical treatment, which often falls on the shoulders of the general public by way of insurance and welfare payments.

Prevalence of Smoking

The CDC noted that, in 1998, 22.9 percent of Americans over the age of 18 reported that they smoked cigarettes. Prevalences ranged from 14.2 percent in Utah to 30.8 percent in Kentucky. Men were somewhat more likely to smoke cigarettes than were women. (See Table 8.7.)

The median prevalence of cigar smoking was 39 percent. State prevalences ranged from 14.8 percent in Arizona to 52 percent in Alaska. Men were far more likely to smoke cigars than were women. (See Table 8.8.) In June 2000 the EPA proposed adding cigars to the list of substances, such as cigarettes, to be labeled with health warnings.

According to the CDC's 1999 *National Youth Tobacco Survey*, 34.8 percent of high-school students and 12.8 percent of middle-school students used tobacco of some kind. Cigarettes were the most commonly used form of tobacco, used by 9.2 percent of middle schoolers and 28.4 percent of high schoolers. Males were more likely to use tobacco than females of either high-school or middle-school age. About 6.6 percent of high schoolers used smokeless tobacco, compared with 2.7 percent of middle schoolers. Students used bidis and kreteks, newer forms

FIGURE 8.8

Ten Great Public Health Achievements in the U.S., 1900–1999

- Vaccination
- Motor-vehicle safety
- Safer workplaces
- Control of infectious diseases
- Decline in deaths from coronary heart disease and stroke
- Safer and healthier foods
- Healthier mothers and babies
- Family planning
- Fluoridation of drinking water
- Recognition of tobacco use as a health hazard

SOURCE: *Ten Great Public Health Achievements—United States, 1900-1999.* Morbidity and Mortality Weekly Report, vol. 48, no. 12, April 2, 1999

TABLE 8.7

Prevalence of Current Cigarette Smoking* Among Adults, by State and Sex, 1998

State	Men %	Men (95% CI†)	Women %	Women (95% CI)	Total %	Total (95% CI)
Alabama	27.2	(±3.5)	22.3	(±2.5)	24.6	(±2.1)
Alaska	28.3	(±3.9)	23.5	(±3.4)	26.0	(±2.6)
Arizona	24.7	(±4.0)	19.2	(±3.3)	21.9	(±2.6)
Arkansas	28.6	(±3.0)	23.7	(±2.2)	26.0	(±1.8)
California	21.9	(±2.2)	16.6	(±1.7)	19.2	(±1.4)
Colorado	26.4	(±3.6)	19.5	(±2.6)	22.8	(±2.2)
Connecticut	21.7	(±3.3)	20.6	(±2.3)	21.1	(±2.0)
Delaware	27.3	(±4.1)	21.9	(±2.8)	24.5	(±2.4)
District of Columbia	24.5	(±4.4)	19.0	(±3.1)	21.6	(±2.6)
Florida	23.5	(±2.2)	20.6	(±1.6)	22.0	(±1.4)
Georgia	28.0	(±3.4)	19.7	(±2.3)	23.7	(±2.0)
Hawaii	22.3	(±3.6)	16.7	(±2.7)	19.5	(±2.3)
Idaho	21.9	(±2.2)	18.8	(±1.7)	20.3	(±1.4)
Illinois	26.0	(±2.7)	20.6	(±2.3)	23.1	(±1.8)
Indiana	29.6	(±3.2)	22.7	(±2.4)	26.0	(±2.0)
Iowa	25.8	(±2.7)	21.1	(±2.0)	23.4	(±1.7)
Kansas	23.0	(±2.5)	19.5	(±1.9)	21.2	(±1.5)
Kentucky	33.3	(±2.8)	28.5	(±2.0)	30.8	(±1.7)
Louisiana	28.2	(±3.9)	23.1	(±3.0)	25.5	(±2.4)
Maine	21.2	(±3.5)	23.5	(±3.2)	22.4	(±2.4)
Maryland	24.3	(±3.2)	20.6	(±2.4)	22.4	(±2.0)
Massachusetts	22.5	(±2.5)	19.5	(±1.9)	20.9	(±1.6)
Michigan	30.3	(±3.1)	24.8	(±2.4)	27.4	(±2.0)
Minnesota	19.7	(±1.9)	16.4	(±1.7)	18.0	(±1.3)
Mississippi	26.9	(±3.4)	21.7	(±2.4)	24.1	(±2.0)
Missouri	29.4	(±3.2)	23.6	(±2.3)	26.3	(±2.0)
Montana	21.5	(±3.0)	21.5	(±2.9)	21.5	(±2.1)
Nebraska	25.2	(±2.8)	19.1	(±2.1)	22.1	(±1.8)
Nevada	32.6	(±4.6)	28.1	(±4.7)	30.4	(±3.2)
New Hampshire	25.7	(±4.0)	21.0	(±3.3)	23.3	(±2.5)
New Jersey	20.9	(±3.0)	17.6	(±2.2)	19.2	(±1.9)
New Mexico	25.1	(±2.4)	20.2	(±2.0)	22.6	(±1.5)
New York	25.9	(±3.1)	22.9	(±2.5)	24.3	(±2.0)
North Carolina	27.4	(±3.6)	22.3	(±2.6)	24.7	(±2.2)
North Dakota	21.8	(±3.1)	18.3	(±2.6)	20.0	(±2.0)
Ohio	29.7	(±3.6)	23.0	(±2.7)	26.2	(±2.3)
Oklahoma	26.7	(±3.2)	21.1	(±2.3)	23.8	(±2.0)
Oregon	21.6	(±3.4)	20.6	(±2.7)	21.1	(±2.2)
Pennsylvania	24.0	(±2.5)	23.6	(±2.1)	23.8	(±1.6)
Rhode Island	24.1	(±2.5)	21.5	(±1.9)	22.7	(±1.6)
South Carolina	29.8	(±3.0)	20.2	(±2.0)	24.7	(±1.8)
South Dakota	36.5	(±3.6)	18.5	(±2.4)	27.3	(±2.3)
Tennessee	30.3	(±3.2)	22.4	(±2.2)	26.1	(±1.9)
Texas	25.3	(±2.4)	18.9	(±1.6)	22.0	(±1.4)
Utah	15.9	(±2.5)	12.5	(±2.0)	14.2	(±1.6)
Vermont	23.6	(±2.7)	21.0	(±2.3)	22.3	(±1.8)
Virginia	25.8	(±3.1)	20.2	(±2.4)	22.9	(±1.9)
Washington	22.4	(±2.4)	20.3	(±2.1)	21.4	(±1.6)
West Virginia	29.6	(±3.3)	26.4	(±2.5)	27.9	(±2.0)
Wisconsin	24.0	(±3.4)	22.9	(±3.2)	23.4	(±2.3)
Wyoming	23.9	(±3.1)	21.7	(±2.3)	22.8	(±1.9)
Range	*15.9–36.5*		*12.5–28.5*		*14.2–30.8*	
Median	*25.3*		*21.0*		*22.9*	

* Persons aged ≥18 years who reported having smoked ≥100 cigarettes and who reported smoking every day and some days.

† Confidence interval.

SOURCE: *State-Specific Prevalence of Current Cigarette and Cigar Smoking Among Adults—United States, 1998.* Morbidity and Mortality Weekly Report, vol. 48, no. 45, Nov. 19, 1999

---

of tobacco, at about the same frequency as they used smokeless tobacco. (See Table 8.9.)

Tobacco and the Legal System

The tobacco industry possesses one of the strongest lobbying groups in the United States. As a result the U.S. government has always, to some degree, accommodated the sale and use of tobacco. For example, the U.S. Department of Agriculture (USDA) administers a price-support system to protect American tobacco farmers. For decades, however, the tobacco industry has had to fight against legislation that would strictly regulate the sale of tobacco products, and lawsuits seeking compensation from the tobacco industry for the damage that their products have done to people's health. For the most part, the

TABLE 8.8

Prevalence of Cigar Smoking Among Adults, by State and Sex, 1998

State	Ever cigar smoking*						Past month cigar smoking†					
	Men		Women		Total		Men		Women		Total	
	%	(95% CI§)	%	(95% CI)	%	(95% CI)	%	(95% CI)	%	(95% CI)	%	(95% CI)
Alabama	65.8	(±3.9)	18.4	(±2.5)	40.8	(±2.5)	11.2	(±2.6)	2.0	(±0.9)	6.3	(±1.3)
Alaska	75.4	(±4.0)	26.0	(±3.6)	52.0	(±3.1)	9.9	(±2.8)	2.0	(±1.2)	6.1	(±1.6)
Arizona	23.1	(±3.7)	6.9	(±2.1)	14.8	(±2.1)	2.9	(±1.6)	0.1	(±0.1)	1.4	(±0.8)
Arkansas	60.9	(±3.2)	13.0	(±1.8)	35.6	(±2.0)	9.8	(±2.2)	1.4	(±0.7)	5.4	(±1.1)
California	63.0	(±2.5)	20.7	(±1.8)	41.7	(±1.7)	10.1	(±1.5)	1.8	(±0.6)	5.9	(±0.8)
Colorado	66.9	(±3.8)	22.4	(±2.9)	44.2	(±2.6)	8.2	(±2.0)	0.9	(±0.6)	4.4	(±1.0)
Connecticut	56.8	(±3.6)	13.0	(±2.0)	33.8	(±2.3)	9.7	(±2.2)	1.2	(±0.6)	5.2	(±1.1)
Delaware	52.3	(±4.4)	9.0	(±1.8)	29.6	(±2.6)	9.8	(±3.3)	0.5	(±0.3)	4.9	(±1.6)
District of Columbia	32.3	(±4.8)	10.5	(±2.4)	20.6	(±2.6)	7.1	(±2.5)	1.0	(±0.8)	3.8	(±1.2)
Florida	59.4	(±2.6)	15.8	(±1.6)	36.6	(±1.6)	10.8	(±1.7)	2.1	(±0.6)	6.2	(±0.9)
Georgia	64.7	(±3.9)	19.0	(±2.4)	40.9	(±2.4)	10.5	(±2.2)	1.8	(±1.0)	5.9	(±1.2)
Hawaii	53.6	(±4.3)	11.6	(±2.1)	32.8	(±2.6)	6.6	(±1.9)	0.8	(±0.6)	3.7	(±1.0)
Idaho	64.5	(±2.4)	18.3	(±1.6)	40.9	(±1.6)	7.2	(±1.3)	1.6	(±0.6)	4.3	(±0.7)
Illinois	68.9	(±4.2)	18.4	(±3.1)	41.8	(±2.9)	13.1	(±2.9)	2.0	(±1.6)	7.1	(±1.6)
Indiana	72.6	(±3.1)	18.3	(±2.2)	44.2	(±2.2)	13.2	(±2.4)	2.0	(±0.8)	7.3	(±1.2)
Iowa	73.5	(±2.7)	18.0	(±1.9)	44.4	(±1.9)	9.7	(±1.9)	1.3	(±0.5)	5.2	(±1.0)
Kansas	49.8	(±2.9)	12.5	(±1.6)	30.5	(±1.8)	5.4	(±1.2)	0.5	(±0.3)	2.8	(±0.6)
Kentucky	67.5	(±2.8)	11.7	(±1.4)	38.2	(±1.9)	10.4	(±2.1)	1.1	(±0.6)	5.5	(±1.1)
Louisiana	57.6	(±4.4)	12.4	(±2.4)	33.8	(±2.7)	7.8	(±2.2)	0.8	(±0.6)	4.1	(±1.1)
Maine	56.9	(±4.3)	14.2	(±2.8)	34.6	(±2.7)	7.3	(±2.4)	1.3	(±1.2)	4.1	(±1.3)
Maryland	53.7	(±3.6)	15.5	(±2.1)	33.7	(±2.2)	8.8	(±2.2)	1.6	(±1.0)	5.0	(±1.2)
Massachusetts	60.8	(+2.9)	17.1	(±2.1)	37.8	(±1.9)	11.2	(±1.8)	1.2	(±0.6)	5.9	(±0.9)
Michigan	74.5	(±3.0)	23.6	(±2.4)	47.9	(±2.2)	12.1	(±2.2)	2.2	(±0.8)	6.9	(±1.2)
Minnesota	45.3	(±2.4)	16.1	(±1.7)	30.3	(±1.5)	7.5	(±1.3)	1.3	(±0.5)	4.3	(±0.7)
Mississippi	66.1	(±3.6)	14.3	(±2.0)	38.6	(±2.3)	9.5	(±2.4)	1.0	(±0.6)	5.0	(±1.2)
Missouri	69.0	(±3.0)	18.2	(±2.1)	42.2	(±2.2)	10.9	(±2.3)	2.1	(±1.0)	6.2	(±1.2)
Montana	68.7	(±3.4)	16.9	(±2.5)	42.1	(±2.5)	8.2	(±2.0)	0.2	(±0.2)	4.1	(±1.0)
Nebraska	70.4	(±3.5)	20.0	(±2.2)	44.2	(±2.2)	9.5	(±2.0)	1.3	(±0.6)	5.2	(±1.0)
Nevada	71.1	(±4.3)	25.6	(±4.5)	48.6	(±3.3)	11.9	(±2.9)	2.9	(±1.4)	7.4	(±1.6)
New Hampshire	66.8	(±4.0)	15.9	(±3.0)	40.6	(±2.9)	10.7	(±3.2)	1.5	(±1.0)	5.9	(±1.6)
New Jersey	54.3	(±3.7)	15.1	(±2.2)	33.8	(±2.2)	12.5	(±2.4)	1.3	(±0.7)	6.6	(±1.2)
New Mexico	68.6	(±2.6)	20.0	(±1.9)	43.6	(±1.8)	7.7	(±1.5)	0.9	(±0.4)	4.2	(±0.8)
New York	54.4	(±3.5)	15.2	(±2.1)	33.6	(±2.2)	12.1	(±2.4)	1.0	(±0.5)	6.2	(±1.2)
North Carolina	61.0	(±4.3)	16.2	(±2.5)	37.6	(±2.6)	7.6	(±2.2)	1.6	(±1.0)	4.5	(±1.2)
North Dakota	68.1	(±3.6)	15.7	(±2.6)	41.5	(±2.6)	7.0	(±1.9)	1.0	(±0.8)	4.0	(±1.0)
Ohio	65.7	(±3.7)	14.8	(±2.2)	39.0	(±2.5)	10.0	(±2.5)	1.8	(±1.0)	5.7	(±1.3)
Oklahoma	35.4	(±3.4)	12.7	(±1.9)	23.6	(±2.0)	3.5	(±1.4)	1.2	(±0.7)	2.3	(±0.8)
Oregon	72.5	(±3.6)	22.3	(±2.7)	46.7	(±2.6)	8.8	(±2.3)	1.1	(±0.6)	4.8	(±1.2)
Pennsylvania	60.0	(±2.9)	14.3	(±1.7)	35.8	(±1.8)	11.9	(±2.0)	1.9	(±0.7)	6.5	(±1.0)
Rhode Island	59.3	(±2.9)	15.1	(±1.7)	36.0	(±1.8)	10.8	(±1.9)	1.0	(±0.5)	5.5	(±0.9)
South Carolina	60.6	(±3.1)	15.7	(±2.0)	37.1	(±2.0)	10.0	(±1.9)	1.6	(±0.7)	5.6	(±1.0)
South Dakota	66.2	(±3.5)	14.2	(±2.2)	39.5	(±2.4)	9.7	(±2.3)	1.0	(±0.7)	5.2	(±1.2)
Tennessee	46.2	(±3.5)	11.3	(±1.7)	27.8	(±2.0)	7.4	(±1.8)	0.8	(±0.4)	3.9	(±0.9)
Texas	62.9	(±2.6)	16.7	(±1.4)	39.2	(±1.7)	7.5	(±1.1)	1.6	(±0.6)	4.5	(±0.6)
Utah	47.8	(±3.8)	13.4	(±2.0)	30.2	(±2.3)	3.9	(±1.2)	1.1	(±0.7)	2.5	(±0.7)
Vermont	66.8	(±3.0)	17.4	(±2.1)	41.3	(±2.2)	9.6	(±3.1)	0.9	(±0.5)	5.1	(±1.6)
Virginia	65.4	(±3.6)	15.4	(±2.3)	39.6	(±2.5)	10.5	(±2.0)	1.3	(±0.6)	5.7	(±1.0)
Washington	69.7	(±2.6)	22.4	(±2.2)	45.6	(±1.9)	9.0	(±1.7)	1.4	(±0.5)	5.1	(±0.9)
West Virginia	65.9	(±3.3)	15.0	(±2.0)	39.0	(±2.2)	7.1	(±1.8)	1.0	(±0.6)	3.8	(±0.9)
Wisconsin	76.7	(±3.1)	24.6	(±3.1)	49.7	(±2.6)	11.8	(±2.5)	1.6	(±1.0)	6.5	(±1.3)
Wyoming	71.9	(±3.3)	21.6	(±2.3)	46.5	(±2.3)	5.9	(±1.5)	1.2	(±0.8)	3.5	(±0.8)
Range	23.1–76.7		6.9–26.0		*14.8–52.0*		2.9–13.2		0.1–2.9		*1.4–7.4*	
Median	64.7		15.8		*39.0*		9.7		1.3		*5.2*	

* Persons aged ≥18 years who reported having ever smoked a cigar, even just a few puffs.

† Persons aged ≥18 years who reported smoking a cigar within the previous month.

§ Confidence interval.

SOURCE: *State-Specific Prevalence of Current Cigarette and Cigar Smoking Among Adults—United States, 1998.* Morbidity and Mortality Weekly Report, vol. 48, no. 45, Nov. 19, 1999

tobacco industry was successful in these battles, but in the 1990s this began to change.

When individuals sue tobacco companies they claim that cigarette smoking contributed to their ill health in a variety of ways. Some of the issues raised in these suits include:

• The cost to society for medical expenses related to tobacco use

TABLE 8.9

Middle School and High School Students Currently* Using Tobacco Products, by Type of Product, Sex, and Race/Ethnicity, 1999

	Sex				Race/Ethnicity									
	Male		Female		White		Black		Hispanic		Total			
Type of tobacco product	%	(95% CI†)	%	(95% CI)	%	(95% CI)	%	(95% CI)	%	(95% CI)	%	(95% CI)		
Any use§														
Middle school	14.2	(± 2.2)	11.3	(± 2.2)	11.6	(± 2.3)	14.4	(± 2.7)	15.2	(± 5.2)	**12.8**	**(± 2.0)**		
High school	38.1	(± 3.2)	31.4	(± 3.1)	39.4	(± 3.2)	24.0	(± 4.2)	30.7	(± 4.4)	**34.8**	**(± 2.7)**		
Cigarette														
Middle school	9.6	(± 1.7)	8.8	(± 1.7)	8.8	(± 2.0)	9.0	(± 1.8)	11.0	(± 4.1)	**9.2**	**(± 1.6)**		
High school	28.7	(± 2.8)	28.2	(± 3.3)	32.8	(± 3.1)	15.8	(± 3.8)	25.8	(± 4.7)	**28.4**	**(± 2.7)**		
Smokeless														
Middle school	4.2	(± 1.3)	1.3	(± 0.5)	3.0	(± 1.1)	1.9	(± 0.9)	2.2	(± 0.9)	**2.7**	**(± 0.7)**		
High school	11.6	(± 2.8)	1.5	(± 0.6)	8.7	(± 2.1)	2.4	(± 1.3)	3.6	(± 1.6)	**6.6**	**(± 1.6)**		
Cigar														
Middle school	7.8	(± 1.3)	4.4	(± 1.3)	4.9	(± 1.0)	8.8	(± 2.3)	7.6	(± 2.9)	**6.1**	**(± 1.1)**		
High school	20.3	(± 1.9)	10.2	(± 1.6)	16.0	(± 1.6)	14.8	(± 3.5)	13.4	(± 2.9)	**15.3**	**(± 1.4)**		
Pipe														
Middle school	3.5	(± 0.8)	1.4	(± 0.6)	2.0	(± 0.6)	2.0	(± 0.9)	3.8	(± 1.7)	**2.4**	**(± 0.5)**		
High school	4.2	(± 0.9)	1.4	(± 0.5)	2.6	(± 0.6)	1.8	(± 0.9)	3.8	(± 1.4)	**2.8**	**(± 0.5)**		
Bidi														
Middle school	3.1	(± 0.8)	1.8	(± 0.6)	1.8	(± 0.5)	2.8	(± 1.3)	3.5	(± 1.6)	**2.4**	**(± 0.6)**		
High school	6.1	(± 1.0)	3.8	(± 1.0)	4.4	(± 0.9)	5.8	(± 2.1)	5.6	(± 2.1)	**5.0**	**(± 0.8)**		
Kretek														
Middle school	2.2	(± 0.6)	1.7	(± 0.7)	1.7	(± 0.7)	1.7	(± 0.8)	2.1	(± 0.6)	**1.9**	**(± 0.5)**		
High school	6.2	(± 1.1)	5.3	(± 1.5)	6.5	(± 1.5)	2.8	(± 1.5)	5.5	(± 1.9)	**5.8**	**(± 1.2)**		

* Used tobacco on one or more of the 30 days preceding the survey.

† Confidence interval.

§ Use of cigarettes, smokeless tobacco, cigars, pipes, bidis, or kreteks.

SOURCE: *Tobacco Use Among Middle and High School Students—United States, 1999.* Morbidity and Mortality Weekly Report, vol. 49, no. 3, January 28, 2000

- Advertising and selling tobacco products to youth

- The location and placement of tobacco products

- Taxes on tobacco products

- The harmful effects of secondhand smoke

- The addictive nature of nicotine

- The failure of tobacco companies to divulge information about the harmful and addictive nature of smoking, and their concealment of evidence to that fact

- Smoking in public buildings

Tobacco companies argue that smokers choose to smoke and that they must assume the risk for doing so. In a landmark 1994 case, *Cipollone v. Liggett Group Inc.* (60 LW 4703), the Supreme Court ruled that smokers may sue cigarette companies for concealing facts about smoking and that the tobacco company was at least partially liable in the death of Rose Cipollone. However, the court also determined that Cipollone herself was partially responsible for her cigarette use and subsequent death.

In 1991 the American Medical Association (AMA) publicly charged R.J. Reynolds (RJR) with targeting children through its Joe Camel advertising campaign. Later that year Janet Mangini, a California attorney, brought suit to end the Joe Camel campaign, becoming the first person to legally challenge the tobacco industry for targeting minors in its advertising. Six years of "discovery," taking depositions and obtaining records, followed. As the trial date neared in May 1997, RJR offered to halt the ad campaign to stop the trial, and also agreed to provide for public release its documents about youth marketing and the Joe Camel campaign.

In 1997, in *Broin v. Philip Morris*, a flight attendant sued claiming that second-hand smoke in airplanes had harmed her health. The suit was settled out of court with Philip Morris agreeing to pay some $300 million to establish a medical foundation.

Nevertheless, most legal action against "Big Tobacco" prior to 1998 was unsuccessful because most of the suits were filed as "class action" suits where large numbers of complainants unite to sue. A 1995 case, *Castono v. The American Tobacco Company Inc.* (85 F 3rd 734, 5th Cir. 1996), filed in Louisiana, included people who had purchased and smoked cigarettes and had become nicotine-dependent. The U.S. District Appeals Court in Louisiana "decertified" the suit, ruling that individual issues predominated over common issues, thus making the case not a proper candidate for a class action.

That ruling—that individual issues were greater than common issues in suits against the tobacco industry—has been handed down both by state and federal courts in at least 30 suits. In 1996, however, in *Howard A. Engle, MD, v. R.J. Reynolds Tobacco, Philip Morris, Brown and Williams, Lorillard Tobacco, the American Tobacco Company, et al.*, filed in Dade County, Florida, the state appellate court ruled that the suit could proceed to trial, although it was reduced to include only Florida residents. In 1999 the jury ruled for the plaintiffs and, in July 2000, assessed penalties of $145 billion in punitive damages against the country's five largest tobacco companies. The money was to be split among Florida residents who could prove they became ill from smoking. Philip Morris was ordered to pay $73.96 billion, R.J. Reynolds $36.28 billion, Brown and Williamson $17.59 billion, Lorillard Tobacco $16.25 billion, and Liggett Group $790 million.

The trial, which lasted for two years, was the longest civil trial in the history of tobacco litigation, and the penalty was the largest ever levied in any case. Rather than building their case around the dangers of smoking, the plaintiffs focused on the negligent conduct of the tobacco companies that they claimed had covered up smoking risks for more than four decades. While some observers suggested that the jury's decision would encourage other class actions, tobacco industry executives contend that the ruling will be reversed on appeal. They claim that the case will be overturned because it should not have been allowed as a class action, and that the fine is outrageous and will put the tobacco companies out of business. In any case, legal experts predicted the verdict was many years away from being final.

In the mid-1990s a number of state governments initiated suits against tobacco companies to recoup state Medicaid spending on tobacco-related illnesses. With the pressure of state and private lawsuits building up, the tobacco industry has begun to seek settlements with the states. In November 1998, in what was termed a "Master Settlement Agreement" between the major tobacco manufacturers and 46 state attorneys general (Texas, Florida, Minnesota, and Mississippi settled independently), the tobacco companies agreed to accept a number of limitations on how they marketed and sold their products. These included: ceasing youth-targeted advertising, marketing, and promotion by stopping the use of cartoon characters in advertising; limiting brand-name sponsorship of events with significant youth audiences; terminating outdoor advertising; banning youth access to free samples; and setting the minimum cigarette package size at 20. In addition to these limits on their business practices, the tobacco industry agreed to pay more than $200 billion to the states.

Furthermore, in his 1999 State of the Union address, President Bill Clinton announced his intent to sue the tobacco industry to recover money spent by the federal government to treat illnesses caused by smoking. Accord-

ingly, the U.S. Department of Justice filed suit in September 1999 (*United States of America v. Philip Morris Inc., et al*) in the U.S. District Court for the District of Columbia.

In the 1990s testimony before Congress claimed that the tobacco industry deliberately manipulated the amount of nicotine in its cigarettes, and that cigarettes were nothing more than a delivery system for nicotine, a drug now widely recognized as physiologically addictive. The industry feared that if cigarettes were perceived as a delivery system for nicotine, tobacco products might fall under the control of the FDA. In 1995 the FDA ruled that nicotine was indeed a drug, and liable to its regulation—the first time the tobacco industry had been regulated. In March 2000, however, the Supreme Court ruled 5–4 that the FDA does not have jurisdiction to regulate tobacco products or cigarette-company marketing practices under existing law.

Most adult smokers claim they became addicted as teenagers. An aggressive advertising campaign, financed by the tobacco companies largely as a concession to improve their image in the eyes of the general public, has stressed the unlawfulness of vendors selling tobacco products to minors. With public health care costs already overwhelming government budgets, the public-health issue has increasingly become a fiscal issue. Some countries, such as Sweden, have attempted to change the smoking habits of residents by imposing high taxes on cigarettes.

ENDOCRINE DISRUPTORS—ENVIRONMENTAL HORMONES

Medical and scientific researchers are increasingly linking chemical compounds known as organochlorines to the endocrine systems of humans and wildlife. The endocrine system—also called the hormone system—is made up of glands located throughout the body, hormones that are synthesized and secreted by the glands into the bloodstream, and receptors in the various target organs and tissues. The receptors recognize and respond to the hormones. The function of the system is to regulate the many bodily processes, including control of blood sugar, growth and function of the reproductive systems, regulation of metabolism, brain and nervous system development, and development of the organism from conception through adulthood and old age.

Substances that interfere with these processes are called "endocrine disruptors." Although some occur naturally—for example, plant-derived hormones—most appear to be man-made. Disruption of the endocrine system can produce certain genetic, reproductive, and behavioral abnormalities in humans and wildlife; increases in several types of cancer not related to smoking or age; malformations; and nervous system disorders.

Endocrine disruptors are sometimes referred to as "environmental estrogens" because they are so widely dis-

FIGURE 8.9

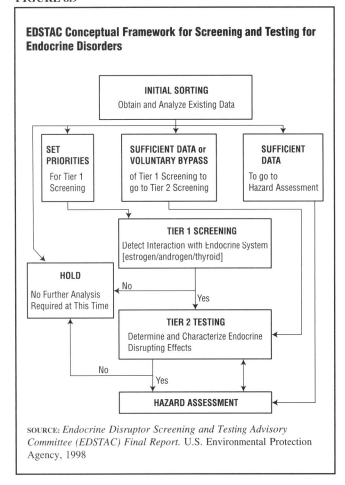

EDSTAC Conceptual Framework for Screening and Testing for Endocrine Disorders

SOURCE: *Endocrine Disruptor Screening and Testing Advisory Committee (EDSTAC) Final Report.* U.S. Environmental Protection Agency, 1998

industrial cleaning compounds. Many such compounds have been banned in the United States. Nonetheless, they persist in the food chain for many years and accumulate in animal tissue. Moreover, many of these chemicals continue to be used in developing countries.

Researchers in North America and Europe are studying the possibility of a link between environmental estrogens and the occurrence of birth abnormalities, Alzheimer's disease, sterility in both males and females, hyperactivity in children, neurological illnesses, and many cancers. Both men and women appear to be susceptible to endocrine disruption. In men, some studies show that estrogenic compounds affect the development of the Sertoli cells in the testicles. These cells secrete masculinizing hormones that regulate sperm production, the descent of the testicles, and the development of the urethra.

Because of the potentially serious consequences of human exposure to endocrine-disrupting chemicals, in 1996 Congress included specific language on endocrine disruption in the Food Quality Protection Act (PL 104-170) and Safe Drinking Water Act Amendments (PL 104-182, 1996). The first mandated the EPA to develop an endocrine disruptor-screening program while the latter authorized the EPA to screen endocrine disruptors found in drinking water. The laws required the EPA to develop a screening program by 1998, to implement the program by 1999, and to report to Congress on the program's progress by August 2000. Accordingly, in 1996, the EPA formed the Endocrine Disruptor Screening and Testing Advisory Committee (EDSTAC) to design a protocol for the program. (See Figure 8.9.) That committee's report was due out in late 2000.

THE GULF WAR SYNDROME

Many veterans of the Persian Gulf War (1991) returned with physical complaints that have baffled medical experts. "Gulf War syndrome" is the name given to an array of symptoms—fatigue, skin rashes, memory loss, headaches—experienced by men and women who served in the war. Some sources have claimed these conditions may have resulted from exposure to poison gas that was inadvertently released when American troops destroyed caches of the gas following the actual conflict. Some theories contend the troops were heavily exposed to fumes of burning oil during the war, while others suggest that the immunizations given to the soldiers may have caused such responses.

The Department of Veterans Affairs (VA) offers Gulf War veterans eligibility for medical treatment. More than 67,000 service members have responded to the VA's Gulf War Registry program, which offers physical examinations to all eligible service members. Most veterans are diagnosed and treated; for some, however, symptoms have been chronic. Data from about 10,000 claimants'

persed in the environment that they turn up in rain water, well water, lakes, and oceans, as well as in foods consumed by birds, fish, animals, and humans. Some scientists worry that organochlorines mimic or block the action of natural estrogen, thereby disrupting the endocrine system.

Some effects of certain estrogenic compounds have been well known for some time. Among these are the eggshell thinning and cracking that led to the population decline of the American bald eagle; the reproductive abnormalities of women exposed *in utero* to diethylstilbestrol (DES), a synthetic estrogen prescribed between 1948 and 1971 to prevent miscarriages; and reported declines in the quantity and quality of sperm in humans. In 1996 researchers at the National Biological Service, in a study of Columbia River otters, found a direct correlation between the level of chemicals and pesticides in the otters' livers and the size of the males' genitalia.

Only recently, however, have researchers begun to realize how many compounds in the environment are estrogenic. More than 50 of these endocrine-disrupting chemicals have been observed to disrupt the hormone or reproductive system, but the remainder of the 85,000 chemicals currently in use remain to be studied. Among them are many herbicides, pesticides, insecticides, and

health exams found unexplained illness in approximately 12 percent of the cases. If a veteran's symptoms defy diagnosis, the veteran can be referred to one of the nation's four Gulf War Referral Centers for treatment.

Since 1995 the federal government has committed $115 million for 121 research projects related to Gulf War illnesses. The VA has provided compensation payments to chronically disabled Gulf War veterans with undiagnosed illnesses. (A disability is considered chronic if it lasts more than six months.)

The federal response to the health consequences of Gulf War service is led by the Persian Gulf Veterans Coordinating Board composed of the VA, the U.S. Department of Defense, and the U.S. Department of Health and Human Services. The Presidential Advisory Committee on Gulf War Veterans' Illnesses was formed by President Bill Clinton in 1995. In 1999 he signed into law the Veterans Millennium Health Care and Benefits Act (PL 106-117) which extends medical care to veterans, their spouses, and their children to December 31, 2003.

THE TOXICS RELEASE INVENTORY

In 1984 a deadly cloud of chemicals was released from the Union Carbide pesticide plant in Bhopal, India, following an explosion in the plant. The methyl isocyanate gas killed approximately 3,000 people and injured 200,000 others. Shortly after, a similar chemical release occurred in West Virginia. Such incidents fueled the demand by workers and the general public for information about hazardous materials in their areas. As a result Congress passed the Emergency Planning and Community Right-to-Know Act of 1986 (EPCRA; PL 99-499).

The EPCRA established, among other things, the Toxics Release Inventory (TRI), a publicly available database (http://www.epa.gov/tri) that contains information on toxic chemical releases by various facilities. Under the TRI program emissions of nearly 650 toxic chemicals must be reported to the EPA each year. Facilities that must report include electric utilities, chemical distributors, petroleum bulk terminals, hazardous waste treatment and disposal plants, solvent recovery services, and facilities in the manufacturing, metal mining, and coal mining industries. Companies must report if they meet certain criteria based on the size of the company and whether it exceeds certain thresholds for the 650 toxic chemicals. In 1998, 23,000 facilities reported to the EPA. Companies that fail to report face civil penalties—fines of up to $27,500 a day per violation. In addition, private citizens may file lawsuits to force a company to obey the law.

According to the 1998 TRI (the latest data available) facilities reported a total of 7.3 billion pounds of chemical releases. On-site releases accounted for 93.9 percent of all releases in 1998, while off-site releases (when a

FIGURE 8.10

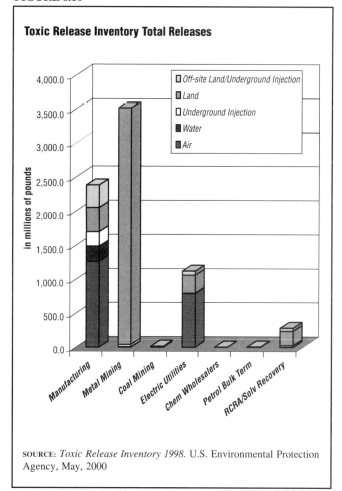

SOURCE: *Toxic Release Inventory 1998*. U.S. Environmental Protection Agency, May, 2000

facility sends toxic chemicals to another facility where they are then released) accounted for the remainder. Most (66 percent) off-site releases were to landfills or surface impoundments.

Metal mining (48 percent) accounted for the largest amount of total emissions, followed by manufacturing (33 percent), electric utilities (15 percent), and solvent recovery and hazardous waste disposal facilities (4 percent combined). (See Figure 8.10.) Because 1998 was the first year in which most of the categories reported, no trends can yet be derived from the data. Emissions from the manufacturing industry, which has reported since 1988, declined 45 percent over that time period.

INDOOR AIR QUALITY

Indoor pollution has become a serious problem in America. Although most people think of outdoor air when they think of air pollution, studies now reveal that indoor environments are not safe havens from air pollution. Improvements in home and building insulation and the widespread use of central air conditioning and heating systems have largely ensured that any contaminant present indoors will not be diluted by outside air and, there-

FIGURE 8.11

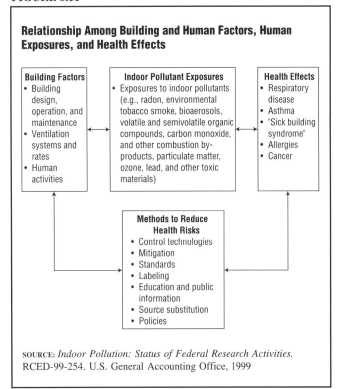

Relationship Among Building and Human Factors, Human Exposures, and Health Effects

Building Factors
• Building design, operation, and maintenance
• Ventilation systems and rates
• Human activities

Indoor Pollutant Exposures
• Exposures to indoor pollutants (e.g., radon, environmental tobacco smoke, bioaerosols, volatile and semivolatile organic compounds, carbon monoxide, and other combustion by-products, particulate matter, ozone, lead, and other toxic materials)

Health Effects
• Respiratory disease
• Asthma
• "Sick building syndrome"
• Allergies
• Cancer

Methods to Reduce Health Risks
• Control technologies
• Mitigation
• Standards
• Labeling
• Education and public information
• Source substitution
• Policies

SOURCE: *Indoor Pollution: Status of Federal Research Activities.* RCED-99-254. U.S. General Accounting Office, 1999

fore, will become more concentrated. This is a particular problem with radon, lead, asbestos, fungus and molds, carbon monoxide, cigarette smoke, or any of the various chemicals used in cleaning products or carpeting and insulation materials. (See Figure 8.11.)

Indoor pollution, however, is not just a product of modern, well-insulated homes and buildings. In 1995 doctors in the Bronx, New York, began observing an emerging epidemic of asthma, citing hospitalization rates as high as 17.3 per 1,000 people and death rates as high as 11 per 100,000. Both rates were eight times the national average and the rate of incidence among children was twice the national rate. The Bronx, an area with many dilapidated and neglected buildings, is among the worst places in the country for asthma. Among the causes cited by area physicians were many factors associated with indoor pollution: dust mites (minute insects that live on house dust and human skin residue), cockroaches, smoking, dust, and respiratory viruses and bacteria (which spread easily in crowded quarters). Scientists suspected that stress and the startup of a waste incinerator in the Bronx three years earlier were factors as well.

Modern indoor environments contain a variety of pollution sources, including synthetic building materials, consumer products, and dust mites. People, pets, and indoor plants also contribute airborne pollution. Efforts to lower energy costs by tightly sealing and insulating buildings have increased the likelihood that pollutants will accumulate.

Reports of illness and allergy among building occupants have become commonplace. Scientific evidence suggests that respiratory disease, allergies, mucous membrane irritation, nervous system defects, cardiovascular symptoms, reproductive problems, and lung cancer may be linked to exposure to indoor air pollution. Scientists rank indoor air pollution among the top five environmental health risks, although public opinion polls report that most Americans do not perceive the risks of indoor pollution to be great.

Providing good air quality is not only a complex scientific and technical issue, but also a complicated issue of public policy in determining the proper role of the government in safeguarding people's health. For example, the EPA estimates that exposure of nonsmokers to cigarette smoke may cause thousands of lung cancer deaths a year in the United States. Cigarette smoke also contributes to a wide range of diseases and incurs enormous expense in work-time loss and medical and insurance costs. Many states and communities now have levied restrictions on smoking in public places. As discussed earlier in this chapter lead, radon, and asbestos have also posed serious and costly problems in indoor settings.

Poorly ventilated buildings sometimes become the cause of "sick-building syndrome." The term, first employed in the 1970s, describes a spectrum of specific and nonspecific complaints reported by building occupants. Such symptoms might include headache, fatigue, or difficulty breathing, and begin soon after entering a building and subside after leaving that building. When 20 percent of a building's occupants report complaints, the World Health Organization (WHO) calls that building "sick." Experts generally believe there are many people who do not complain, even when they experience symptoms, and the WHO and the EPA estimate that 30 percent of all buildings worldwide are unfit for human occupation. Surveys and assessments by private corporations and federal agencies estimate that sick-building syndrome goes undetected in another 10–15 percent of structures. Many billions of dollars in income and productivity are lost annually because of employees falling ill from problems linked to sick-building syndrome.

The first legislation to deal specifically with indoor air quality was Title IV of the Superfund Amendments and Reauthorization Act of 1986, which called for the EPA to establish an advisory committee to conduct research and disseminate information. In October 1991 the General Accounting Office (GAO) reported on progress of the legislation in *Indoor Air Pollution: Federal Efforts are Not Effectively Addressing a Growing Problem.* The GAO concluded not only that the EPA's emphasis on indoor pollution was not commensurate with the health risks posed by the problem, but also that research had been, and would likely continue to be, constrained by lack of fund-

FIGURE 8.12

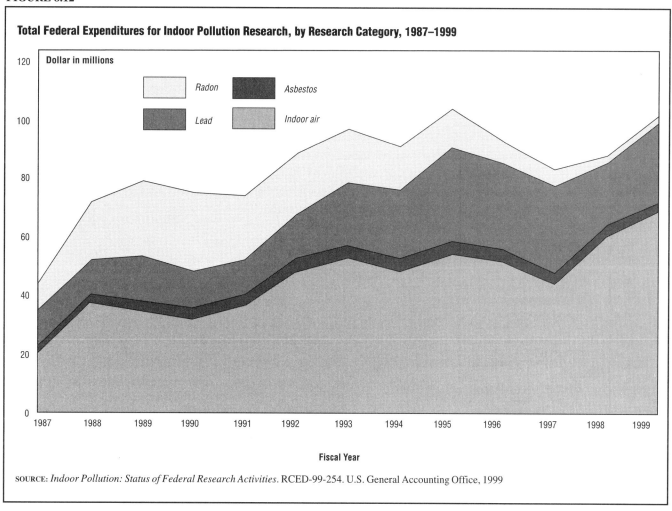

Total Federal Expenditures for Indoor Pollution Research, by Research Category, 1987–1999

SOURCE: *Indoor Pollution: Status of Federal Research Activities*. RCED-99-254. U.S. General Accounting Office, 1999

ing. Accordingly, the proposed Indoor Air Quality Act of 1991 was not enacted by Congress.

The EPA and other sources continue to identify indoor pollution as one of the most serious environmental risks to human health because:

- Concentrations of pollutants in indoor air can be more than two to five times higher than those in outdoor air.

- People spend an average of 80–90 percent of their time indoors, even more for certain groups, such as the elderly, the chronically ill, and the very young.

- The indoor environment is unique in that it contains materials and surfaces that act as emitters and reservoirs of pollutants.

Federal agencies reported that they spent a total of almost $1.1 billion on indoor pollution-related research from 1987 to 1999. Most of that amount went toward indoor air research, followed by studies of lead, asbestos, and radon. (See Figure 8.12.) The National Institute of Environmental Health Sciences (NIEHS) spent the most (almost $400 million), with the National Heart, Lung, and Blood Institute spending $175.2 million, the EPA

$140.4 million, the DOE $136.5 million, and the National Institute of Allergy and Infectious Diseases $93.7 million. (See Figure 8.13.)

In 1999 the GAO once again reviewed the status of indoor air quality activities in its *Indoor Pollution: Status of Federal Research Activities*. It found that significant strides have been made in understanding the risks posed by chemicals and other contaminants commonly found in homes, offices, and schools. Nonetheless, it concluded that "many gaps and uncertainties remain in the assessment of exposures to known indoor pollutants." These gaps include specific sources of exposures; the magnitude of exposures; the relative role of specific exposures such as inhalation, ingestion, and skin contact; the nature, duration, and frequency of human activities that contribute to exposures; and the geographic distribution of exposures to certain pollutants for the U.S. population as a whole.

NOISE POLLUTION

Noise is unwanted sound. The word is derived from the Latin word *nausea*, meaning "seasickness." Experts agree that noise pollution is bad and getting worse in

FIGURE 8.13

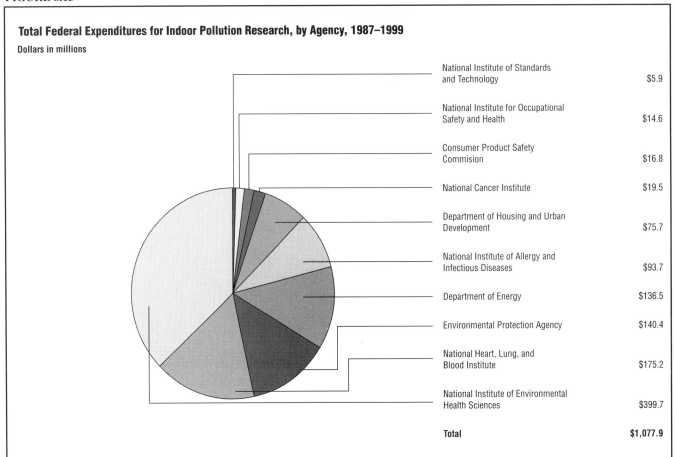

Total Federal Expenditures for Indoor Pollution Research, by Agency, 1987–1999

Dollars in millions

National Institute of Standards and Technology	$5.9
National Institute for Occupational Safety and Health	$14.6
Consumer Product Safety Commision	$16.8
National Cancer Institute	$19.5
Department of Housing and Urban Development	$75.7
National Institute of Allergy and Infectious Diseases	$93.7
Department of Energy	$136.5
Environmental Protection Agency	$140.4
National Heart, Lung, and Blood Institute	$175.2
National Institute of Environmental Health Sciences	$399.7
Total	**$1,077.9**

NOTES: Amounts are expressed in constant 1999 dollars. Amounts for fiscal year 1999 represent planned expenditures. Spending amounts do not add to total because of rounding.

SOURCE: *Indoor Pollution: Status of Federal Research Activities*. RCED-99-254. U.S. General Accounting Office, 1999

America. Noise from road traffic, airplanes, jet skis, garbage trucks, construction equipment, manufacturing processes, lawn mowers, subways, and leaf blowers are just a few of the unwanted sounds that are routinely broadcast into the air. Physicists, audiologists, engineers, architects, and physicians report that permanent hearing loss caused by amplified music is a widespread affliction in the United States. Although hearing loss is the most dramatic effect of noise pollution, even smaller amounts of noise can negatively affect health and well-being. Besides hearing loss, some other problems related to noise include:

• Stress

• High blood pressure

• Sleep loss

• Distraction and lost worker productivity

• A general decline in quality of life

Noise levels are measured in decibels (db). A noise level of less than 65 db is considered acceptable from an environmental standpoint, although several studies have found that levels of 60–65 db are annoying to 9 percent of the public. Soft whispers have a decibel level of 30. An air conditioner at 20 feet measures 60 db. The noise level of a vacuum cleaner or a crowded restaurant is about 70 db. Average city traffic, garbage disposals, or alarm clocks at two feet could be 80 db. The subway, a motorcycle, or a lawn mower would be approximately 90 db; a basketball arena 108 db. A rock concert or thunderclap (120 db), a gunshot blast or jet plane (140 db), or a rocket launching pad (180 db) can be dangerous to those under constant exposure. Researchers for the EPA have found that 20 percent of the population is "highly annoyed" if sound levels reach 55 db.

Many cities have pressed the Federal Aviation Administration (FAA) to steer airplane flight paths around metropolitan areas in order to reduce noise over residential areas. The airline industry has responded by beginning to build jet engines with noise levels in mind.

Legislation Against Noise

The air into which noise is emitted is a "commons," or a public space. It belongs to no one person but to everyone. People, organizations, and businesses, therefore, do not have unlimited rights to broadcast noise. The United States has been slow to confront the issue of noise. At a time when European nations were addressing the issue of noise abatement, in 1981 Congress eliminated the EPA's previously allocated funds for noise programs.

The Office of Noise Abatement and Control (ONAC) of the EPA was established by the Noise Control Act of 1972 (NCA; PL 92-574). During President Reagan's administration, noise pollution came to be viewed as a local problem because noise pollution does not travel very far and quickly dissipates. Some legislators believed that state and local regulation was more efficient than federal regulation since local governments could more easily gauge and respond to noise situations in their area. Consequently, in 1981, Congress eliminated all funding for ONAC, although it did not repeal the NCA. And while many of the provisions of that original law have become outdated and obsolete, others still—technically—could be invoked, although they have not generally been.

Thus, noise pollution has fallen to state and local governments to define and regulate. Much like the federal government, most states have been slow to do so. Increasingly, however, citizens are filing lawsuits based on noise issues. People regularly file noise complaints against airports and road builders, and police often respond to noise-related neighborhood conflicts.

Noisy Highways

In 1968 California built what are believed to be the first noise barriers along modern federal highways to shield neighboring residential neighborhoods from the sound of heavy traffic. Within four years the federal government followed California's lead requiring that, wherever a state builds or expands a federally funded highway, an attempt must be made to curb excessive highway noise. Federal regulations require that when noise levels on a road approach 67 db, a state must bring about a "substantial" reduction in sound level. Since then most states have built highway noise barriers, with the primary cost paid by the U.S. Department of Transportation (DOT). Many states have voluntarily built noise barriers along existing highways. The DOT reported that, by 1997 (the latest data available), noise barriers had been erected along 1,300 miles of federal highway and were being added at a rate of approximately 100 miles per year. California led the way with 438 miles of barriers.

One type of barrier is a natural wall created by mounding earth and planting grass or other vegetation. Walls are also built of concrete, brick, and other construction materials. Typically they are 12 feet in height, though in the Northeast they rise from 16 to 20 feet and in Minnesota, which has been especially anxious to reduce highway noise, to 30 feet. Noise barriers cost an average of a million dollars a mile, and elaborate ones cost considerably more.

FOOD SAFETY—CHANGING MARKET, CHANGING RISKS

According to federal officials the U.S. food supply is among the safest in the world. Nevertheless, episodes of food poisoning and diseases such as salmonella are becoming increasingly common in the United States. The National Centers for Disease Control and Prevention estimates that as many as 76 million illnesses and 6,000 deaths annually are caused by foodborne hazards. Most of these cases can be traced to microbial contamination, although health effects from chronic pesticide exposure in the diet, particularly among children, are causing growing concern.

Americans have changed their eating habits, and with such changes have come additional risks. Some explanations for the increase in the number of foodborne illnesses include:

- Americans are eating out more than in the past.

- More and more foods are being imported from foreign countries; imports have more than doubled in the past seven years. For example, in 1996 and 1997, outbreaks of illness linked to the *Cyclospora* parasite in raspberries from Guatemala affected nearly 2,500 people in the United States. Also, in 1997, a hepatitis A outbreak was traced to contaminated strawberries from Mexico.

- New pathogens and strains of organisms are turning up in the food supply.

- Many people are careless about food preparation in the home.

More than 200 known illnesses associated with bacteria, parasites, or viruses are transmitted through food, ranging from relatively minor episodes of diarrhea or vomiting to serious and chronic diseases such as gastroenteritis and meningitis. Individuals may also experience long-term health effects from pesticide or drug residues in food. According to the USDA's Economic Research Service, medical treatment costs and productivity losses resulting from these illnesses and deaths range from $7 billion to $37 billion a year.

The USDA and FDA have primary responsibility for ensuring the safety of imported foods. These agencies, however, inspect and test less than 2 percent of the nearly three million shipments of imported foods each year. They rely on selecting and testing samples taken at food inspection sites such as ports of entry, warehouses, and businesses. Inspectors rely, in part, on their own senses of sight, smell, and touch to assess the food. In addition they send samples to laboratories for testing, which may take several days or a week. Before these agencies can complete such tests, food has sometimes already been released into the market and consumed. As the number of

TABLE 8.10

Reported Foodborne-Disease Outbreaks, Cases, and Deaths in the U.S.*, by Food Type, 1996†

Vehicle of transmission	Outbreaks No.	Outbreaks (%)	Cases No.	Cases (%)	Deaths No.	Deaths (%)
Beef	7	(1.5)	227	(1.0)	0	(0.0)
Ham	4	(0.8)	89	(0.4)	0	(0.0)
Pork	2	(0.4)	115	(0.5)	0	(0.0)
Chicken	6	(1.3)	315	(1.4)	0	(0.0)
Turkey	3	(0.6)	187	(0.8)	0	(0.0)
Other/unknown meat	1	(0.2)	59	(0.3)	0	(0.0)
Shellfish	5	(1.0)	514	(2.3)	0	(0.0)
Other fish	24	(5.0)	105	(0.5)	0	(0.0)
Milk	2	(0.4)	48	(0.2)	0	(0.0)
Eggs	3	(0.6)	66	(0.3)	0	(0.0)
Ice cream	6	(1.3)	183	(0.8)	0	(0.0)
Other/unknown dairy	2	(0.4)	31	(0.1)	0	(0.0)
Baked foods	6	(1.3)	81	(0.4)	0	(0.0)
Fruits and vegetables	13	(2.7)	1,807	(8.0)	1	(25.0)
Mushrooms	3	(0.6)	10	(0.0)	0	(0.0)
Potato salad	1	(0.2)	12	(0.1)	0	(0.0)
Poultry, fish, and egg salads	7	(1.5)	789	(3.5)	0	(0.0)
Other salad	18	(3.8)	628	(2.8)	0	(0.0)
Mexican food	3	(0.6)	196	(0.9)	0	(0.0)
Nondairy beverage	6	(1.3)	140	(0.6)	0	(0.0)
Multiple vehicles	38	(8.0)	12,692	(56.1)	0	(0.0)
Known vehicle	160	(33.5)	18,294	(80.9)	1	(25.0)
Unknown vehicle	317	(66.5)	4,313	(19.1)	3	(75.0)
Total 1996	**477**	**(100.0)**	**22,607**	**(100.0)**	**4**	**(100.0)**

*Includes Guam, Puerto Rico, and the U.S. Virgin Islands.
†Totals might vary by <1% from summed components because of rounding.

SOURCE: *Surveillance for Foodborne-Disease Outbreaks—United States, 1993–97.* Morbidity and Mortality Weekly Report, vol. 49, no. SS-1, March 17, 2000

imported food shipments has increased, inspection coverage has fallen from an estimated 8 percent of shipments in 1992 to 1.7 percent in 1997.

In its latest study, the CDC reported that, in 1996, there were 477 outbreaks of food-borne disease resulting in 22,607 illnesses and four deaths. In 317 of these outbreaks the cause went undetermined. In 38 incidents more than one cause ("vehicle") was found. Fish were the largest single source of transmission (24 incidents), followed by salad (18) and fruits and vegetables (13). (See Table 8.10.)

In 1997 the CDC found that, among the factors that contributed to the transmission of food-borne disease, improper handling temperatures caused the most cases, followed by poor personal hygiene of handlers, contaminated equipment, inadequate cooking, and food from unsafe sources. The most commonly reported diseases were bacterial—most often salmonella. (See Table 8.11.)

Contamination from Produce

The per capita consumption of fresh produce has increased in the United States in recent years. Some Americans may be eating more produce for health reasons and, because of growing commerce between nations

all over the globe, a wider variety of fruits and vegetables are available today. In 1992 Americans consumed about 107 pounds of vegetables per person. By 1996 that figure rose to 145 pounds. In 1975 Americans ate, on average, 55 pounds of non-citrus fruit compared with about 70 pounds in 1995. Consumption of citrus fruit declined from about 30 pounds per person in 1975 to approximately 23 pounds in 1995. (See Figure 8.14.)

In 1997 researchers Larry R. Beuchat and Jee-Hoon Ryu, in a study for the CDC ("Produce Handling and Processing Practices," *Emerging Infectious Diseases*, vol. 3, no. 4), studied factors associated with produce contamination. They determined that contamination of produce can occur in the field or orchard, during harvesting or processing, in transport or marketing, or in the home or restaurant. The scientists categorized sources of contamination as preharvest or postharvest.

Preharvest sources include feces in soil or fertilizer, organisms in the soil, pollution of water used to irrigate or spray crops, dust and air, animals (including birds), insects, and human handling. Postharvest factors include feces, handling by workers or consumers, farm equipment, transport containers, insects, and animals. Other possible postharvest sources are air and dust, wash water, processing equipment, ice, transport vehicles, improper storage or

TABLE 8.11

Reported Foodborne-Disease Outbreaks in the U.S.*, by Etiology and Contributing Factors, 1997

Etiology	Contributing factors						Outbreaks in which factors reported	Total
	Improper holding temperatures	Inadequate cooking	Contaminated equipment	Food from unsafe source	Poor personal hygiene	Other		
Bacterial								
Bacillus cereus	4	1	—	—	1	—	4	4
Campylobacter	—	—	2	—	1	—	2	2
Clostridium botulinum	—	1	—	1	—	—	1	1
Clostridium perfringens	5	2	—	—	—	1	5	6
Escherichia coli	1	2	—	—	—	—	2	8
Salmonella	32	23	16	2	17	7	46	60
Shigella	3	—	1	—	4	1	6	10
Staphylococcus aureus	3	1	—	1	2	1	5	9
Streptococcus, group A	1	—	—	—	1	—	1	1
Vibrio parahaemolyticus	1	1	1	1	—	—	2	4
Total bacterial	**50**	**31**	**20**	**5**	**26**	**10**	**74**	**105**
Chemical								
Ciguatoxin	—	—	—	3	—	8	9	17
Mushroom poisoning	—	—	—	1	—	1	2	3
Scombrotoxin	4	—	1	—	1	1	6	15
Total chemical	**4**	**—**	**1**	**4**	**1**	**10**	**17**	**35**
Parasitic								
Giardia lamblia	—	—	1	—	1	—	1	1
Other parasitic	—	—	—	3	—	2	4	10
Total parasitic	**—**	**—**	**1**	**3**	**1**	**2**	**5**	**11**
Viral								
Hepatitis A	—	—	—	—	1	—	1	3
Other viral	1	—	1	1	5	2	6	14
Total viral	**1**	**—**	**1**	**1**	**6**	**2**	**7**	**17**
Confirmed etiology	**55**	**31**	**23**	**13**	**34**	**24**	**103**	**168**
Unknown etiology	**99**	**17**	**63**	**6**	**66**	**19**	**163**	**336**
Total 1997	**154**	**48**	**86**	**19**	**100**	**43**	**266**	**504**

*Includes Guam, Puerto Rico, and the U.S. Virgin Islands.

SOURCE: *Surveillance for Foodborne-Disease Outbreaks—United States, 1993–97.* Morbidity and Mortality Weekly Report, vol. 49, no. SS-1, March 17, 2000

packaging, inappropriate temperatures, cross-contamination from other foods, and incorrect handling. Beuchat and Ryu concluded, "Control or elimination of pathogenic microorganisms from fresh fruit and vegetables can be achieved only by addressing the entire system—from the field, orchard, or vineyard to the point of consumption."

Organic Foods—A Booming Industry

Some consumers are also concerned about technologies being applied to the food supply such as irradiation, the use of hormones in milk production, and genetically engineered crops. These consumers may seek out costlier, organically grown foods, although it is uncertain whether they are, indeed, healthier.

The USDA defines organic agriculture as an "ecological production management system that promotes and enhances biodiversity, biological cycles, and soil biological activity. It is based on minimal use of off-farm inputs and on management practices that restore, maintain, and enhance ecological harmony." Organic agriculture is both an approach to food production based on biological methods that avoid the use of synthetic crops or livestock prod-

ucts, and a broadly defined philosophical approach to farming that puts value on ecology, conservation, and nonintensive animal breeding practices. Some conventional practices not accepted in organic agriculture include:

• Synthetic fertilizer and pesticides.

• Confinement livestock operations such as feed lots or cages where animals are fattened before slaughter.

• Routine use of growth-enhancing animal drugs such as hormones or antibiotics.

• Genetically modified crops.

• Irradiation of foods for preservation or decontamination.

While organic methods of farming emerged in the United States and Europe in the early 1900s, it was not until the late 1980s that research groups and consumers began to express widespread interest in such practices. Beginning in 1989 sales of organically produced products began to climb growing, on average, 20 percent per year. According to the USDA, by 1998 organic agriculture was

FIGURE 8.14

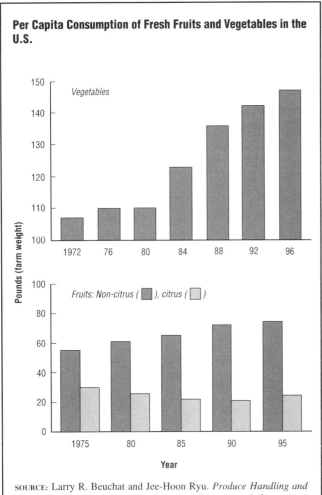

Per Capita Consumption of Fresh Fruits and Vegetables in the U.S.

SOURCE: Larry R. Beuchat and Jee-Hoon Ryu. *Produce Handling and Processing Practices.* Emerging Infectious Diseases, vol. 3, no. 4, Oct.-Dec., 1997

a $756 billion industry, the fasting-growing segment of the food industry.

The Organic Consumers Association, an industry trade group, predicts that, by 2010, organic products will account for 10 percent of U.S. agriculture. Industry sources estimate that the United States exports about $200 million of organic products, primarily to Europe and Japan. The USDA estimates there are 6,000 organic farming operations in the United States.

For most of its history the organic food industry established its own organizations and standards, with approximately 33 private certification operations. However, there is no consistency in labeling and no guarantee that foods labeled as organic are actually grown and processed in a purely organic fashion.

In 1990 Congress passed the Organic Foods Production Act (Title 21 of PL 10-624) to regulate the organic food industry. The act authorized the National Organic Program (NOP) to be administered by the USDA. The program would define standard practices and certify that operations meet those standards. It would be illegal for anyone to use the word "organic" on a product if it does not meet the defined criteria. The program has been hampered by a lack of funds since the beginning but, in 1997, the NOP drafted a proposal on the regulations and set up a "comment period" to end in 1998. The response was overwhelming and positive, primarily from within the organic farming industry. The proposal was revised and was expected to be implemented by the end of 2000.

CHAPTER 9

DEPLETION AND CONSERVATION OF NATURAL RESOURCES

Human activity on Earth has always altered the land. When populations were small enough, and productive, accessible land was abundant, people could abandon land that had been damaged by overuse and move on. While some countries still have excess land available, if population growth continues at the expected rate, virtually all arable land will be in use.

THE ECONOMIC VALUE OF THE WORLD'S ECOSYSTEMS—HOW MUCH IS NATURE WORTH?

Nature performs valuable, practical, measurable functions. Without them the human economy could not exist. Many experts contend that, as human activity gradually consumes or destroys this natural capital, the monetary value of the ecosystem to the economy must be calculated and considered. Thirteen economists, ecologists, and geographers studied 16 different biomes (ecological areas such as lakes, urban areas, and grasslands) to estimate the economic value of 17 ecosystem services. To do this they assigned dollar values to services performed by nature that are considered necessary to the human economy. Their report, published in the journal *Nature* (May 1997) estimated that ecosystems perform at least $33 trillion worth of services annually. Marine systems contribute about 63 percent of the value, mainly from coastal systems ($10.6 trillion). Terrestrial systems account for 37 percent of the value, mainly from forests ($4.7 trillion) and wetlands ($4.9 trillion). This total was 1.8 times the 1997 global gross national product (GNP) of approximately $18 trillion. In other words the services performed by nature were 180 percent as valuable as all of mankind's economic activities.

Most experts, including the authors of the study, recognize the figure is a crude, conservative, "starting point" estimate of what the environment does for humans. Others contend nature's value is incalculable. Virtually everyone agrees that, without nature's ecological contribution

human life could not exist. Furthermore, as ecosystems become more stressed and scarce in the future, their value will increase. If significant, irreversible thresholds are passed, the valuable services of these ecosystems may become irreplaceable.

THE ROLE OF FORESTS AND HABITAT

For millennia, humans have left their mark on the world's forests, although it was difficult to see. By the twenty-first century, however, forests that humans once thought were endless were shrinking before their eyes. Forests are not only a source of timber; they perform a wide range of social and ecological functions. They provide a livelihood for forest dwellers; they protect and enrich soils, regulate the hydrologic cycle, affect local and regional climate through evaporation, and help stabilize the global climate. Through the process of photosynthesis they absorb carbon dioxide and release the oxygen humans and animals breathe. They provide habitat for half of all known plant and animal species, are the main source of wood for industrial and domestic heating, and are widely used for recreation.

Forests are attractive and accessible sources of natural wealth. They are not, however, unlimited. Deforestation is caused by farmers, ranchers, logging and mining companies, and fuel wood collectors. Governments have often encouraged the settlement of land through cheap credit, land grants, and the building of roads and infrastructure. Much of these activities led to the destruction of forests and some governments have begun to reverse their policies.

Forests play a particularly crucial role in the global cycling of carbon. The earth's vegetation contains two trillion tons of carbon, roughly triple the amount stored in the atmosphere. When trees are cleared the carbon they contain is oxidized and released into the air, adding to the atmospheric store of carbon dioxide. Many scientists believe that carbon dioxide contributes to global warming. This release happens slowly if the trees are used to manufacture lumber or are allowed to decay naturally. If they are burned as fuel, however, or in order to clear forestland for farming, almost all of their carbon is released rapidly. The clearing for agriculture in North America and Europe has largely stopped but the burning of tropical forests has taken over the role of producing the bulk of carbon dioxide added to the atmosphere by land use changes.

A 1995 National Oceanic and Atmospheric Administration study found that about half the carbon dioxide emitted by burning fossil fuels is absorbed by plants in the Northern Hemisphere. The finding showed that plants play a role about equal to that of oceans, to which most of the absorption had previously been attributed. The study showed that plants absorb carbon dioxide that is rich in the carbon isotope 12, or C_{12}, while ocean water absorbs C_{12} and C_{13} equally. By determining the ratios of the isotopes, scientists can determine the relative effects of oceans and plants on the atmosphere's carbon dioxide concentration.

Rising Pressures on Forests

Worldwatch Institute, an independent, non-profit environmental research organization, reported in *Taking a Stand: Cultivating a New Relationship with the World's Forests* (Janet Abramovitz, 1998) that, between 1980 and 1995 alone, at least 494 million acres of forests vanished—an area larger than Mexico.

According to the United Nations Food and Agriculture Organization (FAO) approximately half the wood cut worldwide is used for fuel and charcoal. Most fuel wood is used in developing countries. In dry countries such as India, the majority of trees cut are for fuel; in moist tropical areas such as Malaysia, most trees are cut for industrial timber.

Tropical Rain Forests

Tropical forests are the most "alive" places on Earth. Although they cover less than 2 percent of the earth's surface they are home to as many as 30 million species of plants and animals—more than half of all life forms on the planet. A single acre of tropical rain forest supports 60 to 80 tree species and an enormous number of vines and mosses.

Rain forests also play an essential role in the weather. They absorb solar energy which affects wind and rainfall worldwide. Regionally, they reduce erosion and act as buffers against flooding. Tropical trees contain huge amounts of carbon which, when they are destroyed, is released into the atmosphere as carbon dioxide.

Of roughly 3,000 plants identified as having cancer-fighting properties, 70 percent grow in the rain forests. One of every three species of birds in the world nest there. In addition to wildlife more than 1,000 indigenous tribes still survive in tropical forests, just as they have for thousands of years.

The Natural Resources Defense Council (NRDC), a private organization that supports environmental health, reports that, every year, about 78 million acres of tropical forests and woodlands are destroyed for agriculture or logging—an area larger than Poland. That equals 2.47 acres per second (two football fields) and 214,000 acres per day (an area larger than New York City).

The now-defunct U.S. Office of Technology Assessment, in *Combined Summaries: Technologies to Sustain Tropical Forest Resources and Biological Diversity* (Washington, D.C., 1992), concluded that the major underlying cause of deforestation and species extinction is the lack of alternative employment for the growing populations of tropical countries. Logging and the conversion of forest-land to short-term, usually unsustainable, agricultural use results in destruction of the land, declining fisheries, erosion, and flooding.

Species in tropical rain forests possess a high degree of *mutuality* in which two species are completely dependent upon one another for survival; for example, a species of wasp and a species of fig tree. Such relationships are believed to evolve as a result of the relatively constant conditions in the tropics. Any species dependent upon trees therefore becomes imperiled when a tree is cut down.

THE AMAZON—AN EXAMPLE. The Amazon rain forests are the most famous of the earth's tropical forests. They serve as a good example of the controversies surrounding rainforests worldwide. This controversy generally centers on the interest of environmentalists (often from developed countries) in stabilizing the environment, and the developing world's basic need to cut down its forests for fuel and livelihood. Most developing nations claim that these needs are too great to be set aside for the sake of the environment. They also resent the industrialized world's disdain of practices the developed countries once followed themselves in building their own nations. These poorer, developing countries also wonder why they are expected to pay for the cleanup of a world that they did not contaminate.

There are also international incentives for continuing to cut down the rainforests. Foreign countries, especially Asian nations, as a source of ancient trees to make plywood, ornamental moldings, and furniture, are increasingly eyeing the Amazon forests. Desperate for money countries such as Surinam, located on the northern coast of South America, are considering granting several large Asian companies the rights to selectively cut its old forests.

Based on studies of satellite photographs taken over the Amazon, researchers believe that as much as 10 percent of the original Amazon forest has been destroyed, mainly through "slash and burn" methods of clearing land that are used to convert the land to farming use. The cleared land's productivity usually decreases within a few years, and farmers often have to abandon the fields and move on—slashing and burning a new area.

Although some observers expected a decline in forest burning after the 1992 Earth Summit held in Rio de Janeiro, Brazil, the burning of the Amazon actually increased from 4,296 square miles in the 1990–91 season to about 5,750 square miles per year since.

The Role of Logging in Deforestation

U.S. government scientists report that the forests of the northwest United States are being depleted by "clear-cutting" practices—the method of logging in which all trees in an area are cut—as opposed to "selective management" techniques, in which only certain trees are removed from an area. The lumber industry continually battles with environmentalists and the U.S. Forest Service over the right to clear-cut ancient forests. "The 'old growth' forests (stands of old, large trees) of North America store more carbon, acre for acre, than any other ecosystem on Earth," claims Tom Kuhnle of the Natural Resources Defense Council.

NASA scientists report that satellite pictures show a high level of damage to the evergreen forests of the Pacific Northwest. They attribute the damage to clear-cutting and claim the region has been so fragmented by clear-cutting that the overall health of the forest is at risk. Many observers believe that the biggest threat from this logging technique is loss of diversity of species in the area. The logging industry contends that restrictions on logging devastate rural communities by causing loss of thousands of jobs and leading to an increase in retail prices for lumber nationwide.

Logging roads are increasingly blamed for contributing to landslides, floods, and changes in rivers and streams. More than 383,000 miles of logging roads criss-cross the nation's countryside, enough to circle the globe nearly 15 times. The cost of building and maintaining those roads is often cited as a reason many national forests lose money on timber sales. In 1999 the U.S. Forest Service placed an 18-month moratorium on new road building within 191 million acres of national forests. The effort to halt road building until environmental impact studies can be completed was part of President Clinton's agenda to reduce environmental damage to the forests. The off-limits forests are located almost exclusively in the West, mostly in northern California.

REPLANTING. In an effort to counteract tree loss, forests are often "replanted" or replaced. Most experts contend that, when a natural forest (that has been replanted after clear-cutting) is replanted with commercially valuable trees, the plot becomes a tree farm, not a forest, and the biological interaction is damaged. Primary forests represent centuries, perhaps a millennium, of undisturbed growth. Trees will rebound after clear-cutting within 70 to 150 years but, researchers have found, the plants and herbs of the understory (growth under the canopy of the trees) never regain the richness of species diversity and complexity of their predecessors.

Pollution and the Forests

Many biologists believe that regional air pollution is a serious anthropogenic (made by humans) threat to temperate forest ecosystems. The most dangerous impact on forests comes from ozone, heavy metals, and acid deposition. Numerous studies suggest that both photosynthesis and growth decline significantly after one or two weeks of ozone at levels of 50 to 70 parts per billion (ppb), more than twice the normal background level of 20 to 30 ppb. During growing seasons average ozone levels are highest in the West (California, Nevada, Utah, and Arizona) and on the East Coast south of Pennsylvania.

In 1995, in a four-year study of a widespread timber species called loblolly pine, researchers at the Oak Ridge National Laboratory in Tennessee determined that ground level ozone levels that frequently occur in the eastern United States caused growth to slow, especially under drier soil conditions. The loblolly pine, covering approximately 60 million acres, contributes billions of dollars to the economy of the South.

The Role of Fire in Forest Health

The U.S. Forest Service manages about 155 national forests covering 192 million acres of land. About 70 percent of those lands are located in the dry, interior areas of the western United States. (See Figure 9.1.) Management practices in the past called for the Forest Service to put out all wildfires in the national forests. Scientists have recently put forward the idea that wildfires are necessary, however. They point out that wildfires are natural occurrences that serve to remove flammable undergrowth without greatly damaging larger trees.

Before pioneers settled the West, fires occurred about every five to 30 years. Those frequent fires kept the forest clear of undergrowth, fuels seldom accumulated, and the fires were generally of low intensity, consuming undergrowth but not igniting the tops of large trees. Disrupting this normal cycle of fire has produced an accumulation of vegetation capable of feeding an increasing number of large, uncontrollable, and catastrophic wildfires. Thus, the number of large wildfires has increased over the past decade, as have the costs of attempting to put them out. Figure 9.2 shows the number of acres burned from 1910

FIGURE 9.1

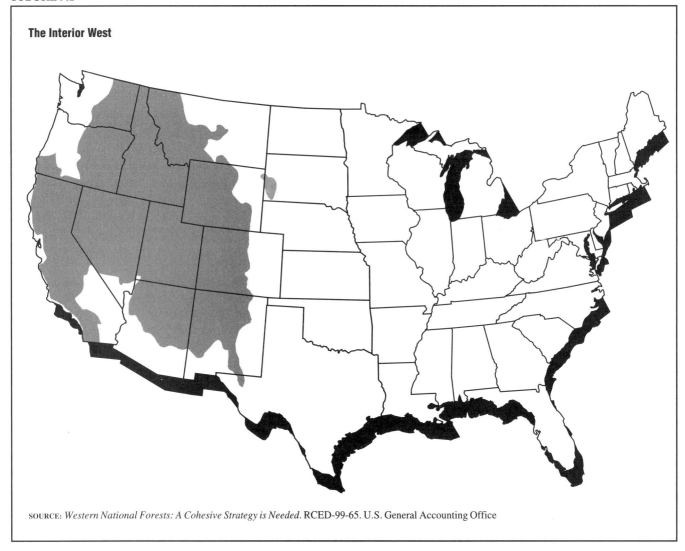

The Interior West

SOURCE: *Western National Forests: A Cohesive Strategy is Needed.* RCED-99-65. U.S. General Accounting Office

to 1997, while Figure 9.3 shows the costs of suppressing those fires.

Because the national forests are attractive for recreation and enjoyment, human population has grown rapidly in recent years along the boundaries scientists refer to as the "wildland/urban interface." (See Figure 9.4.)

According to a 1999 General Accounting Office (GAO) study, *Western National Forests—A Cohesive Strategy is Needed to Address Catastrophic Wildfire Threats,* this combination of rising population and increased fire risk poses a catastrophic threat to human health and life along the "wildland/urban interface" areas. In addition to the risk the fire poses to nearby inhabitants, smoke from such fires contains substantial amounts of particulate matter which contaminates the air for many hundreds of miles. In addition forest soils then become subject to erosion and mudslides after fires, further threatening the ecosystem and those who live near the forests.

In 1997 the Forest Service began an attempt to improve forest health by reducing, through "controlled

burns," the amount of accumulated vegetation, a program to be completed by 2015. The GAO found that lack of funding and inadequate preparedness may render the program "too little, too late." The National Commission on Wildfire Disasters concluded:

> Uncontrollable wildfire should be seen as a failure of land management and public policy, not as an unpredictable act of nature. The size, intensity, destructiveness and cost of...wildfires...is no accident. It is an outcome of our attitudes and priorities.... The fire situation will become worse rather than better unless there are changes in land management priority at all levels.

Ironically, in May 2000, a "controlled burn" near Los Alamos, New Mexico, raged out of control, sweeping hundreds of acres of land, destroying homes and businesses for miles.

THE BATTLE OVER PUBLIC LANDS

In much of the West ranchers have petitioned Congress to loosen restrictions on grazing on thousands of

FIGURE 9.2

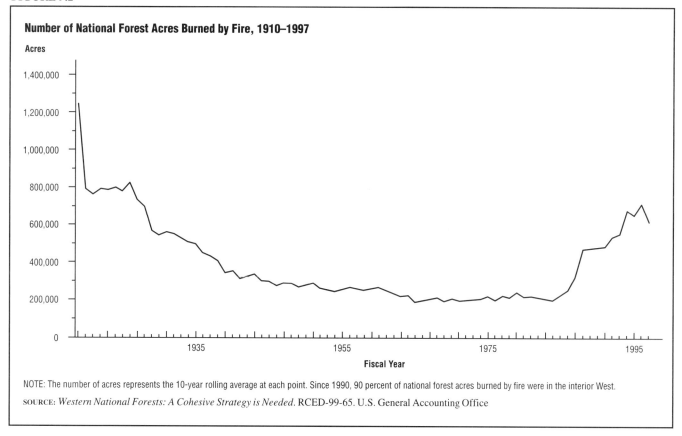

Number of National Forest Acres Burned by Fire, 1910–1997

NOTE: The number of acres represents the 10-year rolling average at each point. Since 1990, 90 percent of national forest acres burned by fire were in the interior West.

SOURCE: *Western National Forests: A Cohesive Strategy is Needed.* RCED-99-65. U.S. General Accounting Office

acres of federally owned ranch land. Environmental groups strongly oppose the proposal, claiming that grazing imperils land conservation, wildlife, and recreation. Grazing, they charge, is especially destructive to stream banks and sensitive wildlife habitat. Such concern for the soil and wildlife also lies at the heart of the dispute between oil companies and environmentalists over control of public lands such as the Arctic National Wildlife Refuge in Alaska.

A Land Grab

Unfortunately, love of the land has led Americans and developers into isolated, undeveloped areas in record numbers, threatening to destroy the very beauty they enjoy. Increasingly, developers are trying to build in choice, remote locations, such as the Sonoran Desert in Arizona.

In Phoenix and Scottsdale, some government leaders are concerned that developers will transform the desert into a sea of asphalt. Residents have indicated a willingness to invest tax money to protect mountain preserves from encroaching development and have declared several mountain areas off limits to developers. The state of Arizona and those cities involved have undertaken to purchase as many as 700,000 acres of land with tax revenues with the purpose of doing nothing with it and simply allowing it to remain in a natural state.

In 1995 South Carolina's Office of Ocean and Coastal Resource Management denied permission to the owners of 7,000 acres of Sandy Island to build a bridge connecting the island to the mainland. Although the owners claimed the bridge would be used to transport timber harvested there to the mainland, opponents believed the bridge would, in fact, lead to construction of homes, condominiums, and golf courses on the island.

In Texas, in an effort to balance development with wildlife preservation, the city of Austin invited the Nature Conservancy, a non-profit environmental group, to develop a plan to protect the environment while enabling building. The Endangered Species Act allows such regional arrangements. The result is the Balcones Canyonlands Conservation Plan, a 70,000- to 75,000-acre preserve in the Texas Hill Country, home to a number of endangered species.

In 1997 the Nature Conservancy finalized one of its many negotiations to keep family ranches in environmentally sensitive areas of the country, especially the West, from being broken up. The organization purchased the Dugout Ranch in Utah, which consists of 5,167 acres of privately owned pastureland and 250,000 acres of grassland leased from the government for grazing, with $4.6 million donated by individuals, foundations, and corporations. The ranch was a favored setting for movies and commercials. The Conservancy will maintain it as a

FIGURE 9.3

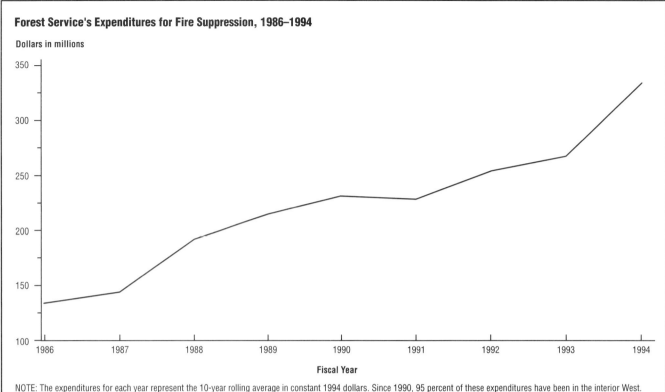

Forest Service's Expenditures for Fire Suppression, 1986–1994

Dollars in millions

Fiscal Year

NOTE: The expenditures for each year represent the 10-year rolling average in constant 1994 dollars. Since 1990, 95 percent of these expenditures have been in the interior West.

SOURCE: *Western National Forests: A Cohesive Strategy is Needed.* RCED-99-65. U.S. General Accounting Office

FIGURE 9.4

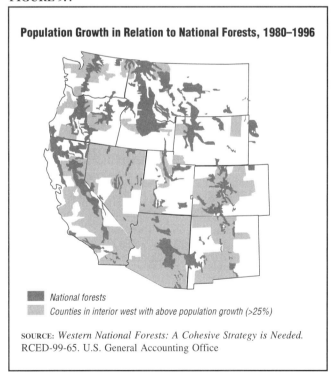

Population Growth in Relation to National Forests, 1980–1996

■ National forests
■ *Counties in interior west with above population growth (>25%)*

SOURCE: *Western National Forests: A Cohesive Strategy is Needed.*
RCED-99-65. U.S. General Accounting Office

working ranch, ecological preserve, and model for how cattle grazing and conservation—at odds throughout the West—can go hand in hand. The Nature Conservancy has a total of 1,340 properties in the United States, Latin America, and the South Pacific—the largest private network of nature sanctuaries in the world.

WETLANDS—NEW FRONTIERS?

Marshes, swamps, bogs, estuaries, and bottomlands comprise about 5 to 9 percent of the contiguous (48) United States and about 40 percent of Alaska. Although these terms refer to specific biosystems with sometimes very distinctive characteristics, they are commonly grouped together under the name "wetlands." Wetlands provide a vivid example of the dynamic, yet fragile, interactions that create, maintain, and repair the world's ecological system. Unfortunately, the fate of many wetlands can also offer concrete evidence of the harmful consequences of human activities that are carried out without regard for, and often without knowledge of, the relationship of each part of the ecosystem to the whole.

Once regarded as useless swamps, good only for breeding mosquitoes and taking up otherwise valuable space, wetlands have become the subject of increasingly heated debate. Many people want to use them for commercial purposes such as agricultural and residential development. Others want them left in their natural state because they believe that wetlands and their inhabitants are indispensable parts of the natural cycle of life on Earth.

TABLE 9.1

Locations of Various Wetland Types in the United States

Wetland type	Primary regions	States
Inland freshwater marsh	Dakota-Minnesota drift and lake bed (8); Upper Midwest (9); and Gulf Coastal Flats (4)	North Dakota, South Dakota, Nebraska, Minnesota, Florida
Inland saline marshes	Intermontane (12); Pacific Mountains (13)	Oregon, Nevada, Utah, California
Bogs	Upper Midwest (9); Gulf-Atlantic Rolling Plain (5); Gulf Coastal Flat (4); Atlantic Coastal Flats (3)	Wisconsin, Minnesota, Michigan, Maine, Florida, North Carolina
Tundra	Central Highland and Basin; Arctic Lowland; and Pacific Mountains	Alaska
Shrub swamps Carolina,	Upper Midwest (9); Gulf Coastal Flats (4)	Minnesota, Wisconsin, Michigan, Florida, Georgia, South North Carolina, Louisiana
Wooded swamps	Upper Midwest (9); Gulf Coastal Flats (4); Atlantic Coastal Flats (3); and Lower Mississippi Alluvial Plain (6)	Minnesota, Wisconsin, Michigan, Florida, Georgia, South Georgia, South Carolina, North Carolina, Louisiana
Bottom land hardwood	Lower Mississippi Alluvial Plain (6); Atlantic Coastal Flats (3); Gulf-Atlantic Rolling Plain (5); and Gulf Coastal Flats (4)	Louisiana, Mississippi, Arkansas, Missouri, Tennessee, Alabama, Florida, Georgia, South Carolina, North Carolina, Texas
Coastal salt marshes	Atlantic Coastal Zone (1); Gulf Coastal Zone (2); Eastern Highlands (7); Pacific Mountains (13)	All Coastal States, but particularly the Mid- and South Atlantic and Gulf Coast States
Mangrove swamps	Gulf Coastal Zone (2)	Florida and Louisiana
Tidal freshwater wetlands	Atlantic Coastal Zone (1) and Flats (3); Gulf Coastal Zone (2) and Flats (4)	Louisiana, Texas, North Carolina, Virginia, Maryland, Delaware, New Jersey, Georgia, South Carolina

SOURCE: *Wetlands: Their Use and Regulation.* Office of Technology Assessment, 1984

What Are Wetlands?

Wetlands is a general term used to describe areas that are always or often saturated by enough surface or groundwater to sustain vegetation that is typically adapted to saturated soil conditions—cattails, bulrushes, red maples, wild rice, blackberries, cranberries, and peat moss. The Florida Everglades and the coastal Alaskan salt marshes are examples of wetlands, as are the sphagnum-heath bogs of Maine. Because they are rich in minerals and nutrients and provide many of the advantages of both land and water environments, wetlands are dynamic systems that teem with a diversity of species, including many insects—a basic link in the food chain.

Wetlands are generally located along sloping areas between uplands and deep-water basins such as rivers, although they may also form in basins far from large bodies of water. Of the 90 million acres of wetlands in the lower 48 states, almost all (95 percent) are inland, freshwater areas; the remaining 5 percent are coastal saltwater wetlands. (See Table 9.1.) Alaska is estimated to have more than 200 million acres of wetlands.

There are several distinct forms of wetlands, each with its own unique characteristics. The main factors that distinguish each type of wetland are location (coastal or inland), salinity (freshwater or saltwater), and the dominant type of vegetation (marsh, swamp, or bog). Wetlands are a contin-

uum in which plant life changes gradually from predominantly aquatic to predominantly upland species. The difficulty in setting a definition of the exact point at which a wetland ends and upland begins results in much of the confusion as to how wetlands should be regulated.

The Many Roles of Wetlands

Experts have understood some of the functions of wetlands for many years. (See Figure 9.5.) Other purposes have come to light more recently.

FOOD AND HABITAT. Wetlands are a source of food and habitat for numerous game and non-game animals. For some species of waterfowl and freshwater and saltwater fish, wetlands are essential for nesting and breeding. About one in five plant and animal species listed as endangered by the U.S. government depend on wetlands for their survival. Two-thirds of the species of Atlantic fish and shellfish that humans consume depend upon wetlands for some part of their life cycle, as do nearly half of all species listed as endangered or threatened.

Coastal marshes and some inland freshwater wetlands boast some of the highest rates of plant productivity of any natural ecosystem, thus supporting abundant animal populations within the food chain. After a plant dies nearly 70 percent of it breaks down and is flushed into

FIGURE 9.5

Ecological Value of Wetland Processes

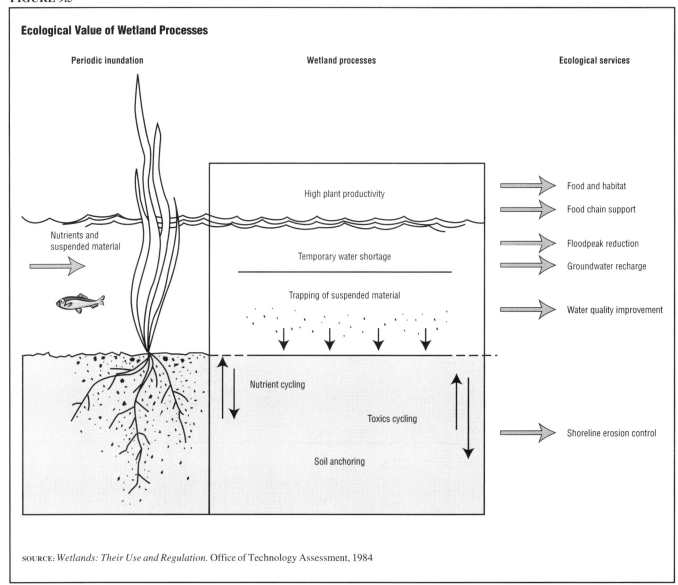

SOURCE: *Wetlands: Their Use and Regulation.* Office of Technology Assessment, 1984

adjacent waters where it can be consumed by fish and shellfish.

Inland wetlands also serve as way stations for migrating birds. The 30,000-acre region in the northcentral United States and southcentral Canada, for example, provides a resting place and nourishment for more than 400 of some 800 species of protected migratory birds (which individually number in the millions) during the migration season. Without this stopover corridor the flight to their Arctic breeding grounds would be impossible.

IMPROVING WATER QUALITY. Wetlands can temporarily or permanently trap pollutants such as excess nutrients, toxic chemicals, suspended materials, and disease-causing microorganisms—thus cleansing the water that flows over and through them. Some pollutants that become trapped in wetlands are biochemically converted to less harmful forms; other pollutants remain buried there; still others are absorbed by wet-

land plants and either recycled through the wetland or carried away from it.

COMMERCIAL FISHING. Between 60 and 90 percent of the United States' commercial fish species spawn in coastal wetlands. More than one-half of the country's seafood catch depends on wetlands during some part of their life cycles.

FLOODWATER REDUCTION. Isolated and floodplain wetlands can reduce the frequency of flooding in downstream areas by temporarily storing runoff water. For example the Cache River watershed in southern Illinois retains about 8.4 percent of the watershed's total runoff during flooding.

SHORELINE STABILIZATION. Because of their density of plant life wetlands can dramatically lessen shoreline erosion caused by large waves and major flooding along rivers and coasts.

RECREATION. Many popular recreational activities, including fishing, hunting, and canoeing, occur in wetlands. In addition wetland areas provide open space, an important but increasingly scarce commodity.

History of Wetlands Use

Early Americans considered wetlands nature's failure, a waste in nature's economy. They sought not to preserve nature in its original form but to increase the efficiency of natural processes. In an agricultural economy land unable to produce crops or timber was considered worthless. Many Americans began to think of draining these lands, an undertaking requiring government funds and resources.

In the nineteenth century state after state passed laws to drain (reclaim) wetlands by the formation of drainage-districts and statutes. Coupled with an agricultural boom and technological improvements, reclamation projects multiplied in the late nineteenth and early twentieth centuries. The farmland under drainage doubled between 1905 and 1910 and again between 1910 and 1920. By 1920 state drainage districts in the United States encompassed an area larger than Missouri.

After the Great Depression of the 1920s and 1930s programs such as the Works Progress Administration (WPA) and the Reconstruction Finance Corporation encouraged wetland conversion to form land for urban development. In 1945, at the end of the Second World War, the total area of drained farmland increased sharply.

In the final few decades of the twentieth century, however, conservationists and the courts have challenged reclamation. If drainage once improved the look of the land, today it is more likely to be seen as degrading it. Wetlands turned out not to be wastelands but the conservationist's ideal—systems efficient in harnessing the sun's rays to feed the food chain. Studies have shown that wetlands have value for flood protection far greater than their potential value for agriculture.

A LOSS IN RECENT YEARS. When the first Europeans arrived in America there were an estimated 215 million acres of wetlands. By the beginning of the twenty-first century only approximately 99 million acres remained. In the 200 years since the birth of the United States more than 50 percent of the wetlands in the lower 48 states have been taken over for agriculture, mining, forestry, oil and gas extraction, and urbanization. Some loss resulted from natural causes such as erosion, sedimentation (the buildup of soil by the settling of fine particles over a long period of time), subsidence (the sinking of land because of diminishing underground water supplies), and a rise in the sea level. However, 95 percent of the losses over the past 25 years have been caused by humans, especially by the conversion of wetlands to agricultural land. (See Table 9.2.)

TABLE 9.2

Major Causes of Wetland Loss and Degradation

Human Impacts

Drainage
Dredging and stream channelization
Deposition of fill material
Diking and damming
Tilling for crop production
Grazing by domestic animals
Discharge of pollutants
Mining
Alteration of hydrology

Natural Threats

Erosion
Subsidence
Sea level rise
Droughts
Hurricanes and other storms
Overgrazing by wildlife

SOURCE: *National Water Quality Inventory: 1992 Report to Congress.* U.S. Environmental Protection Agency

Eighty percent of the wetland conversions were for agricultural purposes; 8 percent for the construction of impoundments (water confined within an enclosure) and large reservoirs; 6 percent for urbanization; and 6 percent for other purposes such as mining, forestry, and road construction.

More than half (56 percent) the losses of coastal wetlands resulted from dredging for marinas, canals, port development, and, to some extent, from natural shoreline erosion. Twenty-two percent accompanied urbanization, 14 percent was from creating beaches, 6 percent from natural or human-made transitions of saltwater wetlands to freshwater wetlands, and only 2 percent from agriculture.

California, Ohio, Iowa, Indiana, and Missouri have lost almost all their wetlands. (See Figure 9.6.) The Fish and Wildlife Service estimates that the United States is losing more than 250,000 acres each year—about 30 acres every hour. Developers have discovered that many of the most tempting sites for new housing or shopping centers are wetlands.

The conversion of wetlands causes the loss of natural pollutant sinks. As water floods into wetlands from rivers and streams, the loss in velocity causes sediments and their absorbed pollutants to settle out in the wetland before they can enter other water bodies. In the United States artificial wetlands have been proposed as a means of controlling pollution from non-point sources. The dramatic decline in wetlands globally suggests not only loss of habitat but decreases in water quality.

Concern over Property Rights

The dispute over wetlands regulation reflects Americans' ambivalence when private property and public

FIGURE 9.6

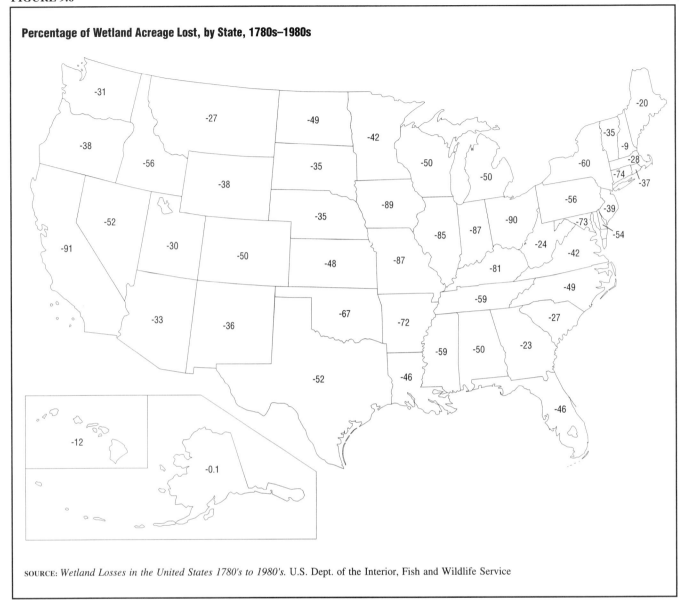

Percentage of Wetland Acreage Lost, by State, 1780s–1980s

SOURCE: *Wetland Losses in the United States 1780's to 1980's.* U.S. Dept. of the Interior, Fish and Wildlife Service

rights intersect, especially since three-fourths of the nation's wetlands are owned by private citizens. In recent years many landowners have complained that wetland regulation devalued their property by blocking its development. They argue that efforts to preserve the wetlands have gone too far, citing instances where a small wetland precludes the use of much larger surrounding areas. Some large landowners have long opposed any federal (and state and local) powers to protect resources such as wetlands that might limit their land use options.

Another policy question of concern to the public is the right of the federal government to take property without compensation. The "takings" clause of the Constitution (Fifth Amendment) provides that, when private property is taken for public use, just compensation must be paid to the owner. Owners claim that, when the government—through its laws—eliminates some uses for

their land, the value is decreased and they should, therefore, be paid for the loss.

In 1986 homebuilder David Lucas bought two residential lots on a South Carolina barrier island, planning to build and sell single-family homes similar to those on nearby lots, completely within the law at the time. In 1988, however, South Carolina declared that parcel of land in danger of erosion and, under the new Beachfront Management Act, prohibited building on the area. Lucas went to court claiming the act was taking his property without just compensation—since he could no longer build on the property, it had no value. In 1992 the United States Supreme Court, in *Lucas v. South Carolina Coastal Council* (505 US 1003) concurred and ruled that, when a regulation as in those preventing owners from developing their wetlands causes land to be "off limits" to economic uses, compensation must be paid.

MOUNTAINS

Mountains are one of Earth's most important features. They span one-fifth of the landscape and house one-tenth of Earth's population. An additional two billion people live downstream from mountains and depend on the water, hydropower, grassland, timber, and mineral resources generated by those mountains.

What Is a Mountain?

A mountain is a land-mass that projects conspicuously above its surroundings and is higher than a hill, generally at least 300 meters in height. An additional criterion is that a mountain's rise creates climates, soils, and vegetation distinct from those in surrounding lowlands. Mountains share common physical attributes of steepness, instability, and ecology that create natural hazards, microclimates, niches of biodiversity, and inaccessibility. The collision of tectonic plates produces mountain uplift and numerous physical hazards, such as earthquakes, volcanic eruptions, landslides, avalanches, and floods. The slope and altitude of mountains create variations in climate—temperature, radiation, wind, and moisture—over very short distances.

Mountains function as the earth's water towers by attracting much of its precipitation—mountains are the predominant and most dependable source of fresh water for humanity. A diversity of wild plants makes mountain ecosystems vital to mankind's future food and pharmacological security.

A Naturally Vulnerable Resource

A distinguishing feature of mountains is vulnerability to disturbance, largely due to the vertical dimension (height and slope). Because a doubling in water speed magnifies the size of materials that water can transport, the erosive power of rapid runoff from mountains is immense. Unlike lowland environments mountain ecosystems are typically less able to recuperate from disruptions such as soil erosion or loss of vegetation; soils are usually thin and poorly anchored, and gravity-powered erosion speeds silt and sediment movement. Also, many mountains are still growing and are less geographically stable than flatter landmasses. Seven of the world's 14 tropical "hotspots" of plants threatened by destruction have at least half their area in mountains, and 131 of the world's 247 bird habitats are in tropical mountains.

As studies seem to confirm the warming of Earth's environment (see Chapter 2), experts predict that such warming will proceed too fast for many ecosystems to adapt. Austrian researchers have found that nine plant species typical of the nival zone (above alpine grasslands) were migrating to higher altitudes at one meter per decade, but would have to move at eight to ten meters per decade to keep up with the current rate of warming.

Impacts on Mountain Ecology

Mountains face threats from poor land use patterns, resource extraction, and mass tourism and recreation. Half of U.S. rangeland, most of it in the mountainous West, is now considered severely degraded, with its livestock-supporting ability reduced by 50 percent. The FAO reports that hill and mountain forests are more susceptible to ecological damage from excessive population densities than are lowland forests. Because their slopes permit gravity to increase the power of flowing waters, mountains attract most of the earth's hydroelectric projects and irrigation reservoirs.

Of all the economic activities in the world's mountains, nothing rivals the destructive power of mining. Environmental impacts include habitat destruction, erosion, air pollution, acid drainage, and metal contamination of water bodies. The result is often denuded forests, eroded hillsides, and dammed or polluted rivers.

TOURISM AND RECREATION. Many lowlanders feel that mountains are a refuge from modern life, but the problems they hoped to leave behind are following them up the slopes. In slick ads and commercials, images of pristine mountain wilderness lure multitudes for respite and sport. In industrial countries mass tourism and recreation are fast becoming the largest threats to mountain environments.

Since 1945 visits to the ten most popular mountainous national parks in the United States have increased twelvefold. Infrastructure for leisure and recreation in the mountains can be exorbitant. In 1990 there were 100 Alpine golf courses; by the end of 1996, 500 existed. Thundering helicopters bring skiers to untracked slopes not only in the American Rockies but worldwide. The populations of many small ski towns, like Vail, Colorado, have more than doubled since 1980, causing new home construction and retail sales to grow at double or triple the national average. White-water rafters and mountain climbers prowl the slopes in numbers some consider dangerous. A generation ago conquering Mount Everest was considered an unimaginable feat; today dozens of climbers reach the peak every month leaving behind trash (including bodies of climbers that die on the dangerous climb) that has become a serious problem on the mountain.

EROSION

Erosion is the process in which the materials of the earth's crust are worn and carried away by wind, water, and other natural forces. Demands on the earth to feed growing populations and changes in the earth's landscape caused by human activities have speeded up soil erosion. Anson Bertrand, of the U.S. Department of Agriculture, describes the situation as "epidemic." Soil erosion has increased to the point where it far exceeds the natural formation of new soil. The Worldwatch Institute, an environmental advocacy research organization, estimates that excessive erosion is occurring on approximately 44 percent of U.S. cropland.

In 1995 the U.S. Department of Agriculture (USDA) reported that "soil erosion in the United States does not pose an immediate threat to the nation's ability to produce food and fiber, but it does reduce the productivity of some soils, and it also causes offsite damages." The USDA estimates that soil productivity damages are about one to three billion dollars annually, while water quality and dust damages are each several times higher.

Erosion of beaches on the East Coast is becoming a more serious problem as development inches closer to the ocean. The Army Corps of Engineers has been rebuilding eroded beaches since the 1950s. The federal government pays 65 percent of the cost of beach rebuilding, with states and local governments paying the remaining 35 percent. Many experts, however, believe that beach replenishment is a futile effort and that funds could be better spent elsewhere.

Farming Practices

Agriculture depends primarily upon the top six to eight inches of topsoil. Fields planted in rows, such as corn, are susceptible to soil runoff; cover crops such as hay provide more soil cover to hold the land. The destruction of forests and native grasses has allowed water and wind greater opportunity to erode the soil. Changes in river flow and seepage from human technology have shifted the runoff patterns of water and the sediment load of rivers that, in turn, deposit into lakes and oceans. Historically, when most of the topsoil was lost, farmers would abandon the land. Now, however, farmers continue to plow the soil, even when it consists of as much subsoil as topsoil. It costs more money to produce food on such land than on land where topsoil is present. Farmers often use more fertilizer to make up for the decreasing productivity of the soil, and that, in turn, adds to environmental pollution.

Agricultural lands are the principal source of eroded soil. The GAO estimates that about 28 percent of the nation's cropland is highly erodible. The states with the highest percentage of highly erodible cropland are New Mexico (90 percent), Arizona (81 percent), and Colorado (77 percent). In absolute terms Texas and Montana have the most erodible land. The amount of erosion has declined in past decades. The USDA attributes the decline to the Conservation Reserve Program which pays farmers to take land out of production for 10 years, and to the Conservation Compliance Program. As part of the 1985 Farm Act (PL 99-198) the Conservation Compliance Program was initiated as a major policy tool. To be eligible for agricultural program benefits, farmers must meet minimum levels of conservation on highly erodible land.

ENVIRONMENTAL QUALITY INCENTIVES PROGRAM (EQIP). Congress, under the 1996 Farm Bill (PL 104-127), authorized the Environmental Quality Incentives Program (EQIP) to address agriculture's natural resource and environmental problems. It is a flexible, voluntary, effective conservation program that allocates $200 million per year to farmers' conservation efforts. Under EQIP the USDA provides assistance to family-sized farms and ranches for up to 75 percent of the cost of certain environmental protection practices, such as grassed waterways, filter strips, manure management facilities, capping abandoned wells, and wildlife habitat enhancement. The USDA may also offer incentive payments to encourage producers to apply such land management practices as nutrient, manure, irrigation water, wildlife, and integrated pest management.

In 1997, the first year of the program, the USDA approved 23,000 contracts with farmers and ranchers. Nearly 58,000 applications for assistance were received, with requests totaling more than three times the amount of available funding.

IRRIGATION. Human beings have survived in deserts or arid areas only because they have been able to increase the quantity of water available to meet their needs. An elaborate system of dams, reservoirs, irrigation pipelines, aqueducts, and canals allows residents of the American West, for example, and especially California, to ignore the fact that they live in a naturally dry climate. That fact has made California's Central Valley the most productive agricultural region in the world on only 3 percent of U.S. farmland.

Constant irrigation, however, is not a miracle solution as once was thought. According to the United Nations Environmental Programme (UNEP), 90 percent of the land in Egypt, 68 percent in Pakistan, 50 percent in Iraq, 38 percent in Peru, 30 percent in the United States, and 20 percent each in India, Russia, and Australia, are suffering salinization (saltiness) caused by irrigation. Sodium in the soil or irrigation water accumulates at the root level of soils or turns into a sterile, rock-hard crust. An estimated two million hectares (about five million acres) of irrigated land are pulled from production each year because of waterlogging and salinization, the result of poor land management. In addition, irrigated land is often paved over for housing, factories, and roads, especially in the United States and Asia, further reducing the productive use of land for agriculture.

BIODIVERSITY

Biological diversity, or biodiversity, refers to the full range of plant, animal, and microbial life and the ecosystems that house them. Environmentalists began using the term during the 1980s when biologists increasingly warned that human activities were causing a loss of plant and animal species.

Studies of deforestation have supported the concerns about declining biodiversity, showing that tropical rain forests have dwindled from 3.5 billion acres before the industrial era to fewer than two billion acres. Deforestation has meant extinction for hundreds of species of plants and animals each year. The exact number of species in the remote forests is unknown, although it is generally accepted that they house the greatest number of species on the planet.

Extinction Rates

No one knows how many species of plants and animals exist in the world. By the beginning of the twenty-first century scientists had named and documented 1.4 million species. Educated guesses range from five million to 100 million. Just as the health of a nation is promoted by a diverse economy, so the health of the biosphere is promoted by a diverse ecology.

Widespread extinctions have occurred infrequently in Earth's history and are generally believed to have been due to major geological and astronomical events. Scientists call the disappearance of only a few species over the period of a million years a "background rate." When that background rate doubles for many different groups of plants and animals at the same time, a mass extinction is taking place.

THE SIXTH EXTINCTION? At least five times in the last 600 million years planet-wide cataclysms, such as drastic climate change or colliding asteroids, have wiped out whole families of organisms. Because of these losses scientists believe that more than 95 percent of all species that have ever existed are extinct. Researchers predict that, as tropical ecosystems are converted to farms and pasture, the extinction rate will approach several hundred extinctions per day in the next 20 to 30 years—millions of times higher than background levels. The Worldwatch Institute believes that more species of flora and fauna may disappear in our lifetimes than were lost in the mass extinction that included the disappearance of the dinosaur 65 million years ago.

The loss of diversity leads to problems beyond the simple loss of animal and plant variety. When local populations of species are wiped out the genetic diversity within that species that provides the ability to adapt to environmental change is diminished, resulting in a situation of "biotic impoverishment." Those organisms that do survive are likely to be hardy, "opportunistic" organisms tolerating a wide variety of conditions—characteristics often associated with pests. Experts suggest that, as some species dwindle, their places may be taken by a disproportionate number of pest or weed species that, while a natural part of life, will be less beneficial to human beings.

Most living species have never been identified. Mammals, including humans, make up barely three-tenths of 1 percent of all known organisms. Approximately 1,205 species in the United States alone were listed as endangered in 1999, up from 281 in 1980 and 596 in 1990. (See Table 9.3.) States with the highest number of endangered species as of July 2000 were Hawaii (312), California (275), and Florida (100). (See Figure 9.7.)

In April 1998 the World Conservation Union listed nearly 34,000 plant species on its growing "red list" of imperiled plants. (See Chapter 3.) This was one of every eight known plant species in the world, and nearly one in three in the United States.

Scientists participating in the "Global Biodiversity Strategy," an international team of 500 researchers, believe that the extinction of species will deprive future generations of new medicines and new strains of food crops. With as many as 50 plant species disappearing daily the researchers calculate that the planet's diversity could be reduced by 10 percent by the year 2015. One-fourth of all medical prescriptions in the United States contain active ingredients from plants. (See Table 9.4.) Among the medically useful species are some used in the treatment of cancer, HIV/AIDS, circulatory disorders, bacterial infections, anxiety, inflammatory diseases, and for the prevention of organ rejection in transplants.

Species Loss—Crisis or False Alarm?

As with most environmental questions, not all experts agree. Some observers believe the threat to species diversity has not been proven and claim that, while wild habitats are, indeed, disappearing because of human expansion, the seriousness of the extinction has been exaggerated and is unsupported by scientific evidence. They point to the fact that the total number of species and their geographic distribution are unknown. How, they ask, can forecasts be made based on such sketchy data?

Other observers contend that extinctions, even mass ones, are inevitable and occur as a result of great geological and astronomical events that humans cannot affect. They do not believe that disruptions caused by human activity are enough to create the mega-extinction prophesied by people they consider "alarmists."

Furthermore, some critics of the environmental movement believe that the needs of humans are being made secondary to those of wildlife. They contend that the Endangered Species Act protects wildlife regardless of the economic cost to human beings. Sometimes, as in the case of the spotted owl of the Pacific Northwest forests, that cost is jobs for people. The bird's presence halted logging there—following protests of environmental groups—at considerable economic loss to communities and families in the area. Furthermore, critics contend that halting development because it threatens a species

TABLE 9.3

Number of U.S. Listed Endangered and Threatened Species, By Major Group, 1980–1999
Total Listed (Endangered & Threatened)

CY	Mam-mals	Birds	Rep-tiles	Amphi-bians	Fish	Crusta-ceans	Snails	Insects	Arach-nids	Clams	Plants	CY Total
1980	36	61	25	8	47	1	7	14	0	23	59	281
1981	36	61	25	8	47	1	8	13	0	23	61	283
1982	36	61	26	8	49	3	8	13	0	23	67	294
1983	39	61	26	8	49	4	8	13	0	23	69	300
1984	42	69	26	8	51	4	8	13	0	23	82	326
1985	48	72	26	8	64	4	8	13	0	23	118	384
1986	49	75	28	8	70	5	8	15	0	23	141	422
1987	52	82	32	9	74	6	8	15	0	28	174	480
1988	56	81	32	9	77	9	8	18	4	31	201	526
1989	58	81	32	11	82	9	9	19	4	34	217	556
1990	61	83	32	11	86	10	9	21	4	39	240	596
1991	64	83	32	11	88	10	13	23	4	42	302	672
1992	65	84	33	11	91	11	18	25	4	42	369	753
1993	65	88	33	11	98	13	19	26	4	53	403	813
1994	66	90	33	12	105	17	22	28	4	54	510	941
1995	66	91	33	12	105	17	22	29	5	57	525	962
1996	66	90	33	13	107	17	22	29	5	57	614	1,053
1997	66	93	36	16	108	19	22	37	5	62	668	1,132
1998	69	93	36	16	119	20	28	37	5	69	702	1,194
1999	69	89	38	17	112	20	28	37	5	69	721	1,205

Note: Totals are not additive. Number of species listed fluctuate between years because of new listings, reclassifications, delistings, new information on taxonomy, and other reasons.

SOURCE: U.S. Dept. of the Interior, Fish and Wildlife Service

whose whole population occupies only a few acres and numbers only in the hundreds is simply nonsense.

Out of Sight, Out of Mind?

When it comes to the environment biologists and the public sometimes see the world differently. According to a 1998 poll conducted for the American Museum of Natural History by Louis Harris & Associates, biologists overwhelmingly view the loss of biodiversity as a serious problem. The general public, on the other hand, is mostly unfamiliar with the concept of biodiversity and views human pollution as more important.

According to the survey most scientists agreed that, if trends continue, the loss of species will have a very negative effect on the earth's ability to recover from both natural and human-made disasters. Most of the general public claimed they did not believe a mass extinction is occurring: 62 percent of the respondents believed pollution was the single most important environmental problem. Biologists, on the other hand, felt the earth was more threatened by population growth (39 percent) and loss of biodiversity (22 percent). "Pollution is obvious—you can see it and smell it," one responding biologist concluded. "But pollution can be reversed, while extinction is irreversible."

The Earth Summit Biodiversity Treaty

At the 1992 Earth Summit in Rio de Janeiro, 156 nations signed a pact to conserve species, habitats, and ecosystems. This Biodiversity Treaty is regarded as one of two main achievements of the United Nations Con-

ference on Environment and Development, the other being a treaty on global warming. The treaty makes nations responsible for any environmental harm in other countries produced by companies headquartered in their country.

One provision of the treaty concerns "biotechnology," a term referring to the ownership of genetic material. Plants, seeds, and germ plasm have historically been in the public domain (belonging to the general public) rather than belonging to any particular government. Therefore, anyone could exploit or use them without compensation to the country of origin. For example, the rosy or Madagascar periwinkle, a plant found only in the tropical rain forests of Madagascar, is used as a base for medication to treat Hodgkins' disease and childhood leukemia. Madagascar receives no compensation for use of the plant. The biotechnology treaty drafted in Rio called for compensation to be paid for the use of those genetic materials.

The United States did not sign the treaty at the time. The Bush administration, while agreeing with many provisions of the pact, believed the economic requirements for accomplishing those goals were unacceptable to American businesses because they would be forced to compensate for the use of these species. President Bill Clinton signed the treaty in 1994.

The Endangered Species Act

The 1973 Endangered Species Act (ESA; PL 93-205), passed into law during the administration of Presi-

FIGURE 9.7

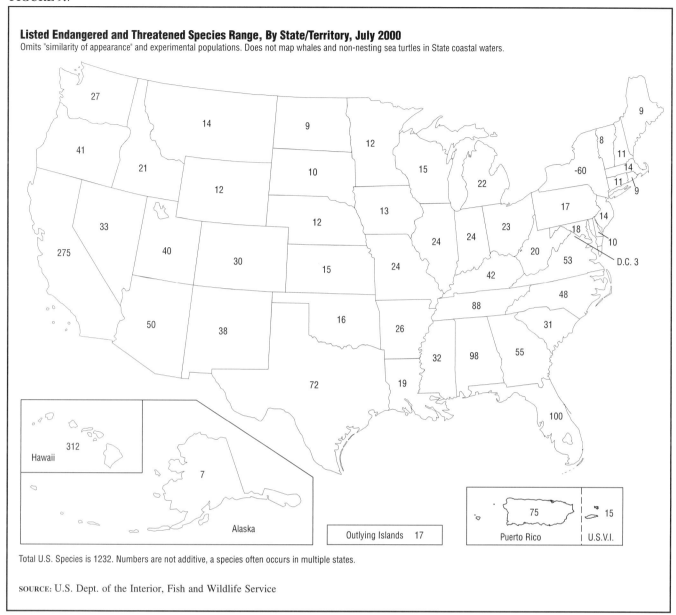

Listed Endangered and Threatened Species Range, By State/Territory, July 2000
Omits "similarity of appearance" and experimental populations. Does not map whales and non-nesting sea turtles in State coastal waters.

Total U.S. Species is 1232. Numbers are not additive, a species often occurs in multiple states.

SOURCE: U.S. Dept. of the Interior, Fish and Wildlife Service

dent Richard Nixon, was originally intended to protect creatures like grizzly bears and whales with whose plight Americans found it easy to identify. In the words of its critics, however, it has become the "pit bull of environmental laws," policing the behavior of entire industries. In almost three decades the Endangered Species Act has gone from being one of the least controversial laws passed by Congress to one of the most contentious.

The law regulates industries that can cause fish and wildlife populations to decline. It also determines the criteria to decide which species are endangered. On average, over the past decade, the government has added about 50 new species to the list annually. Since the act was first passed the pendulum has periodically swung between increased protection and the need to soften the law's economic impact.

The U.S. Supreme Court, in *Manuel Lujan, Jr., Secretary of the Interior v. Defenders of Wildlife et al.* (504 US 55 1992), determined that groups and individuals cannot sue the government solely on behalf of the public interest or on behalf of the flora and fauna they seek to protect. Instead, they must demonstrate harm to themselves. The ruling has been generally regarded as a victory for business interests and a defeat for environmentalists in their efforts to protect endangered species, since immediate harm is often difficult to show in environmental issues.

WILDLIFE

The loss of habitats, the contamination of water and food supplies, poaching, and indiscriminate hunting and fishing, have depleted the population of many species. Most scientists agree that prospects for the survival of many species of wildlife, and hence biodiversity, are wors-

TABLE 9.4

Selected Examples of Medically Useful Species

species	drug & use
	CANCER
Madagascar periwinkle	vincristine & vinblastine: childhood (lymphocyctic) leukemia & Hodgkin's disease
Pacific yew	taxol: ovarian, breast
	HIV/AIDS
Cameroon vine	michellamine B
tree in Malaysia	calanolides
tree in Samoa	prostatin
bush in Australia	conocurvone
mulberry trees	Butyl-DNJ (analogue of natural compound)
	HEART/CIRCULATORY
foxglove	dlgltalls: regulates contractions
ergot (fungus of wheat)	atenolol & metoprolol: blocks adrenaline (especially in coronary disease)
Strophanthus gratus & Strophanthus kombe	cardiac glycoside, G-strophanthin, K- strophan-thin: treatment of acute heart failure hypotension during surgery
snake venom	captopril & enalapril — reduces blood pressure
fungal metabolite	lovastatin — used to reduce cholesterol levels byblocking its biosynthesis
	INFECTIOUS
molds	avermectins: worm killers
molds	penicillin
sewer microbes	cephalosporin; developed into cefaclor, ceftriaxone, & cefoxitin — produces antibiotics
	IMMUNOSUPPRESSANTS (used in organ transplants)
molds	cyclosporin
molds	FK506
	PARKINSON'S DISEASE
Atropa acuminata, Atropa belladonna	hyosyamine: treatment of Parkinson's disease, epilepsy, and gastric ulcers
	TRANQUILIZERS
Indian snake root	reserpine: muscle relaxant, antianxiety; developed alprazolam
	ANTI-INFLAMMATORY
ergot (fungus on wheat)	anti-histimines Terphenadine (H1): used to treat allergies & motion sickness; ranitidine & cimetidine (H2) used to treat gastric ulcers

SOURCE: Endangered Species Coalition, 1993

ening. The expansion of human development into wildlife habitats has resulted in some animals being squeezed into cities and suburbs where encounters between humans and wildlife have become increasingly common.

Species loss and habitat loss are related. Scientists have recognized the connection between the size of an area and the number of species it contains for some 150 years—as large tracts of land are lost, so are some species that make their homes there. Other major causes of animal extinction are hunting and introduced (non-native) species.

Sharing the Planet

In the nineteenth century, miners took parakeets with them into the mines. If a bird died, they knew they were in danger from noxious gases. While more scientific and humane procedures now exist to determine how dangerous the situation is, some scientists believe that plants and animals may still serve as indicators of the safety of the world. When biologists discover toxic amounts of poisons in wildlife, they ask whether human beings are also ingesting these poisons.

Some observers believe that animals should be protected out of an intrinsic respect for life, aside from any market value or use to humans. Others contend that humankind must manage wildlife correctly because biodiversity makes good economic and survival sense. Still others believe that there is no species loss "problem," that species loss is a natural part of evolution. All of these issues are being deliberated as the people of the world struggle to decide how best to live with the other animals and plants that populate the earth.

NEW EVIDENCE. A 1996 study by the Nature Conservancy on more than 20,000 American plant and animal species found that about one-third of species were rare or imperiled, a larger fraction than some scientists had expected. The study, the most comprehensive assessment to date of the state of American plants and animal species, found that mammals and birds were doing relatively well compared to other groups, but that a high proportion of flowering plants and freshwater marine species like mussels, crayfish, and fish, were in trouble. Of the 20,481 species examined, about two-thirds were secure, 1.3 percent were extinct or possibly extinct, 6.5 percent were critically imperiled, 8.9 percent were imperiled, and 15 percent were considered vulnerable. The destruction or degradation of habitat was considered to be the main threat.

City Life Collides with Wilderness

The growth of urban areas has resulted in a collision between city life and wildlife. Increasingly humans are encountering wild animals in their communities. The 600,000-acre Angeles National Park in the Los Angeles outback of the San Gabriel Mountains has been the site of numerous attacks on visitors by snakes and wild animals. In addition humans are using some areas designed for wildlife for undesirable purposes; poachers and some hunters shoot deer out of season and prey upon the endangered Nelson bighorn sheep. During Christmas season trees are cut down. Crowds of picnickers and hikers often swell to music-festival size clogging roads. Toxic-waste outlaws heave garbage and poisons into creeks and abandoned mine shafts. Criminals sell drugs in the forest, even growing small marijuana plantations there. The Angeles National Park has also become a well-known dumping ground for homicide victims—eight bodies were found in the forest in 1995.

In 1996 the National Oceanic and Atmospheric Administration, which controls sanctuaries along a large portion of the California coast, initiated new rules against "chumming," a fishing practice in which fish remains and blood are dumped into the ocean. This practice began because passengers aboard charter boats wanted to view rare white sharks, an endangered species. Captains of those boats found they could lure the sharks by dumping fish debris into the water, leading to dangerous confrontations between sharks, surfers, and the shark enthusiasts.

Some Cases of Threatened Species

Almost daily the decline or threat to some plant or animal is reported. Scientists attribute the decline of salmon on the West Coast to spoiled habitat and disruption of river flow. Erosion of the coastline in Florida has left no place for sea turtles to dig their nests, and they are dying off.

Peregrine falcons, one of the first species to be listed on the Endangered Species List, were dying because they were consuming DDT in the food chain. Following their listing under the Endangered Species Act, and the banning of DDT in 1972, the falcon population has rebounded. In 1999 they were officially removed from the Endangered Species List.

The ivory tusks of African elephants are very valuable as they can be fashioned into jewelry and artwork. In the mid-twentieth century African elephants were so extensively hunted for their ivory that their population dropped to dangerous levels. The international community responded in 1990 by banning trade in African elephant ivory under the Convention on the International Trade in Endangered Species (CITES). Poaching of elephants continued but their population began to rebound. By 1999 there were so many elephants in Zimbabwe, Namibia, and Botswana that those countries (unsuccessfully) requested permission to resume limited trade in ivory.

Dolphins tend to swim with schools of tuna in the Pacific Ocean and nets used by commercial fisheries to catch tuna also entrap dolphins. Since netting began in 1958 an estimated seven million dolphins have been killed. In 1972 Congress passed the Marine Mammal Protection Act (PL 92-522) to reduce the deaths of the dolphins. The law was amended in 1985 and 1988 to regulate tuna imported from other countries. Trade groups have challenged these regulations by pointing to the economic losses of companies and nations who abide by the law. Some companies ignore the law, while other companies have printed a "Dolphin Safe" label on their tuna products to show that they obey the law.

Environmentalists have long argued with government and industry over the question of logging in the Pacific Northwest forests. Environmentalists claim that the biological health of the ecosystem is in decline and more than 100 species of plants and animals are threatened with extinction, while the timber industry responds that the forest provides jobs for thousands of Americans and lumber for millions of people.

The argument came to a head in 1990 when the spotted owl—that lived only in this particular region—was added to the list of endangered species. Logging was halted and a succession of lawsuits was filed against the Forest Service and the U.S. Department of Interior. In 1992 President Bush grudgingly restricted logging in that area but, at the same time, moved to amend the law to allow economic considerations to be taken into account. In 1994 President Clinton worked out what was claimed to be a compromise between environmentalists and business interests, allowing logging to resume with restrictions on the size, number, and distribution of trees to be cut.

Scientists investigating a worldwide decline in frogs and other amphibians have found evidence identifying a number of factors that have contributed to the decline: ultraviolet radiation caused by the thinning of the ozone layer, chemical pollution, and a human taste for frog meat. These species are considered indicator species because their sensitivity makes them early indicators of environmental damage. Butterflies, another creature considered an indicator species, are also disappearing in many areas.

Roads and Wildlife

Almost four million miles of public roads cross the contiguous United States. (See Figure 9.8.) As roadways reach further and further into undeveloped areas, encounters with wildlife are inevitable. The U.S. Department of Transportation, in *Critter Crossing: Linking Habitats and Reducing Roadkill* (2000), found that roads impact wildlife in several ways:

- Roadkill—Vehicles traveling U.S. roads kill millions of birds, mammals, reptiles, and amphibians every year. The ocelot, an endangered cat, now numbers only about 80 animals due in part to highway kills. In addition humans are sometimes killed or injured in animal-vehicle collisions. The insurance industry estimates cost of those fatalities and injuries is about $200 million; motorists pay at least $2,000 in vehicle repair when they hit a large animal.

- Habitat loss—When humans build highways and develop areas they destroy habitat. This forces animals into smaller and smaller areas and into areas inhabited by humans. Some species cannot migrate, and therefore die; others are forced to compete for fewer resources to live and breed.

- Habitat fragmentation—When roads cut through wild areas they divide wildlife populations into smaller, more isolated, and less stable groups. These animals become more vulnerable to predators and given to inbreeding with its resulting genetic defects.

FIGURE 9.8

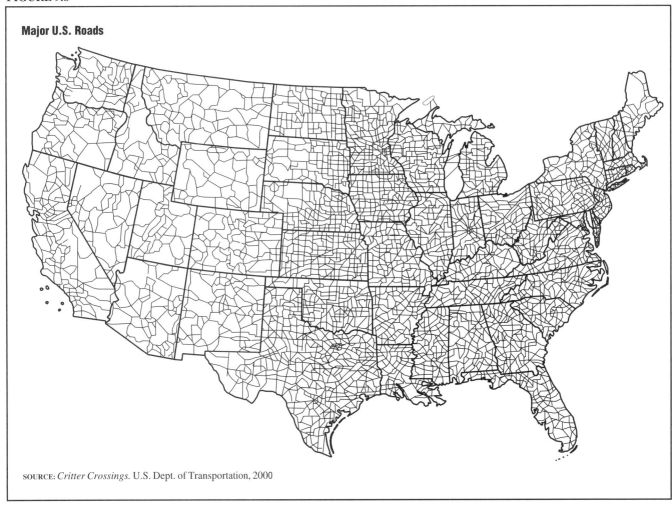

Major U.S. Roads

SOURCE: *Critter Crossings.* U.S. Dept. of Transportation, 2000

Under the 1998 Transportation Equity Act for the 21st Century (PL 105-178), the Federal Highway Administration can provide wildlife crossings—"habitat connectivity measures"—for new and existing roads. Among the strategies used to counteract habitat loss and roadkill are overpasses and underpasses, tunnels, and culverts.

Deep-Sea Harvesting

Worldwide, after centuries of steady growth, the total catch of wild fish peaked in the early 1990s and has declined ever since. The FAO reports that, in six of 11 major fishing regions, more than 60 percent of all commercial fish stocks either have been depleted or are being fished to their limits. The FAO has estimated that more than two-thirds of the world's marine fish stocks were being fished at or beyond their maximum sustainable level by 1993.

A result of the declining catches of fish in shallow fisheries is the recent scouring of the deep seas for other varieties of fish such as the nine-inch long royal red shrimp, rattails, skates, squid, red crabs, orange roughy, orcos, hoki, blue ling, southern blue whiting, and spiny dogfish. Although limited commercial fishing of the deep has been practiced for decades, new sciences and technolo-

gies are making it more practical and more efficient. As stocks of better-known fish shrink and international quotas tighten, experts say the deep ocean waters will increasingly be targeted as a source of seafood. Scientists worry that the rush for deep-sea food will upset the ecology of the ocean.

A Blue Revolution?

The decline in the availability of fish from the deep seas had produced growth of aquaculture, or farmed fish. Worldwide, one in every five fish eaten was raised on a farm, a share that is expected to rise in the years ahead. Aquaculture is one of the fastest growing sectors in world food production. From 12.4 million tons in 1990, farmed fish production nearly doubled—to 23 million tons—by 1996. Worldwatch Institute reports that, for every five kilograms of beef produced worldwide, there are now two kilograms of farm-raised fish.

As a source of animal protein farmed fish are an economical alternative to beef or pork and on par with chicken. While four kilograms of grain are required to produce each kilogram of pork and seven for each kilogram of beef, only two kilograms of grain are needed for a kilogram of chicken or fish. In addition only 40 percent of the

FIGURE 9.9

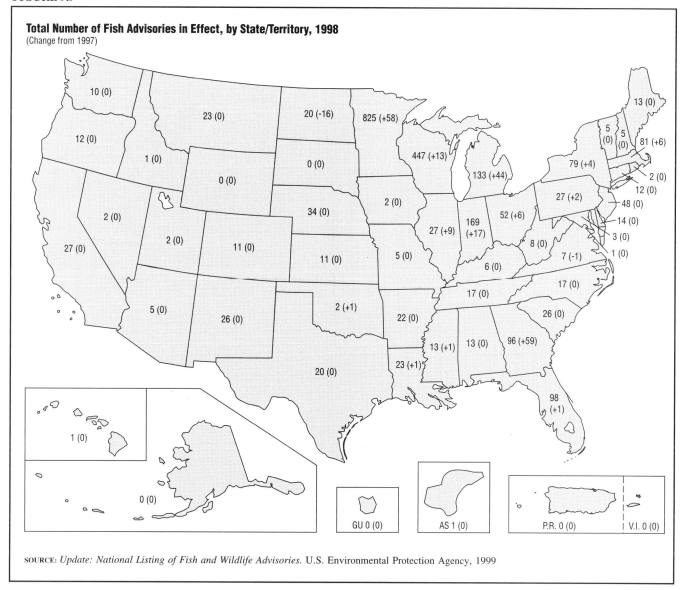

Total Number of Fish Advisories in Effect, by State/Territory, 1998
(Change from 1997)

SOURCE: *Update: National Listing of Fish and Wildlife Advisories.* U.S. Environmental Protection Agency, 1999

weight of sheep is eaten and 50 percent of pigs and chicken, while 65 percent of finfish (opposed to shellfish) is actually consumed. (Because fish are supported by water they have little bone structure; therefore, more of their weight is edible.) Fish are also low in fat and cholesterol, an advantage over other meats.

Contamination of Fish

Noncommercially caught fish and wildlife are sometimes contaminated with chemicals, such as mercury, PCBs, and DDT. In order to protect consumers from health risks associated with consuming such pollutants, the EPA and the states issue consumption advisories to inform the public that high concentrations of contaminants have been found in local specimens. According to the EPA, in *Update: National Listing of Fish and Wildlife Advisories* (1999), in 1998 (the last year for which data are available), 2,506 advisories were issued, a 9 percent increase from 1997. (See Figure 9.9.) However, this was a

98 percent increase from 1993. The trend in the number of advisories issued for various pollutants shows mercury to be—by far—the largest pollutant, followed by PCBs and chlordane. (See Figure 9.10.)

MINERALS

Materials extracted from the earth are needed to provide humans with food, clothing, and housing, and to continually upgrade the standard of living. Some of the materials needed are renewable resources, such as agricultural and forestry products, while others are nonrenewable, such as minerals.

The U.S. Geological Service (USGS), in *Materials Flow and Sustainability* (1998), reported that a significant trend is the decreasing use of renewable resources and the increasing demand for nonrenewable resources. Since 1900 the use of construction materials, such as stone, sand, and gravel, has soared. (See Figure 9.11.)

FIGURE 9.10

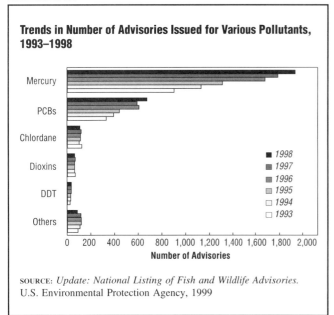

FIGURE 9.11

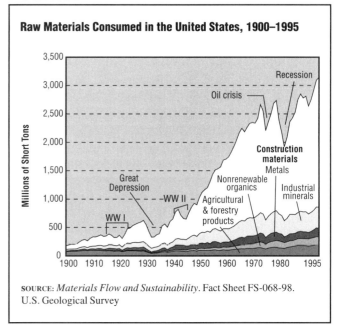

The large-scale exploitation of minerals began in earnest with the industrialization around 1760 in England and has grown rapidly ever since. In a world economy based on fossil fuels, minerals and oil are valuable. The value increases in proportion to demand—which is increasing, and supply—which is decreasing. The result—the search for minerals and fuel sources—has become very aggressive and may be detrimental to the environment.

Mining has always been a dirty industry. As early as 1550, German mineralogist and scholar Georgius Agricola wrote:

> The fields are devastated by mining operations...the woods and groves cut down..., then are exterminated the beasts and birds.... Further, when the ores are washed, the water that has been used poisons the streams, and either destroys the fish or drives them away.

Centuries later mining still pollutes the environment, only on a larger scale. The Clean Air Act (PL 101-576), the Clean Water Act (PL 92-500), and the Resource Conservation and Recovery Act of 1976 (PL 94-580) regulate certain aspects of mining but, in general, the states are primarily responsible for regulation, which varies widely from state to state.

Gold Mining

An extraction process in which cyanide is used for the retrieval of gold from tailings or residues left over from other mining operations has become quite controversial. A series of mishaps around the world have left some communities with polluted lakes, rivers, and streams. A major disaster occurred in Romania on January 30, 2000, when an overflow of polluted mud and wastewater from the Aurul Gold smelter dam sent 100,000 cubic meters of cyanide-tainted wastewater into the adjacent Lapus and Somes Rivers. The cyanide was subsequently carried downstream to the Danube River in Yugoslavia. The spill contaminated drinking water for 2.5 million people and killed more than 100 tons of fish.

Oil in the Arctic

The search for oil has led to the exploration of the Alaskan wilderness. Since the oil supply from the existing North Slope Reserve will steadily decline and then eventually disappear, exploratory oil drillers are focusing their attention on the Arctic wilderness, which includes the 18-million-acre Arctic National Wildlife Refuge in northeast Alaska. Geologists consider the region to be the last great, untapped oil field in North America. Environmental experts fear that oil and gas development will destroy the area.

The future of the refuge lies in the hands of the federal government. Former President George Bush made drilling there a major foundation of his national energy policy. President Clinton reversed this policy. Under Clinton's administration oil and mineral development is prohibited within the wildlife refuge, although most experts expected that to last only as long as President Clinton was in office. The question is how far are Americans willing to go for more domestic oil? Supporters of continued exploration argue that the oil would lessen American dependence on the Middle East, while critics maintain that it is important to preserve one of the earth's largest, undisturbed ecosystems.

Antarctic Resources—Regulate or Prohibit?

THERE ARE MINERALS THERE... Dispute exists over another polar area—Antarctica—as the southern polar

FIGURE 9.12

Antarctic Mineral Occurrences

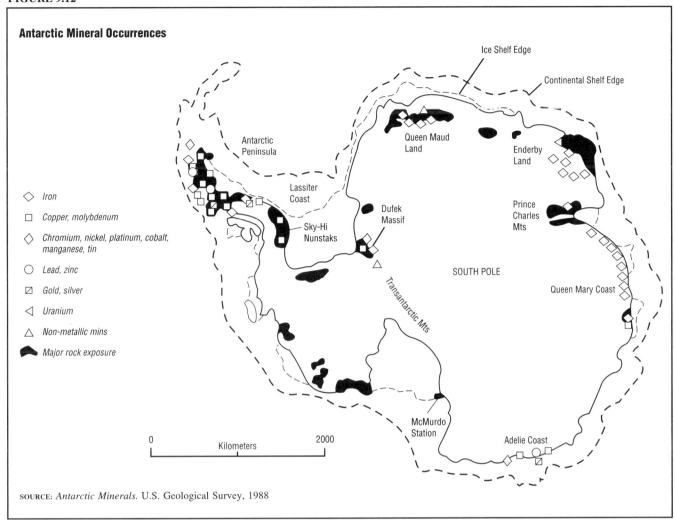

SOURCE: *Antarctic Minerals.* U.S. Geological Survey, 1988

region attracts new interest as a source of petroleum and minerals. Antarctica covers an area of 5.4 million acres—one-tenth of the earth's land surface—and is larger than the United States and Mexico combined. Geologists believe that considerable quantities of mineral deposits probably exist there, as in all large landmasses. Based on the geology of the region, geologists believe they could find base metal (copper, lead, and zinc) and precious metal (gold and silver). There are already some known mineral deposits in Antarctica. (See Figure 9.12.) The huge mass of ice would make recovery difficult, especially in some areas and seasons.

BUT INTERNATIONAL AGREEMENT PROHIBITS THEIR MINING. In 1959, 12 countries (Argentina, Australia, Norway, South Africa, Chile, the United Kingdom, Sweden, France, New Zealand, Belgium, Japan, and the United States) agreed to preserve the region south of 60 degrees south latitude, which includes Antarctica, as an area for scientific research and as a zone of peace. They concluded the Antarctic Treaty, giving equal participation in governance to the signing countries "in the interests of all mankind." The treaty established provisions for new mem-

ber nations; 39 countries representing more than three-fourths of the world's population are party to the treaty.

Seven nations claim territorial sovereignty in Antarctica—Argentina, Australia, Chile, France, Great Britain, New Zealand, and Norway. A 1991 agreement prohibits all mining exploration and development for 50 years, protects wildlife, regulates waste disposal and marine pollution, and provides for increased scientific study of the continent.

Environmentalists want to ban all mining in Antarctica indefinitely. Critics of mining believe the ultimate solution to the problem of mining's destruction of the environment lies in changes in mineral use and a shift from fossil fuels to renewable energy sources. These changes, however, would represent huge transformations in the way people live. Whether these changes are justified, and whether many people are prepared to make them, will be a matter of debate for years to come.

SEEKING GLOBAL SOLUTIONS

The indebtedness and poverty of many developing countries reduce opportunities for conservation. Local, national, and international efforts must be linked to deal effectively with the underlying pressures on the ecosystems that support biological diversity. Some international efforts include "debt for nature" programs and ecotourism.

Since the 1980s a number of debt forgiveness programs have involved "debt for nature" swaps, where governments or conservation groups buy back, or forgive, a portion of a country's debt, usually at a discounted market price, in exchange for the commitment to fund conservation programs.

Another approach to stemming biodiversity-depleting exploitation is to support alternative, less harmful ways for people to earn their livelihoods. Ecotourism—travel oriented around natural sites, native species, and traditional cultural practices—is one alternative economic activity that can use nature, if done carefully, with minimal harm. Ecotourism has spurred communities to protect rare ecological sites and has been modestly successful at generating currency for developing countries.

CHAPTER 10
RENEWABLE ENERGY

WHAT IS RENEWABLE ENERGY?

Imagine an energy source that uses no oil, produces no pollution, cannot be affected by political events and cartels, creates no radioactive waste, and yet is economical. Although that might sound impossible some experts claim that technological advances could make a renewable-energy-based economy achievable by the mid-twenty-first century.

Renewable energy is a term used to describe energy from sources that are naturally regenerated and are, therefore, virtually unlimited. These energy sources include sun, wind, water, vegetation, and the heat of the earth.

Solar energy, wind energy, hydropower, and geothermal power are all renewable, inexpensive, and clean sources of energy. Each of these alternative energy sources has advantages and disadvantages, and many observers hope that one or more of them may someday provide a substantially better energy source than conventional, fossil-fuel-burning methods. As the United States

and the rest of the world continue to expand their energy needs which puts a strain on the environment, alternative sources of energy continue to be explored in the hope that they might provide a higher percentage of America's (and the world's) future energy requirements.

A HISTORICAL PERSPECTIVE

Before the nineteenth century most energy used came from renewable sources. People burned wood for heat, used sails to harness the wind and propel boats, and installed water wheels on streams to grind grain. The large-scale shift to nonrenewable energy sources began in the 1800s with the Industrial Revolution, a period marked by the rise of factories—first in Europe and then in North America. As the demand for energy grew coal replaced wood as the main fuel source. (See Figure 10.1.) Coal was the most efficient fuel for the steam engine, which was perhaps the most important invention of the Industrial Revolution. By the beginning of the twentieth century

FIGURE 10.1

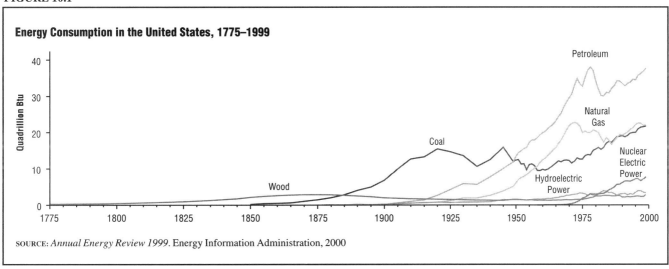

Energy Consumption in the United States, 1775–1999

SOURCE: *Annual Energy Review 1999*. Energy Information Administration, 2000

TABLE 10.1

Energy Consumption by Source, 1949–1999
(Quadrillion Btu)

Year	Coal	Coal Coke Net Imports	Natural Gas[1]	Petroleum[2]	Total Fossil Fuels	Nuclear Electric Power	Hydroelectric Pumped Storage[3]	Conventional Hydroelectric Power[4]	Geothermal[5]	Wood and Waste[6]	Solar	Wind	Total Renewable Energy	Total[7]
1949	11.981	-0.007	5.145	11.883	29.002	0	(8)	1.449	0	1.549	0	0	2.998	32.000
1950	12.347	0.001	5.968	13.315	31.632	0	(8)	1.440	0	1.562	0	0	3.003	34.635
1951	12.553	-0.021	7.049	14.428	34.008	0	(8)	1.454	0	1.535	0	0	2.988	36.996
1952	11.306	-0.012	7.550	14.956	33.800	0	(8)	1.496	0	1.474	0	0	2.970	36.770
1953	11.373	-0.009	7.907	15.556	34.826	0	(8)	1.439	0	1.419	0	0	2.857	37.684
1954	9.715	-0.007	8.330	15.839	33.877	0	(8)	1.388	0	1.394	0	0	2.783	36.660
1955	11.167	-0.010	8.998	17.255	37.410	0	(8)	1.407	0	1.424	0	0	2.832	40.242
1956	11.350	-0.013	9.614	17.937	38.888	0	(8)	1.487	0	1.416	0	0	2.903	41.791
1957	10.821	-0.017	10.191	17.932	38.926	(s)	(8)	1.557	0	1.334	0	0	2.890	41.816
1958	9.533	-0.007	10.663	18.527	38.717	0.002	(8)	1.629	0	1.323	0	0	2.952	41.670
1959	9.518	-0.008	11.717	19.323	40.550	0.002	(8)	1.587	0	1.353	0	0	2.940	43.493
1960	9.838	-0.006	12.385	19.919	42.137	0.006	(8)	1.657	0.001	1.320	0	NA	2.977	45.120
1961	9.623	-0.008	12.926	20.216	42.758	0.020	(8)	1.680	0.002	1.295	0	NA	2.977	45.755
1962	9.906	-0.006	13.731	21.049	44.681	0.026	(8)	1.822	0.002	1.300	0	NA	3.124	47.832
1963	10.413	-0.007	14.403	21.701	46.509	0.038	(8)	1.772	0.004	1.323	0	NA	3.099	49.647
1964	10.964	-0.010	15.288	22.301	48.543	0.040	(8)	1.907	0.005	1.337	0	NA	3.248	51.831
1965	11.581	-0.018	15.769	23.246	50.577	0.043	(8)	2.058	0.004	1.335	0	NA	3.397	54.016
1966	12.143	-0.025	16.995	24.401	53.514	0.064	(8)	2.073	0.004	1.369	0	NA	3.446	57.024
1967	11.914	-0.015	17.945	25.284	55.127	0.088	(8)	2.344	0.007	1.340	0	NA	3.691	58.906
1968	12.331	-0.017	19.210	26.979	58.502	0.142	(8)	2.342	0.009	1.419	0	NA	3.771	62.415
1969	12.382	-0.036	20.678	28.338	61.362	0.154	(8)	2.659	0.013	1.440	0	NA	4.113	65.628
1970	12.265	-0.058	21.795	29.521	63.522	0.239	(8)	2.654	0.011	R1.429	0	NA	R4.094	R67.856
1971	11.598	-0.033	22.469	30.561	64.596	0.413	(8)	2.861	0.012	R1.430	0	NA	R4.303	R69.312
1972	12.077	-0.026	22.698	32.947	67.696	0.584	(8)	2.944	0.031	R1.501	0	NA	R4.476	R72.756
1973	12.971	-0.007	22.512	34.840	70.316	0.910	(8)	3.010	0.043	R1.527	0	NA	R4.579	R75.806
1974	12.663	0.056	21.732	33.455	67.906	1.272	(8)	3.309	0.053	R1.538	0	NA	R4.900	R74.078
1975	12.663	0.014	19.948	32.731	65.355	1.900	(8)	3.219	0.070	R1.497	0	NA	R4.786	R72.041
1976	13.584	(s)	20.345	35.175	69.104	2.111	(8)	3.066	0.078	R1.711	0	NA	R4.855	R76.070
1977	13.922	0.015	19.931	37.122	70.989	2.702	(8)	2.515	0.077	R1.837	0	NA	R4.429	R78.120
1978	13.766	0.125	20.000	37.965	71.856	3.024	(8)	3.141	0.064	R2.036	0	NA	R5.242	R80.122
1979	15.040	0.063	20.666	37.123	72.892	2.776	(8)	3.141	0.084	R2.150	0	NA	R5.375	R81.042
1980	15.423	-0.035	20.394	34.202	69.984	2.739	(8)	3.118	0.110	R2.483	0	NA	R5.710	R78.434
1981	15.908	-0.016	19.928	31.931	67.750	3.008	(8)	3.105	0.123	2.590	0	NA	5.818	76.569
1982	15.322	-0.022	18.505	30.232	64.037	3.131	(8)	3.572	0.105	R2.615	0	NA	R6.292	R73.441
1983	15.894	-0.016	17.357	30.054	63.290	3.203	(8)	3.899	0.129	2.831	0	(s)	6.860	73.317
1984	17.071	-0.011	18.507	31.051	66.617	3.553	(8)	3.800	0.165	2.880	0	(s)	6.845	76.972
1985	17.478	-0.013	17.834	30.922	66.221	4.149	(8)	3.398	0.198	R,92.862	0	(s)	R,976.458	R,976.777
1986	17.260	-0.017	16.708	32.196	66.148	4.471	(8)	3.446	0.219	R,92.840	0	(s)	R,96.506	R,977.065
1987	18.008	0.009	17.744	32.865	68.626	4.906	(8)	3.117	0.229	R2.822	0	(s)	R6.169	R79.633
1988	18.846	0.040	18.552	34.222	71.660	5.661	(8)	2.662	0.217	R,92.940	0	(s)	R,95.819	R,983.071
1989	18.926	0.030	19.384	34.211	72.551	5.677	(8)	R,102.999	R,100.338	R,103.050	R,100.059	R,100.024	R,106.470	R,1084.593
1990	19.101	0.005	19.296	33.553	71.955	R,66.162	-0.036	R,1133.140	R,60.359	R2.665	0.063	R0.032	R6.260	R84.186
1991	18.770	R0.010	19.606	32.845	R71.231	R6.580	-0.047	R3.222	R0.368	R2.679	0.066	R0.032	R6.367	R84.063
1992	1219.158	R0.035	20.131	33.527	R,1272.850	R6.608	-0.043	2.863	0.379	R2.826	0.068	0.030	R6.167	R,1285.512
1993	19.776	R0.027	20.827	33.841	R74.471	R6.520	-0.042	3.147	0.393	R2.782	0.071	0.031	R6.424	87.309
1994	19.960	R0.058	21.288	R34.670	R75.976	6.838	-0.035	2.971	0.395	R2.914	0.072	0.036	R6.387	R89.234
1995	20.024	R0.061	22.163	R34.553	R76.802	7.177	-0.028	3.474	0.339	R3.044	0.073	0.033	R6.963	R90.940
1996	20.940	R0.023	R22.559	R35.757	R79.279	7.168	-0.032	R3.915	0.352	R3.104	0.075	0.035	R7.482	R93.911
1997	21.444	R0.046	R22.530	R36.266	R80.286	6.678	-0.042	R3.940	R0.328	R2.982	0.074	R0.034	R7.358	R94.316
1998	R21.593	R0.067	R21.921	R36.934	R80.515	7.157	-0.046	R3.552	R0.335	R2.991	0.074	R0.031	R6.984	R94.570
1999P	21.698	0.058	22.096	37.706	81.557	7.733	-0.063	3.417	0.327	3.514	0.076	0.038	7.373	96.596

[1]Includes supplemental gaseous fuels.

[2]Petroleum products supplied, including natural gas plant liquids and crude oil burned as fuel.

[3]Represents total pumped storage facility production minus energy used for pumping.

[4]Through 1988, includes all net imports of electricity. From 1989, includes only the portion of net imports of electricity that is derived from hydroelectric power.

[5]Includes electricity imports from Mexico that are derived from geothermal energy.

[6]Values are estimated. For all years, includes wood consumption in all sectors. Beginning in 1970, includes electric utility waste consumption. Beginning in 1981, includes industrial sector waste consumption, and transportation sector use of ethanol blended into motor gasoline. Beginning in 1989, includes expanded coverage of nonutility wood and waste consumption.

[7]From 1989, includes net imported electricity from nonrenewable sources and removes ethanol blended into motor gasoline, which would otherwise be double counted in both petroleum and renewable energy.

[8]Through 1989, pumped storage is included in conventional hydroelectric power.

[9]Not all data were available; therefore, values were interpolated.

[10]There is a discontinuity in this time series between 1988 and 1989 due to the expanded coverage of renewable energy beginning in 1989.

[11]There is a discontinuity in this time series between 1989 and 1990; beginning in 1990, pumped storage is removed and expanded coverage of use of hydroelectric power is included.

[12]Independent power producers' use of coal is included beginning in 1992.

R=Revised. P=Preliminary. (s)=Less than 0.0005 and greater than -0.0005 quadrillion Btu. NA=Not available.

Note: Totals may not equal sum of components due to independent rounding.

SOURCE: *Annual Energy Review 1999.* Energy Information Administration, 2000

The Environment

FIGURE 10.2

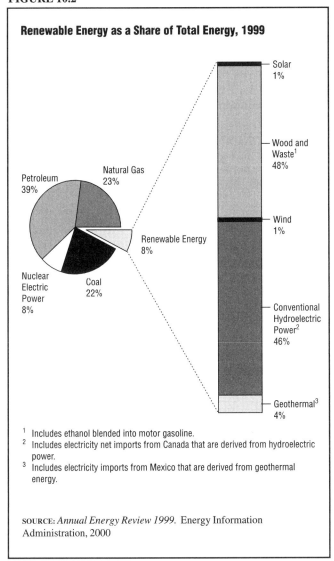

Renewable Energy as a Share of Total Energy, 1999

1 Includes ethanol blended into motor gasoline.
2 Includes electricity net imports from Canada that are derived from hydroelectric power.
3 Includes electricity imports from Mexico that are derived from geothermal energy.

SOURCE: *Annual Energy Review 1999.* Energy Information Administration, 2000

FIGURE 10.3

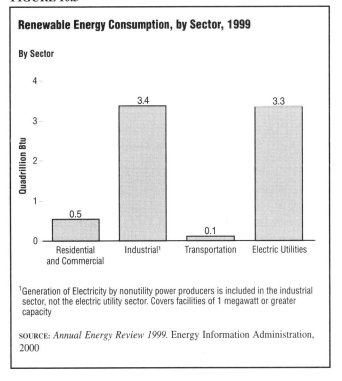

Renewable Energy Consumption, by Sector, 1999

1Generation of Electricity by nonutility power producers is included in the industrial sector, not the electric utility sector. Covers facilities of 1 megawatt or greater capacity

SOURCE: *Annual Energy Review 1999.* Energy Information Administration, 2000

petroleum products such as gasoline and diesel fuel began to replace coal as the fuel of choice.

Until the early 1970s most Americans were unconcerned about the sources of the nation's energy. Supplies of coal and oil, which together provided more than 90 percent of U.S. energy, were believed to be plentiful. The decades preceding the 1970s were characterized by cheap gasoline and little public discussion of energy conservation.

That carefree approach to energy consumption ended in the 1970s. A fuel oil crisis brought on by political events in the Middle East made Americans more aware of the importance of developing alternative sources of energy to supplement and perhaps even replace fossil fuels. In major cities throughout the United States gasoline rationing became commonplace, lower heat settings for offices and living quarters were encouraged, and people waited in line to fill their gas tanks. The crisis was over by the late 1970s and oil prices dropped, but the incident had a lasting impact on American's opinions on oil. In a

country where mobility and personal transportation are highly valued, the oil crisis was a shocking reality for many Americans. As a result President Jimmy Carter's administration (1977–81) encouraged federal funding for research into alternative energy sources.

In 1978 the U.S. Congress passed the Public Utilities Regulatory Policies Act (PURPA; PL 95-617) designed to help the struggling alternative energy industry. The act exempted small producers from state and federal utility regulations and required existing local utilities to buy electricity from the smaller producers. PURPA encouraged the growth of small-scale electric power plants, especially those fueled by renewable sources. The renewable industries responded by growing rapidly, gaining experience, improving technologies, and lowering costs. This act was the single most important factor in the development of the commercial renewable energy market.

In the 1980s President Ronald Reagan decided that private sector financing for the short-term development of alternative energy sources was best. As a result he proposed the reduction or elimination of federal expenditures for alternative energy sources. Although funds were severely cut the U.S. Department of Energy (DOE) continues to support some research and development to explore alternate sources of energy.

How Much of Today's Energy Is Renewable?

Renewable energy at the beginning of the twenty-first century contributed only a small portion of the nation's energy supply, although its importance was expected to

FIGURE 10.4

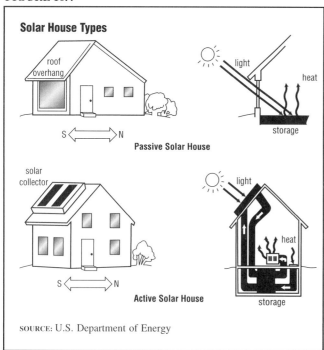

Solar House Types

Passive Solar House

Active Solar House

SOURCE: U.S. Department of Energy

increase. In 1999 the United States consumed an estimated 7.37 quadrillion Btu (British thermal unit—a standard unit of energy measurement) of renewable energy, only 8 percent of the nation's total energy consumption. (See Table 10.1.) Biofuels (mainly wood) and hydroelectric power accounted for, by far, the largest shares (48 percent and 46 percent, respectively) of renewable energy. Geothermal, solar, and wind energy accounted for the remainder. (See Figure 10.2.) Electric utility companies and the industrial sector were the biggest consumers of renewable energy. Residential, commercial, and transportation use accounted for only 0.6 quadrillion Btu combined. (See Figure 10.3.)

SOLAR ENERGY

Ancient Greek and Chinese civilizations used glass and mirrors to direct the sun's rays to start fires. Solar energy (energy from the sun) is a renewable, widely available energy source that generates neither pollution nor hazardous waste. Solar-powered cars have already competed in long-distance races, and solar energy has been used routinely for many years to power spacecraft. Although many people consider solar energy a product of the space age, the Massachusetts Institute of Technology built the first solar powered house in 1939.

Solar radiation is nearly constant outside Earth's atmosphere, but the amount of solar energy, or *insolation*, reaching any point on Earth varies with changing atmospheric conditions such as clouds and dust, and the changing position of Earth relative to the sun. In the United States insolation is greatest in the West and Southwest

regions. Nevertheless, almost all U.S. regions could use solar energy.

Passive and Active Solar Energy Collection Systems

Passive solar systems such as greenhouses, or windows with a southern exposure, use heat flow, evaporation, or other natural processes to collect and transfer heat. It is considered the least costly and least difficult system to implement. (See Figure 10.4.)

Active solar systems use mechanical methods to control the energy process. They require collectors and storage devices as well as motors, pumps, and valves, to operate the systems that transfer heat. Collectors consist of an absorbing plate (solar panel or collector) that transfers the sun's heat to a working fluid (liquid or gas), a translucent cover plate that prevents the heat from radiating back into the atmosphere, and insulation on the back of the collector panel to further reduce heat loss. (See Figure 10.4) Excess solar energy is transferred to a storage facility so it may be used to provide power on cloudy days. In both active and passive systems the conversion of solar energy into a form of power is made at the site where it is used. The most common and least expensive active solar systems are used for heating water.

Solar Thermal Energy Systems

A solar thermal energy system uses intensified sunlight to heat water or other fluids to more than 750 degrees. Mirrors or lenses constantly track the sun's position and focus its rays onto solar receivers that contain fluid. Solar heat (energy) is transferred to the water that, in turn, powers a steam-driven electric generator. In a distributed solar thermal system, the collected energy powers irrigation pumps, provides electricity for small communities, or captures normally wasted heat from the sun in industrial areas. In a central solar thermal system, the energy is collected at a central location and used by utility networks for a large number of customers.

Other solar thermal energy systems include solar ponds and trough systems. Solar ponds are lined ponds filled with water and salt. Because salt water is denser than fresh water, the salt water migrates to the bottom and absorbs the heat, while the fresh water on top keeps the salt water contained and traps the heat. Trough systems use U-shaped mirrors to concentrate the sunshine on water or oil-filled tubes.

Photovoltaic Conversion Systems

The photovoltaic (PV) cell solar energy system converts sunlight directly into electricity without the use of mechanical generators. PV cells have no moving parts, are easy to install, require little maintenance, do not pollute the air, and usually last up to 20 years. PV cells are

FIGURE 10.5

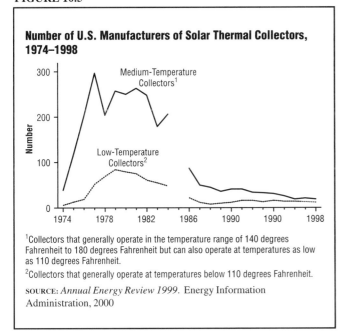

Number of U.S. Manufacturers of Solar Thermal Collectors, 1974–1998

[1]Collectors that generally operate in the temperature range of 140 degrees Fahrenheit to 180 degrees Fahrenheit but can also operate at temperatures as low as 110 degrees Fahrenheit.

[2]Collectors that generally operate at temperatures below 110 degrees Fahrenheit.

SOURCE: *Annual Energy Review 1999.* Energy Information Administration, 2000

FIGURE 10.6

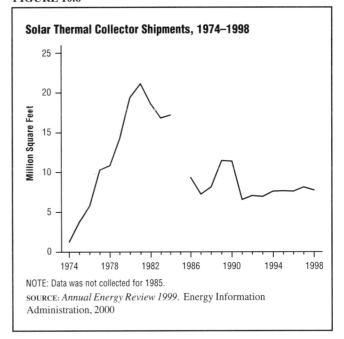

Solar Thermal Collector Shipments, 1974–1998

NOTE: Data was not collected for 1985.

SOURCE: *Annual Energy Review 1999.* Energy Information Administration, 2000

commonly used to power small devices such as watches or calculators. They are also used on a larger scale to provide electricity for rural households, recreational vehicles, and businesses. Solar panels using PV cells have generated electricity for space stations and satellites for many years. Solar panels have also provided electricity for a few major buildings in the United States.

Since PV systems produce electricity only when the sun is shining, a backup energy supply is needed. PV cells produce the most power around noon when sunlight is the most intense. A PV system typically includes storage batteries that provide electricity during cloudy days and at night.

The use of PV technology is expanding both in the United States and abroad. Although PV systems have a higher initial cost than conventional power plants, they have a much lower operating cost.

Using Solar Energy

Because it is difficult to measure solar energy directly, shipments of equipment are often used as an indicator. In 1979 there were 84 low-temperature collector manufacturers. That number dropped to only 22 in 1986 and 12 in 1998. (See Figure 10.5.) Total shipments of solar thermal collectors peaked in 1981 at more than 21 million square feet and declined to approximately 7.76 million square feet by 1998. (See Figure 10.6.)

By 1998 the market for solar energy space heating had virtually disappeared. Most of the solar thermal collector market is for residential purposes (mostly in the Sunbelt states), with only a small proportion for commercial purposes. In 1998 most of the solar thermal collectors shipped were used for heating swimming pools (7.2

FIGURE 10.7

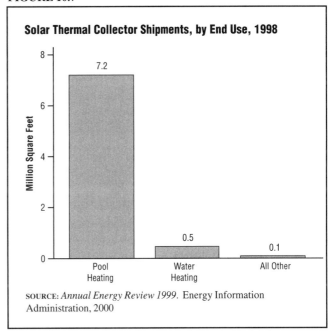

Solar Thermal Collector Shipments, by End Use, 1998

SOURCE: *Annual Energy Review 1999.* Energy Information Administration, 2000

million square feet) while virtually all the rest were used for domestic hot water (0.5 million square feet). (See Figure 10.7.) Some state and municipal power companies have added solar systems as adjuncts to their regular power sources during peak hours.

Advantages and Disadvantages of Solar Energy

The primary advantage of solar energy is its inexhaustible supply, while its primary disadvantage is its reliance on a consistently sunny climate to provide continuous electrical power. Very few areas of the country have enough constant sunshine to make this system an

FIGURE 10.8

Darrieus rotors are grouped together and generate bulk electricity from wind power. (*Corbis Corporation. Reproduced by permission.*)

FIGURE 10.9

Horizontal axis wind turbines are used to create energy on a wind farm in Altamont Pass, California. (*Photograph by Keven Schafer. Corbis Corporation. Reproduced by permission.*)

efficient replacement for conventional methods. In addition a large amount of land area is necessary for the most efficient collection of solar energy for solar thermal units. Experts estimate that a new thermal energy plant would have a 60 percent higher cost of production than a conventional coal-fired plant.

PV cell solar energy systems are probably the most attractive form of solar energy production. A PV cell system is nonpolluting and silent and can be operated by computer. In addition it is less expensive to operate because there are no turbines or other moving parts which makes maintenance minimal. Above all the fuel source (sunshine) is free and plentiful. The disadvantage of a PV cell energy system is the initial cost. Although the price has fallen considerably, PV cells are still too expensive to manufacture and install for widespread use.

Future Development Trends

Interest in PV cell solar energy systems is particularly high in rural and remote areas where it is impractical to

extend traditional electrical power lines. In some remote areas PV cells are used as independent power sources for communications or for the operation of water pumps or refrigerators. This use will most likely increase where the traditional use of an electrical cord is a problem.

Although solar power still costs more than three times as much as fossil fuel energy, utilities could turn to solar energy to provide "peaking power" on extremely hot or cold days. Some people believe that building solar energy systems to provide peak power capacity would be cheaper in the long run than building new and expensive diesel fuel generators. Utility regulators may decide that the price of fossil fuel power must include the hidden cost of fossil fuel damage to the environment caused by acid rain and the greenhouse effect.

WIND ENERGY

Wind energy is really a form of solar energy. Winds are created by the uneven heating of the atmosphere by the sun, the irregularities of the earth's surface, and the rotation of the earth. As a result winds are strongly influenced by local terrain, water bodies, weather patterns, vegetation, and other factors. This wind flow, when "harvested" by wind turbines, can be used to generate electricity.

Wind machines have changed dramatically from those that were common in the 1800s. Early windmills produced mechanical energy to pump water and run sawmills. In the late 1890s Americans began experimenting with wind power to generate electricity. Their early efforts produced enough electricity to light one or two modern light bulbs.

Compared to the pinwheel-shaped farm windmills that can still be seen dotting the American rural landscape, state-of-the-art wind turbines look more like airplane propellers. Their sleek, high-tech fiberglass design and aerodynamics allow them to generate an abundance of electricity, while they also produce mechanical energy and heat. (See Figures 10.8 and 10.9.)

Unlike solar energy systems wind systems produce renewable energy at night as well as during the day. During the 1990s industrial and developing countries alike began using wind power as an adaptable source of electricity to complement their existing power sources and to bring electricity to remote regions. Wind turbines cost less to install per unit of kilowatt capacity than either coal or nuclear facilities. After installing a windmill there are few additional costs, particularly as the fuel (wind) is free.

Wind speeds are generally highest and most consistent in mountain passes and along coastlines. Europe has the greatest coastal wind resources and clusters of wind turbines, or wind farms, are being developed in much of Europe and Asia. Denmark, the Netherlands, China, and India are especially interested in fostering the development of domestic wind industries. In the United States it is estimated that sufficient wind energy is available to provide more than one trillion kilowatt hours of electricity annually. Electricity-producing wind turbines (not windmills used for mechanical energy) operate in 95 countries.

Energy Production by Wind Turbines

Wind is the world's fastest-growing energy source. Although wind power has not been adopted widely in the United States, U.S. companies export turbines to Spain, the Netherlands, Great Britain, India, and China. Following a slow period in the late 1980s when the U.S. government discontinued tax credits for wind installations, the market for wind turbines in the nation has grown.

The wind industry in the United States began with research projects in California. In 1981 the state erected 144 relatively small turbines capable of generating a combined total of seven megawatts of electricity. Within a year the number of turbines had increased 10 times, and by 1986 had multiplied 100-fold. The 1980s saw an explosion of wind technology in California where about 95 percent of the installed wind power capacity in the United States is located.

By the 1990s wind energy facilities began to appear in other states such as Texas, Minnesota, Vermont, Hawaii, and Iowa. Of these states Texas had the most capacity with 43 megawatts in 1997, followed by Minnesota with 25 megawatts. Minnesota and Iowa both have plans for major expansions. In 1995 wind energy generated 3.2 million kilowatts of power across the United States, comprising only a tiny 0.04 percent of all electricity consumed. That year, however, California produced enough wind power to supply all of San Francisco's residents. By 1997 wind energy provided about 1,620 megawatts of power in the United States, up from 1,405

TABLE 10.2

U.S. Wind Net Summer Electric Capacity, 1990–1997

Year	Capacity (megawatts)
1990	1,405
1991	1,653
1992	1,823
1993	1,813
1994	1,745
1995	1,731
1996	1,677
1997	1,620

SOURCE: *Renewable Energy 1998: Issues and Trends.* Energy Information Administration

megawatts in 1990 but less than the 1,823 megawatts achieved in 1992. (See Table 10.2.)

Experts point out that California's dominance has less to do with wind availability than tax credits that were offered by the state until 1999. Studies show that several states, especially the plains states, have wind speeds sufficient to supply electricity to those states. Twelve states—North Dakota, South Dakota, Texas, Kansas, Montana, Nebraska, Wyoming, Oklahoma, Minnesota, Iowa, Colorado, and New Mexico—contain 90 percent of the U.S. wind energy potential. Refinements in wind turbine technology may enable a substantial portion of the nation's electricity to be produced by wind energy.

Problems in the U.S. Wind Industry

For a number of reasons growth in the wind industry has slowed. Crude oil prices fell during the 1990s making oil and gas the lowest-cost fuel sources. Concern about reducing the federal budget resulted in a change in federal policy toward renewable energies. Furthermore, some people are concerned about the uncertainty involving electric utility deregulation. U.S. producers receive around three to four cents per kilowatt hour in price guarantees while wind producers in Germany, Denmark, and India are guaranteed more than twice that amount. As a result investors often find investing in wind energy in the United States too risky.

Development of Wind Energy Throughout the World

During the decade following the 1973 oil crisis, more than 10,000 wind machines were installed worldwide, ranging in size from portable units to multimegawatt turbines. In developing villages small wind turbines recharge batteries and provide essential services. In China small wind turbines allow people to watch their favorite television shows, a major reason for the increased demand for turbines in China. In fact five of the world's ten largest manufacturers of small wind turbines are Chinese.

TABLE 10.3

Wind Electric Capacity Worldwide, 1996 and 1997
(Megawatts)

Country	Year 1996	Year 1997
Europe		
Germany	1,545	1,939
Denmark	857	1,061
Spain	249	406
Netherlands	299	336
United Kingdom	270	330
Sweden	105	108
Italy	71	100
Ireland	11	46
Greece	29	29
Portugal	20	20
Austria	3	20
Finland	8	12
France	10	10
Belgium	7	7
Czech Republic	7	7
Russia	5	5
Ukraine	1	5
Norway	4	4
Poland	1	3
Luxembourg	2	2
Switzerland	2	2
Latvia	1	1
Total	**3,507**	**4,453**
North America		
United States	1,677	1,620
Canada	21	23
Mexico	2	2
Total	**1,700**	**1,645**
Asia		
India	816	845
China	79	166
Japan	14	18
Australia	10	11
New Zealand	4	4
Total	**923**	**1,044**
South and Central America		
Costa Rica	20	20
Argentina	3	9
Brazil	3	3
Total	**26**	**32**
Middle East and Africa		
Iran	9	9
Israel	6	6
Egypt	5	5
Jordan	1	1
Africa	3	3
Total	**24**	**24**
Caribbean	**4**	**4**
Total World	**6,184**	**7,202**

SOURCE: *Renewable Energy 1998: Issues and Trends.* Energy Information Administration

Global wind power generating capacity reached 7,202 megawatts in 1997, up from 2,976 in 1993. Most of the growth has been in northern Europe. In Europe wind energy capacity reached 4,453 megawatts in 1997, followed by North America (1,645 megawatts) and Asia (1,044 megawatts). (See Table 10.3.)

Although wind power supplies less than 0.1 percent of the world's electricity, it is one of the fastest-growing energy sources. The most ambitious wind energy program is planned for India, which plans to provide enough electrical power to serve five million customers. India is expected to be the most rapidly growing market for wind turbines and, if the planned program is successful, wind may supply more energy for India than the country's nuclear program.

Interest in wind energy has been driven in part by the declining cost of capturing wind energy—from more than $.25 a kilowatt-hour in 1980 to $.05 per kilowatt-hour for new turbines in the late 1990s. This makes wind power nearly competitive with gas- and coal-powered plants, even before considering wind's environmental advantages.

Advantages and Disadvantages of Using Wind Energy

The main problem with wind energy is that the wind does not always blow. Some people object to the whirring noise of wind turbines or do not like to see wind turbines clustered in mountain passes and along shorelines because they interfere with scenic views. Some environmentalists have charged that the wind turbines are responsible for the loss of some species of endangered birds that fly into the blades. Finally, as with all types of renewable energy, wind power is more expensive to produce than energy generated through conventional means, at least so long as the price of oil remains low.

On the other hand, generating electricity with wind offers many environmental advantages. Windfarms do not emit climate-altering carbon dioxide, acid-rain-forming pollutants, respiratory irritants, or nuclear waste. Because windfarms do not require water to operate, they are especially well suited to semi-arid and arid regions. Windfarming also offers the added benefit of reducing soil loss on land prone to wind erosion because turbines capture the wind and decrease its potential for downwind destruction.

GEOTHERMAL ENERGY

Since ancient times humans have exploited the earth's natural hot water sources. Although bubbling hot springs became public baths in ancient Rome, using naturally occurring hot water and underground steams to produce power is a relatively modern development. The first electricity to be generated from natural steam was in Italy in 1904. The world's first steam power plant was built in 1958 in a volcanic region of New Zealand. A field of 28 geothermal power plants covering 30 square miles in northern California was completed in 1960.

What Is Geothermal Energy?

Geothermal energy is the natural, internal heat of the earth trapped in rock formations deep within the earth. Only a fraction of this vast storehouse of energy can be extracted, usually where there are large fractures in the earth's crust. Hot springs, geysers, and fumaroles (holes in or near volcanoes from which vapor escapes) are the most easily exploitable sources of geothermal energy.

FIGURE 10.10

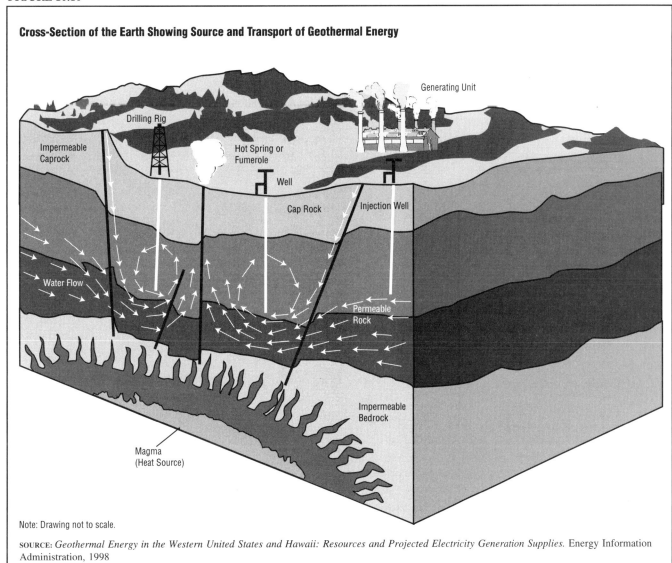

Cross-Section of the Earth Showing Source and Transport of Geothermal Energy

Note: Drawing not to scale.

SOURCE: *Geothermal Energy in the Western United States and Hawaii: Resources and Projected Electricity Generation Supplies.* Energy Information Administration, 1998

(See Figure 10.10.) Geothermal reservoirs provide hot water or steam that can be used for heating buildings, processing food, and generating electricity.

To produce power from a geothermal energy source, pressurized steam or hot water is extracted from the earth and directed toward turbines. The electricity produced by turbines is then fed into a utility grid and distributed to residential and commercial customers. By the late 1990s electricity from this source accounted for almost two-thirds of the world's geothermal energy use.

Types of Geothermal Energy

Like most natural energy sources, geothermal energy is usable only when it is concentrated in one spot, in what is called a "thermal reservoir." The four basic categories of thermal reservoirs are hydrothermal (dry steam and hot, or wet, steam), dry rock, and geopressurized and magma reservoirs. Most of the known areas for geother-mal power in the United States are located west of the Mississippi River. (See Figure 10.11.)

Hydrothermal reservoirs consist of a heat source covered by a permeable formation through which water circulates. Dry steam is produced when hot water boils underground and some of the steam escapes to the surface under pressure. Once at the surface impurities and tiny rock particles are removed and the steam is then piped directly to the electrical generating station. These systems are the cheapest and simplest form of geothermal energy. The Geysers, 90 miles north of San Francisco, California, are the most famous example of this type. (See Figure 10.12.) These Geysers produce enough electricity to meet the needs of about 1.3 million people.

Hot steam systems are created when underground water is heated to more than 700 degrees Fahrenheit by the surrounding hot rock or magma (rock so hot it has

FIGURE 10.11

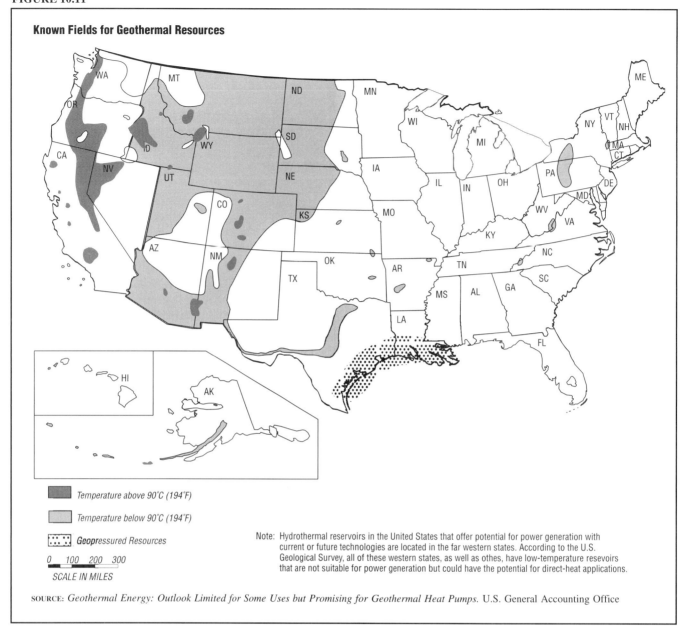

Known Fields for Geothermal Resources

■ Temperature above 90°C (194°F)

▨ Temperature below 90°C (194°F)

⠿ *Geop*ressured Resources

0   100   200   300
SCALE IN MILES

Note: Hydrothermal reservoirs in the United States that offer potential for power generation with
        current or future technologies are located in the far western states. According to the U.S.
        Geological Survey, all of these western states, as well as othes, have low-temperature resevoirs
        that are not suitable for power generation but could have the potential for direct-heat applications.

SOURCE: *Geothermal Energy: Outlook Limited for Some Uses but Promising for Geothermal Heat Pumps.* U.S. General Accounting Office

liquified), but the water remains liquid because of intense pressure. When the water is brought to the surface and the pressure is reduced, a small amount of water becomes steam that is then separated and used to power an electrical generating plant.

Dry rock formations are the most common geothermal source, especially in the West. To tap this source of energy water is injected into naturally hot rock formations to produce steam or water for collection.

Geopressurized reservoirs are sedimentary formations containing hot water and methane gas. Supplies of geopressurized energy remain uncertain, however, and drilling is expensive. Scientists hope that advanced technology will eventually permit the commercial exploitation of the methane content in these reservoirs.

Magma resources are found where molten or partially liquified rock is located from 10,000 to 33,000 feet below the earth's surface. Because magma is so hot, ranging from 1,650 degrees to 2,200 degrees Fahrenheit, it is a good geothermal resource. The process for extracting energy from magma is still in the experimental stages.

Disadvantages of Geothermal Energy

Geothermal plants are expensive because they must be built near the source. Other drawbacks include low efficiency, bad odors from sulfur released in processing, noise, lack of access for most states, potentially harmful pollutants (hydrogen sulfide, ammonia, and radon), and poisonous arsenic or boron often found in geothermal waters. Serious environmental concerns have been raised over the release of chemical compounds, potential water contami-

FIGURE 10.12

The Geysers power plant in California is the largest geothermal development in the world. (*Photograph by Roger Ressmeyer. Reproduced by permisssion.*)

TABLE 10.4

Renewable Energy Consumption, by Source, 1989–1999
(Quadrillion Btu)

Year	Wood and Waste[1]	Geothermal[2]	Conventional Hydroelectric Power[3,4]	Solar[5]	Wind[6]	Total
1989	R3.050	R0.338	R2.999	R0.059	R0.024	R6.470
1990	R2.665	R0.359	R3.140	0.063	R0.032	R6.260
1991	R2.679	R0.368	R3.222	0.066	R0.032	R6.367
1992	R2.826	0.379	2.863	0.068	0.030	R6.167
1993	R2.782	0.393	3.147	0.071	0.031	R6.424
1994	R2.914	0.395	2.971	0.072	0.036	R6.387
1995	R3.044	0.339	3.474	0.073	0.033	R6.963
1996	R3.104	0.352	R3.915	0.075	0.035	R7.482
1997	R2.982	R0.328	R3.940	0.074	R0.034	R7.358
1998	R2.991	R0.335	R3.552	0.074	R0.031	R6.984
1999E	3.514	0.327	3.417	0.076	0.038	7.373

[1]Wood, wood waste, black liquor, red liquor, spent sulfite liquor, pitch, wood sludge, peat, railroad ties, utility poles, municipal solid waste, landfill gas, methane, digester gas, liquid acetonitrile waste, tall oil, waste alcohol, medical waste, paper pellets, sludge waste, solid byproducts, tires, agricultural byproducts, closed looped biomass, fish oil, and straw.

[2]Includes electricity imports from Mexico that are derived from geothermal energy. Includes grid-connected electricity, and geothermal heat pump and direct use energy. Excludes shaft power and remote electrical power.

[3]Hydroelectricity generated by pumped storage is not included in renewable energy.

[4]Includes electricity net imports from Canada that are derived from hydroelectric power.

[5]Includes solar thermal and photovoltaic energy.

[6]Includes only grid-connected electricity.

R=Revised. E=Estimated.

Note: Totals may not equal sum of components due to independent rounding.

SOURCE: *Annual Energy Review 1999.* Energy Information Administration, 2000

nation, the collapse of land surface around the area from which the water is being drained, and potential water shortages resulting from massive withdrawals of water.

American Production of Geothermal Energy

Geothermal energy ranks third in renewable energy production in the United States after hydroelectric and biomass energy. Geothermal energy accounted for 0.327 quadrillion Btu in 1999, down slightly from the mid-1990s. (See Table 10.4.) It represented only 4 percent of renewable energy consumed in the United States. (See Figure 10.2.)

Public sector involvement in the geothermal industry began with the passage of the Geothermal Steam Act of 1970 (PL 91-581), which authorized the U.S. Department of Interior to lease geothermal resources on federal lands. Although the United States is the greatest producer of geothermal power with 44 percent of the world's capacity, the geothermal industry in the United States has become static. As of 1999, the energy market had

excess electrical generating capacity and oil prices were relatively low at a time when most of the easily exploited geothermal reserves had already been developed. In addition utility companies and independent power producers argue over who should build additional generating capacity and what prices should be paid for the power. As a result the rate of growth in U.S. geothermal capacity slowed. Continued growth in the American market depends upon the regulatory environment, oil price trends, and the success of unproven technologies for economically exploiting some of the presently inaccessible geothermal reserves.

World Production of Geothermal Energy

During the oil crisis of the 1970s, when energy was at the forefront of the international agenda, governments scrambled to find domestic alternatives to imported oil. As public interest grew research dollars became available and a large number of geothermal energy plants were built. Although interest has since faded geothermal power's commercial development world-wide has continued at a slow but steady pace.

Since 1979 worldwide geothermal electrical generating capacity has nearly tripled. Nonetheless, in total it is still little more than the energy output of 10 average-size, coal-fired power plants. World geothermal reserves are immense but unevenly distributed. These reserves fall mostly in seismically active areas at the margins or borders of Earth's nine major tectonic plates. Exploited reserves represent only a small fraction of the overall potential—many countries are believed to have in excess of 100,000 megawatts of geothermal energy available.

The United States is the largest geothermal power producer, followed by the Philippines. The Philippine government has committed itself to the development of geothermal power by providing tax incentives and cooperation with the private sector; geothermal energy provides about one-fourth of the nation's electricity. New Zealand and Iceland both use their rich steam reserves to provide significant amounts of power. Italy, Japan, and Mexico are the other major geothermal powers.

A few nations in the developing world—El Salvador, Kenya, Bolivia, Costa Rica, Ethiopia, India, and Thailand—have considerable steam reserves available for power generation. Debt-ridden developing nations that have substantial unexploited geothermal reserves are especially eager to use them instead of relying on costly fossil fuel imports for their energy needs.

HYDROPOWER

Hydropower is the world's largest renewable energy source, accounting for 46 percent of renewable energy. (See Figure 10.2.) Hydropower is the energy that comes from the natural flow of water. Usually, the power is harnessed by taking advantage of gravity when water falls from one level to another. The energy of falling water is converted into mechanical energy. In the past water's energy was harnessed by waterwheels to grind grain or turn saws. Modern technology uses water's energy to turn turbines that create electricity. Hydropower is a renewable, nonpolluting, and reliable energy source. In 1999, in the United States, hydroelectric power generated almost 3.4 quadrillion Btu of energy, down somewhat from the previous two years. (See Table 10.4.)

By the end of the twentieth century hydropower was still the only means of storing large quantities of electrical energy for almost instant use. This is done by holding water in a large reservoir behind a dam with a hydroelectric power plant below. The dam creates a height from which water flows. The fast-moving flow of water from the dam pushes the turbine blades that turn the rotor part of the electric generator. When coils of wire on the rotor sweep past the generator's stationary coil, electricity is produced. Whenever power is needed at peak times water valves are opened and, in a short amount of time, turbine generators produce extra power. The Hoover Dam, located on the Colorado river at the Nevada-Arizona border, is the site of one of the largest U.S. hydroelectric plants.

Advantages and Disadvantages of Hydropower

Small hydropower plants in the United States are costly to build but quickly become cost-efficient because of their low operating costs. One of the disadvantages of small hydropower generators is their reliance on rain and melting snow to fill reservoirs, a problem especially during years with drought conditions. Other concerns include the difficult search for the proper terrain on which to build a hydroelectric power plant; the high cost of construction; and the ecological concern that dams could ruin streams, dry up waterfalls, and interfere with marine life habitats.

Large hydropower plants suffer from the same problems except that they rarely lack sufficient water since they can be built only on very large rivers. There is little potential to build new, large, hydropower plants in the United States because plants have already been built at all of the best sites.

New Directions in Hydropower Energy

Since almost all power sites have already been developed, hydropower's contribution to U.S. energy generation should remain relatively constant—although existing sites can become more efficient as new generators are added.

Most of the new development in hydropower is occurring in developing nations that see it as an effective method of supplying power to their growing populations. Major hydropower development programs are tremendous public works projects requiring huge amounts of money, most of it borrowed from the developed world. Third-world leaders believe that, in the long run, despite threats to the environment, the dams will pay for themselves by bringing cheap electric power to their people.

While the third world has developed only a small portion of its large-scale hydropower potential, the United States and Europe have developed a major proportion of their potential. In addition dams are now less favored because of their harm to the environment. (See Chapter 6.) Large-scale hydropower development has virtually stopped in the United States, with not one new dam being approved for federal funding since the late 1980s.

Dams in the United States were usually constructed entirely with federal monies. Since 1986, however, any new dam proposed in the United States must be built with at least half the money being put up by local governments. Any new major supplies of hydroelectric power for the United States will most likely come from Canada.

OTHER ALTERNATIVES USING WATER

The potential power locked in the world's oceans is unknown. However, since the ocean is not as easily controlled as a river or water that is directed through canals into turbines, unlocking that potential power is far more challenging. Three ideas being considered are tidal plants, wave power, and ocean thermal energy conversion (OTEC).

Tidal Power

A tidal plant uses the power generated by the tidal flow of water as it ebbs (flows back out to sea). A minimum tidal range of three to five yards is generally considered necessary for an economically feasible plant. Canada, for example, has built a small 18-megawatt unit at the Bay of Fundy with its 15-yard tidal range, the largest in the world, and is considering building a larger unit. The largest existing tidal facility is the 240-megawatt plant at the La Rance estuary in northern France. The People's Republic of China (PRC) has a tiny 10-megawatt plant.

Wave Energy

Norway has two operating wave power stations at Tostestallen on its Atlantic coast. The arrival of a wave forces water up a 65-foot tower, displacing the air already in the tower. This air rushes out the top through a turbine, spinning the turbine rotors and generating electricity. When the wave falls back and the water level drops, air is sucked back in through the turbine, again generating electricity.

Another type of wave power plant uses the overflow of high waves. As the wave splashes against the top of a dam, some of the water goes over and is trapped in a reservoir on the other side. The water is then directed through a turbine as it flows back to the sea. These two kinds of plants are experimental. Several projects are underway in Japan and the Pacific region to determine a way to use the potential of the huge waves of the Pacific.

Ocean Thermal Energy Conversion (OTEC)

In 1881 Jacques Arsene d'Arsonval, a French physicist, was the first to propose tapping the thermal energy of the ocean. Not until 1974, however, was a laboratory and test facility for OTEC technologies built at the National Energy Laboratory of Hawaii. In 1980 the U.S. DOE built OTEC-1, a test site on board a converted U.S. Navy tanker.

In 1980 Congress enacted two laws to promote the commercial development of OTEC technology—the Ocean Thermal Energy Conversion Act (PL 96-320) later modified by PL 98-623, and the Ocean Thermal Energy Conversion, Research, and Development and Demonstration Act (PL 96-310).

OTEC uses the temperature difference between the ocean's warm surface water and the cooler water in its depths to produce heat energy that can power a heat engine to generate electricity. OTEC systems can be installed on ships, barges, or offshore platforms with underwater cables that transmit electricity to shore. In addition to providing power, OTEC systems can be used

FIGURE 10.13

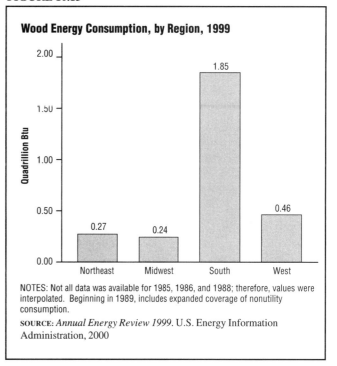

Wood Energy Consumption, by Region, 1999

NOTES: Not all data was available for 1985, 1986, and 1988; therefore, values were interpolated. Beginning in 1989, includes expanded coverage of nonutility consumption.

SOURCE: *Annual Energy Review 1999*. U.S. Energy Information Administration, 2000

to desalinate water, provide air-conditioning and refrigeration, produce methanol, ammonia, hydrogen, aluminum, chlorine, and other chemicals.

BIOMASS ENERGY CONVERSION

The term biomass refers to organic material such as plant and animal waste, wood, seaweed and algae, and garbage. A biofuel is the product of biomass conversion. These raw materials can be converted into liquid or gaseous fuels or used directly to provide heat and electricity. The by-products of biomass conversion can be used for fertilizers and chemicals.

Wood energy was the first energy source in America's industrialization. (See Figure 10.1.) Wood, the most commonly used biofuel, is still used to heat millions of homes every year. Other than hydroelectric power, wood and other biomass resources provide the largest source of renewable electricity produced in the United States. (See Figure 10.2.)

When wood is widely used as a fuel in an area, deforestation can occur, resulting in the possibility of soil erosion and mudslides. Burning wood, as with the burning of fossil fuels, also pollutes the environment.

Types of Biomass Conversion

There are two types of biofuel energy (bioenergy) conversion processes: thermochemical conversion and biochemical conversion. Thermochemical conversion uses heat to produce chemical reactions in biomass. Direct combustion is the easiest and most commonly

FIGURE 10.14

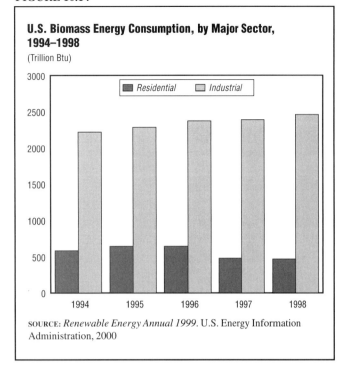

U.S. Biomass Energy Consumption, by Major Sector, 1994–1998

(Trillion Btu)

SOURCE: *Renewable Energy Annual 1999.* U.S. Energy Information Administration, 2000

used method. Materials such as dry wood or agricultural wastes are chopped and burned to produce steam, electricity, or heat for industries, utilities, and homes. Wood burning in stoves and fireplaces is one example. In the United States the number of homes burning wood for fuel—20 million—has remained relatively unchanged since 1980, although these homes are using less fuel.

Homes in the South consume far greater amounts of wood energy than in other parts of the country. (See Figure 10.13.) Industrial-size wood boilers are operating throughout the country and DOE projects that many more will be built over the coming years. The burning of agricultural wastes is also becoming more widespread. In Florida, for example, sugar cane producers use the residue from the cane to generate much of their energy.

Pyrolysis, also called gasification or carbonization, uses heat to break down biomass to yield liquid, gaseous, and solid fuels. Converting wood in to charcoal is an example of this process.

The second type of conversion process, biochemical conversion, uses enzymes, fungi, or other microorganisms to convert high-moisture biomass into either liquid or gaseous fuels. Bacteria convert manure, agricultural wastes, paper, and algae into methane, which is used as fuel. Sewage treatment plants have used anaerobic (without oxygen) digestion for many years to generate methane gas. Small-scale digesters have been used on farms, primarily in Europe and Asia, for hundreds of years. The DOE estimates that many thousands of biofuel

plants are in use today in Korea, and perhaps half a million plants operate in China.

A second type of biochemical conversion process, fermentation, uses yeast to decompose carbohydrates to yield ethyl alcohol (ethanol) and carbon dioxide. Sugar crops, grains (corn, in particular), potatoes, and other starchy crops are common feedstocks that supply the sugar for ethanol production.

Biofuels accounted for 48 percent of renewable energy consumed in 1999. (See Figure 10.2.) Biofuel consumption totaled an estimated 3.51 quadrillion Btu, most of which was wood energy. (See Table 10.4.) Some industries, such as the paper and lumber industries, have ready access to wood and wood by-products and rely heavily on wood as an energy source. Residential use has declined somewhat between 1994 and 1998, while use of biofuels by industry rose slightly over that time period. (See Figure 10.14.)

Ethanol and Methanol—Important Agricultural By-products

Ethanol (ethyl alcohol) is a colorless, nearly odorless, flammable liquid derived from fermenting plant material that contains carbohydrates in the form of sugar. Most of the ethanol manufactured for use as fuel is derived from corn, wood, and sugar. A mixture of 10 percent ethanol and 90 percent gasoline is usable in any internal combustion engine without the need to modify the motor. Although the DOE claims that the demand for alcohol/gasoline blends is increasing because alcohol can substitute for lead as an octane booster, there is little question that the development of ethanol depends more upon the continued support of farm state legislators than any economic benefit.

Ethanol is difficult and expensive to produce in bulk. Methanol-blend fuels have also been tested successfully. (Methanol is methyl alcohol.) Using methanol instead of diesel fuel virtually eliminates sulfur emissions and reduces other environmental pollutants usually emitted from trucks and buses. Burning biofuels in vehicle engines creates a "carbon cycle" in which the earth's vegetation can in turn make use of the products of combustion and, therefore, reduce net greenhouse gases. (See Figure 10.15.) Producing methanol from biofuels, however, is costly.

Some scientists believe ethanol made from wood, sawdust, corncobs, or rice hulls could liberate the alcohol fuel industry from its dependence on food crops such as corn and sugar cane. Worldwide, there are enough corncobs and rice hulls left over from annual crop production to produce more than 40 billion gallons of ethanol.

In just one decade, research was able to cut the cost of wood-derived ethanol from $4.00 per gallon to $1.35 per gallon. Advocates of wood-derived ethanol believe

FIGURE 10.15

The Carbon Cycle

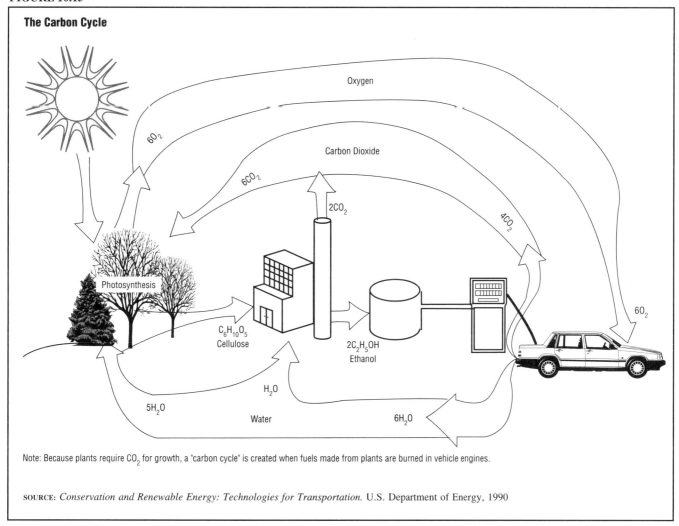

Note: Because plants require CO_2 for growth, a "carbon cycle" is created when fuels made from plants are burned in vehicle engines.

SOURCE: *Conservation and Renewable Energy: Technologies for Transportation.* U.S. Department of Energy, 1990

that the eventual result of wood-to-ethanol conversion research could create a sustainable liquid fuel industry that does not rely on pollution-generating fossil fuels. For instance if new trees were planted to replace those that were cut for fuel, they would be available for later harvesting and, in the meantime, contribute to the prevention of global warming by continuing their carbon-dioxide processing function. Other scientists warn that a huge demand for transportation fuels could create a demand for wood that might accelerate the destruction of old-growth forests and endanger ecosystems. Without careful attention to forestry practices ethanol production might aggravate rather than solve the fuel problem.

MUNICIPAL WASTE RECOVERY

Each year millions of tons of garbage are buried in landfills and city dumps. This method of disposal is not only costly but becoming increasingly difficult as some landfills across the nation near capacity. Many communities have discovered that they can solve both problems at once by constructing waste-to-energy (WTE) plants. Not only is garbage burned and reduced in volume by 90 per-

cent, energy in the form of steam or electricity is generated in a cost-effective way. The potential energy benefit is significant; solid waste generated by the nation's households is equal to more than 200 million barrels of oil per year.

Waste-to-Energy Plants

The two most common WTE plant designs are the mass burn (also called direct combustion) and the refuse derived fuel (RDF) systems.

MASS BURN SYSTEM. Most WTE plants in the United States use the mass burn system. This system's advantage is that the waste does not have to be sorted or prepared before burning, except for removing obviously non-combustible, oversized objects. The mass burn eliminates expensive sorting, shredding, and transportation machinery that may be prone to break down.

Waste is carried to the plant in trash trucks and dropped into a storage pit. Large overhead cranes lift the garbage into a furnace feed hopper that controls the amount and rate of waste that is fed into the furnace. Next the garbage is moved through a combustion zone so that it burns to the greatest extent possible. The burning garbage produces heat, and that heat is used to produce steam. The steam can be used directly for industrial needs or heat or can be sent through a turbine to power a generator to produce electricity.

REFUSE DERIVED FUEL (RDF). RDF systems process waste to remove non-combustible objects and to create homogeneous and uniformly sized fuel. Large items such as bedsprings, dangerous materials, and flammable liquids are removed by hand. The trash is then shredded and carried to a screen to remove glass, rocks, and other material that cannot be burned. The remaining material is usually sifted a second time with an air separator to yield fluff. The fluff is sent to storage bins before being burned, or it can be compressed into pellets or briquettes for long-term storage. This fuel can be used as an energy source by itself in a variety of systems, or it can be used with other fuels such as coal or wood.

PERFORMANCE OF WASTE-TO-ENERGY SYSTEMS. Most WTE systems can produce two to four pounds of steam for every pound of garbage burned. A 1,000-ton-per-day mass burn system will burn an average of 310,250 tons of trash each year and will recover two trillion Btu of energy. In addition the plant will emit 96,000 tons of ash (32 percent of waste input) for landfill disposal. An RDF plant produces less ash but sends almost the same amount of waste to the landfill because of noncombustibles that accumulate in the separation process before burning.

DISADVANTAGES OF WASTE-TO-ENERGY PLANTS. The major obstacle to increasing the use of municipal WTE plants is their effect on the environment. Noise from trucks, fans, and processing equipment at RDF plants can be unpleasant for nearby residents. The emission of particles into the air is controlled by electrostatic precipitators, and most gases can be eliminated by proper combustion techniques. There is concern, however, about the amounts of dioxin (a very dangerous air pollutant) that is often emitted from these plants.

Landfill Gas Recovery

Landfills contain a large amount of biodegradable matter. Gas is created because of the lack of oxygen that helps the growth of methagens—types of bacteria that produce methane gas and carbon dioxide. In the past, as landfills aged, these gases built up and leaked out. This gas leakage prompted some communities to drill holes and burn off the flammable and dangerous methane.

The energy crisis of the 1970s made this methane gas an energy resource too valuable to waste, and efforts were made to find an inexpensive way to tap the gas. The first landfill gas-recovery site was finished in 1975 at the Palos Verdes Landfill in Rolling Hills Estates, California.

Depending on the extraction rates, most existing sites can produce gas for about 20 years.

In a typical operation, garbage is allowed to decompose for several months. When a sufficient amount of methane gas has developed, it is piped out to a generating plant where it is burned to produce electricity. In its purest form methane gas is equivalent to natural gas and can be used in exactly the same way.

The advantages of tapping gas from a landfill go beyond the energy provided by the methane. When internal pressure forces methane gas to seep into the air, it carries very unpleasant odors into the surrounding neighborhoods. The released methane can also be a danger because, if it accumulates and is accidentally ignited, it can explode. Extracting the methane gas for energy eliminates both of these problems.

HYDROGEN—A FUEL OF THE FUTURE?

Hydrogen, the lightest and most abundant chemical element is, from the environmental point of view, the ideal fuel. Its combustion produces only water vapor—it is entirely carbon-free. Three-quarters of the mass of the universe consists of hydrogen such as water, so it is readily available everywhere. However, the elemental, combustible, form of hydrogen is a gas and is not found in the atmosphere. The many compounds of hydrogen, such as water, cannot be converted into pure gas hydrogen without the expenditure of energy. The amount of energy that would be required is about the same as the amount of energy that would be obtained by the combustion of the hydrogen. Therefore, with today's technology, little or nothing could be gained from an energy point of view.

Hydrogen has considerable potential as a clean fuel, however, and, because it is a gas, can be distributed with essentially the same technology as natural gas. Scientists are researching ways to economically produce hydrogen gas.

The possibility of a transition to hydrogen has been considered for more than a century, and many see hydrogen as the logical "third-wave" fuel—hydrogen gas following oil, just as oil replaced coal decades earlier. While advocates note that the world's current energy needs could be met with less than 1 percent of today's fresh water supply and that hydrogen can be produced from seawater, hydrogen as an energy resource is still a long, long way in the future.

THE NEXT ENERGY REVOLUTION?— THE FUTURE OF RENEWABLE ENERGY

In 1998 renewable energy contributed a relatively small and steady 8 percent of the total energy consumed by the nation. The nation's fleet of automobiles has become less energy efficient, and consumers are less interested in energy-saving technology. Many economists believe that only a return of higher energy prices will cause Americans to once again reduce their energy use and consider renewable energy sources. Despite their environmental advantage over fossil fuels, renewable energies have never attracted enough financial support from the government, the public, or energy companies to the point where they can be cost competitive with fossil fuel power.

The Energy Information Administration (EIA) of the DOE annually forecasts energy supply, demand, and prices for the coming two decades. Its projections are based upon business trends and federal, state, and local laws in effect at the time. These forecasts are used by government officials and planners, and decision-makers in the public and private sectors. In its *Annual Energy Outlook 2000* (1999), the EIA predicted that electricity from renewable sources would increase slightly, from 6.7 quadrillion Btu to 8.0 quadrillion Btu between 1998 and 2020. (See Figure 10.16.) Growth will be in the use of geothermal and wind energy, biomass, recovery of methane from municipal solid waste, and use of ethanol. The EIA predicts that use of hydropower and wood in homes will decline.

The EIA predicts that, by the year 2010, renewables could contribute about 10 percent of the nation's total energy consumption and as much as 50 to 70 percent of U.S. energy by 2030 if government supported the effort. Most experts believe that is unlikely.

Abroad —a Grassroots Movement

While the use of renewable energy sources has remained steady in the United States, other countries have shown more interest.

WIND ENERGY. Encouraged by improved technology, falling costs, and government incentives like tax credits and guaranteed prices, wind power is booming across Europe. With windmills springing up from the coasts of Sweden to the tip of Spain, Europe's wind industry already employs 20,000 people. Close to 100,000 Danes own shares in hundreds of small cooperatives that operate 4,700 windmills. Furthermore, they are making a profit. Wind power generates 6 percent of Denmark's electricity, the highest per capita output of wind energy in the world.

One reason for the growth in the industry is that the European Union wants to diversify its energy sources while clamping down on pollution. Almost no country supports the expansion of nuclear power and, in many areas, wind power is becoming economically viable. New wind turbines already generate electricity less expensively than solar panels, biomass, or other nontraditional sources. The International Energy Agency, a research

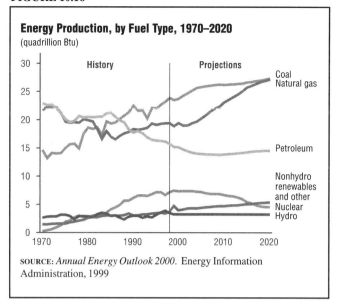

FIGURE 10.16

Energy Production, by Fuel Type, 1970–2020
(quadrillion Btu)

SOURCE: *Annual Energy Outlook 2000.* Energy Information Administration, 1999

organization based in Paris, reported in 2000 that wind energy is now competitive with electricity from Europe's oil- and coal-fired power plants.

In terms of wind power Europe has already overtaken the United States, which led the drive to wind energy in the 1980s. Europe's capacity of 4,453 megawatts is two and one-half times that of the United States, with Germany alone surpassing the output of American wind farms. (See Table 10.3.) Wind energy is produced under entirely different circumstances in the United States and in Europe. American entrepreneurs, seeing wind energy as a potentially profitable business, built large wind farms with huge numbers of turbines. When oil prices fell and tax credits were cut, growth stalled. In Denmark, Germany, Sweden, and the Netherlands, wind energy began as a grass-roots movement, with small groups of politically motivated investors installing one or a few machines at a time. In Spain, Britain, and Greece, the clusters were larger because money was provided by local governments and utilities. The latest trend in Europe is to build wind farms offshore where there is more wind and fewer complaints that they clutter the landscape. The Netherlands, which had approximately 11,000 windmills less than a century ago, has 1,120 modern turbines today, many of them standing beside the old-style windmills.

SOLAR POWER. Rural areas are more expensive to serve than cities, and electrification has been slow to reach many people in rural areas of developing countries. In the United States it was only after the Rural Electrification Administration in 1935 provided low-cost financing to rural electric cooperatives that most farmers received power. Many developing countries have similar programs, but as increasingly remote and mountainous areas have begun receiving electricity, the cost of hooking up new customers has grown greatly. The World Bank estimates that in places such as western China, the Himalayan foothills, or the Amazon basin, the cost of hooking up new rural customers is seven times that in cities. Furthermore, state-owned power systems have been badly managed in many countries. This has left many national power systems all but bankrupt and blackouts common.

In India, blackouts are so common that many factories and other businesses have, at great expense, set up their own private systems based primarily on natural gas, propane, or fuel oil. Although rural families do not have access to those systems, they do have sunlight. In most tropical countries considerable energy falls on rooftops in the form of sunlight. Electricity produced by solar PV cells was initially too expensive, as much as a thousand times more costly than conventional plants. By 1996 the price had fallen to about $1.65 a kilowatt-hour, opening up the global market. Solar energy advocates believe that solar cells can eventually be made so inexpensively that they can provide economical power even for consumers already hooked up to conventional power supplies.

During the 1990's a different approach to solar electrification developed, driven less by government planners and more by the desire of individual families to meet their own need for electricity. In more than a dozen countries solar power is now reaching thousands of families one-by-one, avoiding the delay for government planners to deliver. In Kenya, for example, where the state power company is on the verge of bankruptcy, eight domestic companies have merged to market, install, and maintain solar home systems. With little state or international assistance, those companies managed to electrify 20,000 rural households between 1987 and 1992, 3,000 more than the state power system.

Despite those advances only one of every thousand potential customers has yet been served. Part of the problem is lack of credit. Consumer credit is one of the most momentous financial advances of the twentieth century, leading to wide ownership of homes, automobiles, and appliances that the average person cannot afford outright. According to Neville Williams, president of Solar Electric Light Fund (SELF), a non-profit U.S. energy agency, millions of families in the developing world could afford solar energy if credit were available.

To fill this gap, international agencies are working to set up revolving credit funds to finance solar energy systems in many countries, including Vietnam, India, Indonesia, Uganda, Swaziland, and the Dominican Republic. A number of non-profit agencies, private foundations, investment firms, and the World Bank have been attempting to satisfy this financial need. Many observers hope that newly developing nations can "leapfrog" many of the more damaging environmental practices, including the heavy dependence on fossil fuels that characterized

IMPORTANT NAMES
AND ADDRESSES

American Crop Protection Association
1156 15 St. NW, #400
Washington, DC 20005
(202) 296-1585
FAX (202) 463-0474
E-mail: member_services@acpa.org
URL: http://www.acpa.org

American Lung Association
1740 Broadway
New York, NY10019
(212) 315-8700
E-mail: info@lungusa.org
URL: http://www.lungusa.org

**Biological Resources Division
(BRD)—USGS**
U.S. Dept. of the Interior
12201 Sunrise Valley Dr.
Reston, VA 22092
(301) 317-3819
URL: http://www.doi.gov

Clean Water Fund
4455 U Ave. NW, Suite A300
Washington, DC 20008
(202) 895-0420
FAX (202) 895-0438
E-mail: cleanwaterfund@cleanwater.org
URL: http://www.cleanwater.org

Council on Environmental Quality
722 Jackson Pl. NW
Washington, DC 20503
(202) 395-5750
FAX (202) 456-6546
URL: http://www.whitehouse.gov/ceq

**Environmental Business International
Inc.**
4452 Park Blvd., #306
San Diego, CA 92116
(619) 295-7685
FAX (619) 295-5743
E-mail: moe@ebiusa.com
URL: http://www.ebiusa.com

Environmental Defense Fund
1875 Connecticut Ave., #1016
Washington, DC 20009
(202) 387-3500
FAX (202) 234-6049
E-mail: contact@environmentaldefense.org
URL: http://www.edf.org

Environmental Industries Associations
4301 Connecticut Ave., #300
Washington, DC 20008
(202) 244-4700
FAX (202) 966-4818
E-mail: jleca@envasns.org
URL: http://www.envasns.org

Environmental Protection Agency (EPA)
401 M St. SW
Washington, DC 20460
(202) 260-4700
FAX (202) 260-0279
URL: http://www.epa.gov

Fresh Water Society
2500 Shadywood Rd.
Navarre, MN 55331
(612) 471-9773
FAX (612) 471-7685
E-mail: freshwater@freshwater.org
URL: http://www.freshwater.org

Friends of the Earth
1025 Vermont Ave. NW, #300
Washington, DC 20005-0303
(202) 783-7400
FAX (202) 783-0444
E-mail: foe@foe.org
URL: http://www.foe.org

Greenpeace USA
1436 U St. NW
Washington, DC 20009
(202) 462-1177
FAX (202) 462-4507
E-mail: greenpeace.usa@wdc.greenpeace.org
URL: http://www.greenpeace.org

Izaak Walton League of America
707 Conservation Ln.
Gaithersburg, MD20878-2783
(301) 548-0150
FAX (301) 548-0146
E-mail: general@iwla.org
URL: http://www.iwla.org

**National Aeronautics and Space
Administration (NASA)**
Goddard Space Flight Center
Office of Public Affairs
Greenbolt, MD 20771
(301) 286-8955
E-mail: taday@nasa.gov
URL: http://www.nasa.gov

National Audubon Society
1901 Pennsylvania Ave., #1100
Washington, DC 20006
(202) 861-4290
FAX (202) 861-4290
E-mail: jbianchi@audubon.org
URL: http://www.audubon.org

**National Coalition Against the Misuse of
Pesticides**
701 E St. SE, Suite 200
Washington, DC 20003
(202) 543-5450
FAX (202) 543-4791
URL: http://www.beyondpesticides.org

National Mining Association
1130 17 St. NW
Washington, DC 20036-4677
(202) 861-2800
FAX (202) 463-6152
E-mail: rmaddalena@nma.org
URL: http://www.nma.org

National Parks Conservation Association
1776 Massachusetts Ave. NW, #200
Washington, DC 20036
(202) 223-6722
FAX (202) 659-0650

E-mail: npca@npca.org
URL: http://www.npca.org

National Wildlife Federation
1400 16 St. NW
Washington, DC 20036
(202) 797-6800
FAX (202) 797-6646
URL: http://www.nwf.org

Natural Resources Defense Council
40 West 20 St.
New York, NY 10011
(212) 727-2700
FAX (212) 727-1773
E-mail: nrdcinfo@nrdc.org
URL: http://www.nrdc.org

Nature Conservancy
4245 N. Fairfax Dr., #100
Arlington, VA 22203
(800) 628-6860
URL: http://www.tnc.org

Pesticide Action Network
49 Powell St., Suite 500
San Francisco, CA 94102

(415) 981-1771
FAX (415) 981-1991
E-mail: panna@panna.org
URL: http://www.panna.org

Public Citizen
1600 20 St. NW
Washington, DC 20009
(202) 588-1000
E-mail: member@citizen.org
URL: http://www.citizen.org

Rachel Carson Council
8940 Joans Mill Rd.
Chevy Chase, MD 20815
(301) 652-1877
FAX (301) 951-7179
URL:
http://members.aol.com/rccouncil/ourpage

Sierra Club
408 C St. NE
Washington, DC 20002
(202) 547-1141
FAX (202) 547-6009
URL: http://www.sierraclub.org

Union of Concerned Scientists
2 Brattle Square
Cambridge, MA 02238
(617) 547-5552
FAX (617) 864-9405
E-mail: ucs@ucsusa.org
URL: http://www.ucsusa.org

The Wilderness Society
900 17 St. NW
Washington, DC 20006
(202) 833-2300
FAX (202) 429-3958
URL: http://www.wilderness.org

Worldwatch Institute
1776 Massachusetts Ave. NW
Washington, DC 20036
(202) 452-1999
FAX (202) 296-7365
E-mail: worldwatch@worldwatch.org
URL: http://www.worldwatch.org

World Wildlife Fund
1250 24 St. NW
Washington, DC 20037
(202) 293-4800
FAX (202) 293-9211
URL: http://www.worldwildlife.org

RESOURCES

The now-defunct Office of Technology Assessment (OTA), an analytical support agency of the United States Congress, provided outstanding resource information for committees of Congress and the public on a variety of scientific and technical issues. Among the OTA publications used in the preparation of this book are *Wetlands: Their Use and Regulation* (1984) and *Green Products by Design: Choices for a Cleaner Environment* (1992). The U.S. Congressional Budget Office, which produces studies for Congress on economic and financial issues, published *Federal Options for Reducing Waste Disposal* (1991) and *The Total Costs of Cleaning Up Nonfederal Superfund Sites* (1994).

Other U.S. government publications consulted included *Agriculture and the Environment* (U.S. Department of Agriculture, 1992), *Policy Implications of Greenhouse Warming* (U.S. Forest Service, 1990), and *A Comprehensive Approach to Addressing Potential Climate Change* (U.S. Department of Justice, 1991). Also used were *More than Asphalt, Concrete, and Steel* (U.S. Department of Transportation, 1997). The U.S Bureau of Justice Statistics provided *Federal Enforcement of Environmental Laws, 1997* (1999). The United States Geological Service (USGS), Denver, CO, documents the use of the nation's waters every five years in *Estimated Use of Water in the United States in 1995* (1998). The USGS also produced *Sustainability of Ground Water Resources* (1999), *Fertilizers—Sustaining Global Food Supplies* (1999), *Materials Flow and Sustainability* (1998), and *Water of the World* (undated). The Energy Information Administration (EIA) of the U.S. Department of Energy's *Emissions of Greenhouse Gases in the United States 1998* (1999) was a source of data on global warming. The EIA also published *Renewable Energy 1998: Issues and Trends* (1999), the *Annual Energy Review 1998* (1999), *Annual Energy Outlook 2000* (2000), *Alternatives to Traditional Transportation Fuels 1998* (1999), and *The Waste Isolation Plant* (1999). *Yucca Mountain Studies*

(1990) and bulletins of the Office of Civilian Radioactive Waste Management of the Department of Energy were also helpful.

Data from the U.S. Public Health Service's Centers for Disease Control and Prevention (CDC)'s *Morbidity and Mortality Weekly Report, Surveillance Summaries*, and *Healthy People 2000* was invaluable. The National Institute of Environmental Health Sciences (NIEHS) of the National Institute of Health prepared the *NIEHS Report on Health Effects from Exposure to Power-Line Frequency Electric and Magnetic Fields* (1999) and *Questions and Answers About Electric and Magnetic Fields Associated with the Use of Electric Power* (1995). The CDC also produced their *Journal of Emerging Infectious Diseases* (October–December 1997).

The National Aeronautics and Space Administration (NASA—Goddard Space Flight Center) publishes a variety of materials on environmental and space issues. Useful in this book were *NASA Facts* (1996), *Understanding Our Changing Climate* (1997), and *Looking at the Earth From Space* (1994). The U.S. Department of Transportation produced *Critter Crossings* (2000), a study of the conflicts between humans and wildlife along U.S. roads and highways. The U.S. Government Accounting Office (GAO) has published many useful reports on environmental issues, including energy, nuclear waste, landfills, and pollution. The *Climate Action Report*, prepared by the United Nations Framework Convention on Climate Change (1995) was useful in explaining the scientific community's determinations about global warming.

The Environmental Protection Agency (EPA) monitors the status of the environment. Some of its publications include the *National Water Quality Inventory: 1998 Report to Congress* (2000), *Progress in Ground-Water Protection and Restoration* (1990), *National Air Quality and Emissions Trends Report, 1998* (1999), *Characteriza-*

tion of Municipal Solid Waste in the United States: 1998 Update (1999), *Let's Reduce and Recycle: Curriculum for Solid Waste Awareness* (1990), and *The Plain English Guide to The Clean Air Act* (1993). Also useful were *Municipal Solid Waste Generation, Recycling, and Disposal in the United States: Facts and Figures for 1998* (2000), *National Source Reduction Characterization Report* (1999), *Report on the Supply and Demand of CFC-12 in the United States 1999* (1999), *Light Duty Automotive Technology and Fuel Economy Trends Through 1999* (2000), and *Inventory of U.S. Greenhouse Gas Emissions and Sinks: 1990–1998* (2000). The EPA's Endocrine Screening and Testing Advisory Committee (EDSTAC) published *EDSTAC Final Report* (1998) on endocrine disruptors. The EPA also provided *Update: National Listing of Fish and Wildlife Advisories* (1999) and the *Toxic Release Inventory* (2000).

The National Conference of State Legislatures publishes reports on a wide variety of environmental topics. The Gale Group appreciates especially the *Legislators' Guide to Alternative Fuel Policies and Programs* (1997), *Legisbrief—Lead Hazard Disclosures in Real Estate Transactions* (1997), and *Two Decades of Clean Air: EPA Assesses Cost and Benefits* (1998). The Institute of Chemical Waste Management's *Managing Hazardous Waste: Fulfilling the Public Trust* (1989) provided useful information on modern landfills.

Other important publications include *Trends in the U.S.: Consumer Attitudes and the Supermarket* (Food Marketing Institute, Washington, DC, 1995) and *Policy Implications of Greenhouse Warming* (National Academy of Sciences, Washington, DC, 1991). *Conserving the World's Ecological Biodiversity* (Washington, DC, 1990), prepared by World Resources Institute, World Wildlife Fund, the International Union for Conservation of Nature and Natural Resources, and the World Bank, was most informative regarding biodiversity and the endangerment of species. The World Resources Institute also published *The Greenhouse Trap: What We're Doing to the Atmosphere and How We Can Slow Global Warming* (Beacon Press, 1991). *The U.S. Environmental Industry 1998* (2000), prepared by Environmental Business International (San Diego, CA), an environmental research and consulting group, was the source of data on the status of the environmental industry.

The Gale Group appreciates the use of material from the Gallup Organization's public opinion surveys and to the League of Women Voters for *The Garbage Primer* (New York, 1993). Thanks to the Endangered Species Coalition for information on plant species used in medications. Also helpful was *Garbage Then and Now* (undated) by the National Solid Waste Management Association.

The Gale Group thanks the J.G. Press for the use of its important biennial study, "The State of Garbage in America," from *Biocycle Magazine*. Wirthlin Worldwide, a nationwide public opinion research organization based in McLean, VA, prepared *The Wirthlin Report*. The Gale Group also appreciates use of information from the Environmental Defense Fund, New York, NY, on the leading environmental concerns of Americans and for *Evaluation of Erosion Hazards*, prepared by the H. John Heinz III Center for Science, Economics, and the Environment for the Federal Emergency Management Agency (FEMA; April 2000) on the erosion of U.S. coastlines.

INDEX

by pollutants and contributing sources, 70 (*t*5.2)

relationship among building and human factors, 156*f*

respiratory diseases, 70

waterbourne disease outbreaks, 119–121

Heavy metals, 137

Highway noise barriers, 149

Highways, 180*f*

Hoover Dam, 99

Horizontal axis wind turbines, *190*

Hot steam systems, 193–194

Howard A. Engle, MD v. R. J. Reynolds Tobacco et al (1996), 153

Hunt, Chemical Waste Management v. (1992), 46

Hydrogen, 81, 200

Hydrologic cycle, 93, 94*f*

Hydromodification. *See* Dams

Hydropower, 196–197

I

Idaho, 63

Illinois, 63, 72, 108, 119

In re Agent Orange Product Liability Litigation, 138

In-stream use water, 96

Incineration and incinerators, 51–53, 55

India, air quality, 91

Indiana, 108, 171

Indiana Michigan Power Co. v. Department of Energy (1996), 64

Indoor air quality, 155–157

Indoor pollution. *See* Indoor air quality

Industrial Revolution, 1, 185

Industrial waste, 55

Insolation. *See* Solar energy

Intergovernmental Panel on Climate Change (IPCC) assessments, 28–29, 28*t*

Interior West, U. S., 166*f*

International Convention for the Prevention of Pollution from Ships, 108–109

International environmental protection, 8–9

Iowa, 171, 191

Irrigation, 96–97, 174

Ivy v. Diamond Shamrock (1992), 138

J

James M. Seif v. Chester Residents Concerned Citizens et al, 6–7

Job growth/loss and environmental protection, 4

"Joe Camel" campaign, 152

K

Kansas, 191

Kennedy Heights (Houston, TX neighborhood) contaminated water supply, 6

Kentucky, 149

Kyoto Protocol, 30, 87

L

Landfill gas recovery, 200

Landfills, 42, 44–50, 47*f*, 55

Lead and Copper Rule, 113

Lead and lead poisoning

blood lead levels, by states participating in surveillance, 147*f*

blood lead levels in adults, 146*f*

in drinking water, 112–113

lead, as pollutant, 74, 144–145

lead disclosure laws, by state, 81*f*

long-term ambient lead concentrations, 79*f*

nonattainment designations, 80*f*

Lead-Based Poisoning Prevention Act, 145

Lead Contamination Control Act of 1988, 117

Legislation and international treaties

Antarctic Treaty, 183

Beachfront Management Act, 5–6, 172

Clean Air Act of 1970, 38, 88–89, 133

Clean Water Act, 100

Comprehensive Environmental Response, Compensation, and Liability Act of 1980 (CERCLA), 56–58, 107

Earth Summit Biodiversity Treaty, 176

Emergency Planning and Community Right-to Know Act of 1986, 155

Endangered Species Act, 175, 176–177

Energy Policy Act of 1992, 80, 144

federal environmental and wildlife protection acts, 7*f*

Federal Insecticide, Fungicide, and Rodenticide Act, 107, 137

Food Quality Protection Act, 154

Grand Canyon Protection Act of 1992, 100

International Convention for the Prevention of Pollution from Ships, 108–109

Kyoto Protocol, 30, 87

Lead and Copper Rule, 113

Lead-Based Poisoning Prevention Act, 145

Lead Contamination Control Act of 1988, 117

Low Level Radioactive Waste Policy Act of 1980, 60

Noise Control Act of 1972, 159

North American Free Trade Agreement (NAFTA), 10

Ocean Dumping Act, 109–110

Ocean Thermal Energy Conversion, Research, and Development and Demonstration Act, 197

Oil Pollution Act of 1990, 110

Organic Foods Production Act, 162

Primary Drinking Water Standards, 116

Public Utilities Regulatory Polices Act, 187

Residential Lead-Based Paint Hazard Reduction Act, 145

Resource Conservation and Recovery Act, 47, 54, 107

Safe Drinking Water Act of 1974, 106–107, 116, 117

Taxpayer Relief Act, 59

Toxic Substances Control Act, 145

UN Economic Commission for Europe Convention on Long-Range Transboundary Air Pollution, 131

Water Quality Control Act of 1987, 117

World Trade Organization (WTO), 9–10

Liggett Group, Inc., Cipollone v. (1994), 152

Lindzen, Richard, 29

Litigation and environmental policy, 5–6

See also Court cases

Loblolly pines, 165

Logging, deforestation, and habitat loss, 165, 179

Los Alamos (NM) "controlled burn" disaster, 166

Los Angeles (CA) smog, 72, 86

Low birth-weight babies, 149

Low-level radioactive waste, 60–61

Low Level Radioactive Waste Policy Act of 1980, 60

Lucas, David, 5–6, 172

Lucas v. South Carolina (1992), 6, 172

Lung cancer, 140, 146

M

Magma resources, 194

Maine, 72

Man-made objects, effect of acid rain on, 129–131

Marshes. *See* Wetlands

Maryland, 51, 72

Mass burn system, 199–200

Massachusetts, 72

Mauna Loa atmospheric concentration of carbon dioxide, 19 (*f*2.3)

Melanoma and other skin cancers, 37

Mercury (metal), 114

Mesothelioma, 140

Methane, 23, 23*t*, 46–47, 200

Methanol, 198–199

Methyl chloroform, 37

Mexico forest fires, 91

Michigan, 52

Microbiological organisms, as water contaminants, 114

Middle East water supply, 110

Milwaukee (WI) drinking water contamination, 119

Minerals, 181–183

Mining, 44, 182

Minnesota, 51, 159, 191

Mission to Planet Earth (MTPE), 14, 20
Missouri, 171
Montana, 191
Montreal Protocol, 38
Mount Pinatubo, 19
Mountains, 173
MTBE (Methyl tertiary butyl ether ethanol), 77
Municipal solid waste
generation, imports, and exports, by state, 46–47 (*t*4.2)
historical and projected generation, 53 (*f*4.13)
landfills and incinerators, by state, 45–46 (*t*4.1)
by material, 49 (*f*4.8)
by product, 49 (*f*4.7)
projections of materials generated in the municipal waste stream, 53 (*t*4.4)
recovery and discards, 51*f*
total generation and management, 1960–1997, 45 (*f*4.4)
U. S. production of, 43
waste-to-energy plants, 199–200
Mutagens, 137

N

National Acid Precipitation Assessment Program (NAPAP), 131–132, 133
National Organic Program, 162
National parks, effect of acid rain on, 129
National Public Water System Compliance Report (1998), 118–119
Nature Conservancy Dugout Ranch purchase, 167–168
Nebraska, 191
Nevada
population growth, 98
transuranic waste sites, 63
Yucca Mountain waste site, 63–64, 66*f*, 67
New Hampshire, 72, 108
New Jersey, 51, 72, 108
New Mexico
Los Alamos "controlled burn" disaster, 166
transuranic waste sites, 63
Waste Isolation Pilot Plant (WIPP), 63, 63*f*
waterbourne disease outbreaks, 119
wind power, 191
New York
Adirondack Mountains and pH levels, 126, 133
Bronx asthma epidemic, 156
electric vehicle sales, 83
mountain run-off, 127
New York City smog, 72
pollution credits, 89
Staten Island Fresh Kills landfill, 46
Nissan Altra, 83

Nitrates and nitrites, 114
Nitrogen dioxide, 73, 126*f*
Nitrogen pollution, 125
Nitrous oxide, 23–24, 24*t*, 125 (*f*7.3)
Noise Control Act of 1972, 159
Noise pollution, 157–159
Non-household solid waste, 43–44
See also Industrial waste
Nonattainment areas and designations, 70
carbon monoxide levels, 75
ozone levels, 71 (*f*5.1)
by pollutant, 70 (*t*5.1), 132 *f*(t7.3)
sulfur dioxide levels, 78*f*
North American Free Trade Agreement (NAFTA), 10
North Carolina
beach monitoring, 108
coastal erosion, 108
mountain run-off, 127
population growth, 97
smog, 69
Warren County hazardous waste landfill, 6
North Dakota, 191
Nuclear disarmament, 61–62
Nuclear waste, 60–65
containment canisters, 65
defense transuranic waste generating and storage sites, 64*t*
federal repositories, 62–64
low-level radioactive waste disposal compacts, 61*t*
spent fuel and high-level radioactive waste, by location, 61*t*
Waste Isolation Pilot Plant (WIPP) site, 63*f*, 65

O

Ocean Dumping Act, 109–110
Ocean Thermal Energy Conversion, 197
Ocean Thermal Energy Conversion, Research, and Development and Demonstration Act, 197
Oceans, as carbon sinks, 23
Oceans and coastal waters, 107–110
Off-stream use water, 94–96, 95*t*
Office of Noise Abatement and Control, 159
Ohio, 63, 108, 171
Oil Crisis of 1973 and development of alternative energy sources, 187, 191, 195
Oil drilling, 44, 182
Oil Pollution Act of 1990, 110
Oklahoma, 191
One-time use items, 42–43
Oregon, 50
Organic foods, 138, 161–162
Organic Foods Production Act, 162
Ozone and ozone depletion, 33–40
destruction of ozone, 34*f*

earth's atmosphere, 33*f*
nonattainment designations, 71 (*f*5.1)
ozone as smog component, 71–72
ozone-destroying chemicals, 35 (*t*3.2)
ozone studies, 35 (*t*3.1)
standards revision, 89–90
Ozone hole, 34, 39*f*

P

Pacific Basin, normal conditions, 26 (*f*2.11)
Packaging and waste, 53–54, 54*f*
Particulate matter, 73–74
Partnership for a New Generation of Vehicles (PGNV), 80
Passive smokers, 149
Passive solar energy systems, 188
Pennsylvania
beach monitoring, 108
James M. Seif v. Chester Residents Concerned Citizens et al, 6–7
mountain run-off, 127
Philadelphia (PA) sanitation workers' strike, *42*
smog, 72
Peregrine falcons, 179
Pesticides, 137–138
pH (potential hydrogen) scale, 124, 124 (*f*7.1)
Philadelphia (PA) sanitation workers' strike, *42*
Philip Morris, Broin v. (1997), 152
Philip Morris et al, United States of America v., 153
Phoenix (AZ) smog, 72
Photovoltaic conversion systems, 188–189, 190
Phytoplankton, 37–38
Plants, medically useful species, 178*t*
Point and nonpoint sources of pollution, 102
Pollutant advisories, 182 (*f*9.10)
Pollution credits, 89
Polychlorinated biphenyls (PCBs), 138
Population, human, 2–3, 3*f*
Population growth in relation to national forests, 168 (*f*9.4)
Pratt and Whitney, 84
Primary Drinking Water Standards, 116
Private property v. public rights movement, 172–173
See also Anti-regulatory movement
Produce, consumption per capita, 162*t*
Produce contamination, 160–161
Public health achievments, U. S., 149*f*
Public land development, 167–168
Public opinion, 10–13, 31–32
cost of environmental improvements, 13 (*f*1.5)
drinking water safety, 121 (*t*6.5)
economic growth v. the environment, 12 (*f*1.4)

effect of global warming on unemployment, 31 (t2.10)

energy costs v. global warming, 31 (t2.9)

environmental concerns, by age, 12 (f1.3)

environmental image of industries, 13 (t1.7)

impact of social movements, 11 (t1.6)

importance of issues in presidential campaigns, 13 (t1.8)

seriousness of environmental problems, 11 (t1.5)

seriousness of global warming, 31 (t2.8)

seriousness of social problems, 10t

Public Utilities Regulatory Polices Act, 187

Pyrolysis, 198

R

R. J. Reynolds Tobacco et al, Howard A. Engle, MD v. (1996), 153

Racism, environmental. *See* Environmental justice movement

Radon, 146–148

Raw materials, consumed in the U. S., 182 (f9.11)

Recovery of ecosystems, 136t

Recycling, 41, 50–51, 50 (f4.9)

Reformulated gasoline, 77

Refuse derived fuel, 200

Renewable energy. *See* Energy and renewable energy

Reprocessing spent fuel, 61

Residential Lead-Based Paint Hazard Reduction Act, 145

Residual acid, 127–128

Resource Conservation and Recovery Act, 47, 54, 107

Respiratory diseases, 70

Rhode Island, 51, 72

"Right to know" initiatives. *See* Sector Facility Indexing Project

Roadkill, 179

S

Safe Drinking Water Act amendments, 117–118, 154

Safe Drinking Water Act of 1974, 106–107, 116

San Diego (CA) wastewater treatment, 121

Saturated zone, 104

Saturn EV1, 83, *85*

Sea level and global warming, 24

Sector Facility Indexing Project, 8, 74–75

Shipping hazardous waste, 55–56

Shoreline stabilization, 170

Sick building syndrome, 156

Silent Spring (Carson), 1

Siltation, 101

Sixth Extinction, 175

Skin cancer and melanoma, 37

Smith, Robert Angus, 123

Smog, 33, 69–70

Soil and vegetation, effect of acid rain on, 128–129

Solar cycles, 20

Solar energy, 188–190

Solar houses, 188f

Solar power, 201–202

Solar thermal energy systems, 188, 189 (f10.5), 189 (f10.6), 189 (f10.7)

Sole source aquifers, 117

Source reduction, 49–51, 50 (t4.3)

South Carolina

Beachfront Management Act, 5–6, 172

low-level radioactive waste, 61

nuclear weapons-making plant, 61

public land development, 167

transuranic waste sites, 63

South Carolina, Lucas v. (1992), 6

South Dakota, 191

Spent fuel and high-level radioactive waste, 61, 62f

Sport utility vehicles (SUVs), 79–80

State of the Air: 2000 (American Lung Association), 72

Staten Island (NY) Fresh Kills landfill, 46, 48f

States

cigar smoking prevalence, by gender, 151t

cigarette smoking prevalence, by gender, 150t

compacts for low-level radioactive waste facilities, 60–61, 61t

defense transuranic waste generating and storage sites, 64t

fish advisories in effect, 107f

geothermal energy, known fields for resources, 194f

lead disclosure laws, 81f

locations of spent fuel and high-level radioactive waste, 62f

municipal solid waste generation, imports and exports, 46–47 (t4.2)

municipal solid waste landfills and incinerators, 45–46 (t4.1)

off-stream water use, 96t

percent of land area used for crops, 139f

projected population change, 98f

wetlands acreage lost, 172f

wetlands locations, 169t

See also individual states

Statistical information

air quality pollutant concentrations, decrease in, 1989–1998, 71 (t5.3)

alternative fuel refueling sites, by fuel, 84 (t5.7)

alternative fuel vehicle use, by fuel, 84 (t5.6)

alternative fuel vehicles, new fleet light duty purchases, 83t

anthropogenic volatile organic compound emissions, by source, 72 (f5.3)

anthropogenic volatile organic compound emissions, by year, 72 (f5.2)

asbestosis deaths, by industry, 142 (t8.4)

asbestosis deaths, by occupation, 142 (t8.3)

asbestosis deaths, by sex, race/ethnicity, and age, 140 (t8.1)

asbestosis deaths, by state and age, 141t

automobile market share, by vehicle type, 82 (f5.17)

biomass energy consumption, by major sector, 198f

blood lead levels, by states participating in surveillance, 147f

blood lead levels in adults, 146f

carbon dioxide emissions, by region, 22 (f2.8)

carbon dioxide emissions and sinks, 22 (t2.3)

carbon monoxide, nonattainment designations, 75f

carbon monoxide emissions, by source, 73 (f5.6)

carbon monoxide emissions, by year, 74f

chlorofluorocarbon units in U.S., 36t

chlorofluorocarbons, demand for, 37t

cigar smoking prevalence, by state and gender, 151t

cigarette smoking prevalance, by state and gender, 150t

Clean Air Act compliance costs, 89 (t5.8)

company-selected methods of reducing emissions, 135f

control and no-control scenario emissions, 89 (f5.20)

cost of environmental improvements, public opinion, 13 (f1.5)

drinking water safety, public opinion, 121 (t6.5)

drinking water safety notices, received by public, 121 (t6.6)

economic growth v. the environment, public opinion, 12 (f1.4)

electromagnetic field exposure levels, 24-hour, 144f

electromagnetic fields, by source, 143 (t8.6)

electromagnetic fields, kitchen sources, 143 (t8.5)

endangered species, by major group, 176t

energy consumption, by source, 186t

energy consumption in the U. S., 185f

energy costs v. global warming, public opinion, 31 (t2.9)

energy production, by fuel type, 201f

environmental concerns, by age, public opinion, 12 (f1.3)

T

Tambora volcano, 19

Tampa (FL) wastewater treatment, 121

Taxpayer Relief Act, 59

Temperature, 24–25

Tennessee
    mountain run-off, 127
    nitrogen saturation, 133
    recycling rates, 51
    smog, 69
    transuranic waste sites, 63

Teratogens, 137

Texas
    Kennedy Heights (Houston neighborhood) contaminated water supply, 6
    low-level radioactive waste, 61
    nuclear weapons-making plant, 61
    population growth, 97
    public land development, 167
    smog, 72
    wind power, 191

Thermal and hydrothermal reservoirs, 193

Threatened and endangered species, by major group, 176t

Threatened and endangered species range, by state/territory, 177f

Tidal power, 197

Tobacco and smoking, 149–153

Topography/geography and acid rain, 126

Topsoil erosion, 174

Toxic Substances Control Act, 145

Toxics Release Inventory, 75, 155, 155f

Toxins in everyday life. See Everyday toxins

Toyota RAV-4-EV, 83

Transport and transformation of emissions, 126

Trash sorting process, 43

Tree replanting, 165

Trees, effect of acid rain on, 129

Tropical rain forests, 164–165

U

Ultraviolet radiation, 33, 37

Union Carbide pesticide plant disaster, 155

United Nations Economic Commission for Europe Convention on Long-Range Transboundary Air Pollution, 131

United Nations Environment Program, 8

United Nations global warming treaty, 29–30

United States global warming program, 30–31

United States of America v. Philip Morris Inc. et al, 153

Unsaturated zone, 103–104

Uranium, 61

Utah, 149, 167–168

Utility deregulation, 90

Utility plants, geographic distribution of, 134f

V

Vermont, 72, 191

Victorian Water Treatment Enters the 21st Century (Cohen, Olson), 118

Virginia, 121

Volatile organic compounds, 72, 72 (f5.2), 72 (f5.3)

Volcanoes and climate, 19, 19 (f2.4)

W

Washington, D. C., 72

Washington state
    Hanford radioactive waste leak, 62
    population growth, 97
    recycling rates, 51
    transuranic waste sites, 63

Waste disposal, 41–67
    management of municipal solid waste, 1997, 44 (f4.3)
    materials recovery, 52 (f4.11)
    municipal solid waste, by material, 49 (f4.8)
    municipal solid waste, by product, 49 (f4.7)
    municipal solid waste, historical and projected generation, 53 (f4.13)
    municipal solid waste, recovery and discards, 51f
    municipal solid waste generation, imports and exports, by state, 46–47 (t4.2)
    municipal solid waste generation and management, 1960-1997, 45 (f4.4)
    municipal solid waste landfills and incinerators, by state, 45–46 (t4.1)
    source reduction/expansion, by material and product, 50 (t4.3)
    waste combustion plant with pollution control system, 52 (4.12)

Waste Isolation Pilot Plant (WIPP), 63, 63f

Waste prevention. See Source reduction

Waste-to-energy plants, 199–200
    See also Incineration and incinerators

Water depletion, 98–99

Water desalination, 110

Water issues, 93–122
    average annual erosion rates, by selected counties, 112f
    bacteria sources, 105 (f6.8)
    drinking water, safety notices received by public, 121 (t6.6)
    drinking water safety, public opinion, 121 (t6.5)
    EPA statutory authorities with groundwater protection provisions, 100t
    estimated water use trends, 97t
    excessive nutrients, effect on lake ecosystems, 105 (f6.9)
    fish advisories, by state/territory, 181f
    fish advisories in effect, by state, 107f
    fluoridation, 148–149
    groundwater as drinking water source, 113f
    irrigation, 174
    national groundwater use, 108f
    point and nonpoint sources of pollution, 106
    pollutants/stressors in impaired lakes, reservoirs, and ponds, 101f
    pollutants/stressors in impaired rivers and streams, by pollutant/stressor, 102f
    pollutants/stressors in impaired rivers and streams, by source, 103f
    pollutants/stressors of impaired shoreline waters, 111f
    radon levels in drinking water, 147–148
    siltation in rivers and streams, effects of, 104f
    sources of groundwater contamination, 109f
    water withdrawls, by water-resources region, 96f
    waterbourne disease outbreaks, drinking water, by etiologic agent and month, 120 (f6.19)
    waterbourne disease outbreaks, drinking water, by etiologic agent and type of water system, 120 (t6.4)
    waterbourne disease outbreaks, drinking water, by type of water system and year, 120 (f6.20)
    waterbourne disease outbreaks, recreational water, by etiologic agent and type of exposure, 122f
    waterbourne disease outbreaks, recreational water, by illness and year, 121 (f6.21)
    wetlands, causes of loss and degradation, 171t
    wetlands water quality, 170

Water power. See Hydropower

Water Quality Control Act of 1987, 117

Water table, 104

Water treatment, 114–116, 115f

Waterbourne disease outbreaks, 119–121
    drinking water, by etiologic agent and month, 120 (f6.19)
    drinking water, by etiologic agent and type of water system, 120 (t6.4)
    drinking water, by type of water system and year, 120 (f6.20)
    recreational water, by etiologic agent and type of exposure, 122f
    recreational water, by illness and year, 121 (f6.21)

Wave energy, 197

West Virginia, 133

Wetlands, 168–172
    acreage lost, by state, 172f
    causes of loss and degradation, 171t

ecological value of wetland processes, 170*f*

locations, by state, 169*t*

Wildlife, species and habitat loss, 177–178

Wind power, 190–192, 191*t*, 192*t*, 201

Windmills, 190–191

Wisconsin, 51, 119

Wood energy. *See* Biomass energy conversion

World Trade Organization (WTO), 9–10

Wyoming, 191

Y

You Are What You Drink (National Resources Defense Council), 118

Yucca Mountain (NV) waste site, 63–64, 66*f*, 67